ALGORITHMS IN BIOINFORMATICS

A PRACTICAL INTRODUCTION

T0139125

CHAPMAN & HALL/CRC
Mathematical and Computational Biology Series

Aims and scope:

This series aims to capture new developments and summarize what is known over the whole spectrum of mathematical and computational biology and medicine. It seeks to encourage the integration of mathematical, statistical and computational methods into biology by publishing a broad range of textbooks, reference works and handbooks. The titles included in the series are meant to appeal to students, researchers and professionals in the mathematical, statistical and computational sciences, fundamental biology and bioengineering, as well as interdisciplinary researchers involved in the field. The inclusion of concrete examples and applications, and programming techniques and examples, is highly encouraged.

Series Editors

Alison M. Etheridge
Department of Statistics
University of Oxford

Louis J. Gross
Department of Ecology and Evolutionary Biology
University of Tennessee

Suzanne Lenhart
Department of Mathematics
University of Tennessee

Philip K. Maini
Mathematical Institute
University of Oxford

Shoba Ranganathan
Research Institute of Biotechnology
Macquarie University

Hershel M. Safer
Weizmann Institute of Science
Bioinformatics & Bio Computing

Eberhard O. Voit
The Wallace H. Couter Department of Biomedical Engineering
Georgia Tech and Emory University

Proposals for the series should be submitted to one of the series editors above or directly to:
CRC Press, Taylor & Francis Group
4th, Floor, Albert House
1-4 Singer Street
London EC2A 4BQ
UK

Published Titles

Algorithms in Bioinformatics: A Practical Introduction
Wing-Kin Sung

Bioinformatics: A Practical Approach
Shui Qing Ye

Cancer Modelling and Simulation
Luigi Preziosi

Combinatorial Pattern Matching Algorithms in Computational Biology Using Perl and R
Gabriel Valiente

Computational Biology: A Statistical Mechanics Perspective
Ralf Blossey

Computational Neuroscience: A Comprehensive Approach
Jianfeng Feng

Data Analysis Tools for DNA Microarrays
Sorin Draghici

Differential Equations and Mathematical Biology
D.S. Jones and B.D. Sleeman

Engineering Genetic Circuits
Chris J. Myers

Exactly Solvable Models of Biological Invasion
Sergei V. Petrovskii and Bai-Lian Li

Gene Expression Studies Using Affymetrix Microarrays
Hinrich Göhlmann and Willem Talloen

Glycome Informatics: Methods and Applications
Kiyoko F. Aoki-Kinoshita

Handbook of Hidden Markov Models in Bioinformatics
Martin Gollery

Introduction to BioInformatics
Anna Tramontano

An Introduction to Systems Biology: Design Principles of Biological Circuits
Uri Alon

Kinetic Modelling in Systems Biology
Oleg Demin and Igor Goryanin

Knowledge Discovery in Proteomics
Igor Jurisica and Dennis Wigle

Meta-analysis and Combining Information in Genetics and Genomics
Rudy Guerra and Darlene R. Goldstein

Modeling and Simulation of Capsules and Biological Cells
C. Pozrikidis

Niche Modeling: Predictions from Statistical Distributions
David Stockwell

Normal Mode Analysis: Theory and Applications to Biological and Chemical Systems
Qiang Cui and Ivet Bahar

Optimal Control Applied to Biological Models
Suzanne Lenhart and John T. Workman

Pattern Discovery in Bioinformatics: Theory & Algorithms
Laxmi Parida

Python for Bioinformatics
Sebastian Bassi

Spatial Ecology
Stephen Cantrell, Chris Cosner, and Shigui Ruan

Spatiotemporal Patterns in Ecology and Epidemiology: Theory, Models, and Simulation
Horst Malchow, Sergei V. Petrovskii, and Ezio Venturino

Stochastic Modelling for Systems Biology
Darren J. Wilkinson

Structural Bioinformatics: An Algorithmic Approach
Forbes J. Burkowski

The Ten Most Wanted Solutions in Protein Bioinformatics
Anna Tramontano

Chapman & Hall/CRC Mathematical and Computational Biology Series

ALGORITHMS IN BIOINFORMATICS

A PRACTICAL INTRODUCTION

WING-KIN SUNG

CRC Press
Taylor & Francis Group
Boca Raton London New York

CRC Press is an imprint of the
Taylor & Francis Group, an **informa** business

A CHAPMAN & HALL BOOK

CRC Press
Taylor & Francis Group
6000 Broken Sound Parkway NW, Suite 300
Boca Raton, FL 33487-2742

First issued in paperback 2020

© 2010 by Taylor & Francis Group, LLC
CRC Press is an imprint of Taylor & Francis Group, an Informa business

No claim to original U.S. Government works

ISBN-13: 978-1-4200-7033-0 (hbk)
ISBN-13: 978-0-367-65931-8 (pbk)

Library of Congress Cataloging-in-Publication Data

Sung, Wing-Kin.
    Algorithms in bioinformatics : a practical introduction / Wing-Kin Sung.
        p. cm. -- (CHAPMAN & HALL/CRC mathematical and computational biology
        series)
    Includes bibliographical references and index.
    ISBN 978-1-4200-7033-0 (hardcover : alk. paper)
    1. Bioinformatics. 2. Genetic algorithms. I. Title. II. Series.

QH324.2.S86 2009
572.80285--dc22
                                        2009030738

Visit the Taylor & Francis Web site at
http://www.taylorandfrancis.com

and the CRC Press Web site at
http://www.crcpress.com

Contents

Preface

Bioinformatics is the study of biology through computer modeling or analysis. It is a multi-discipline research involving biology, statistics, data-mining, machine learning, and algorithms. This book is intended to give an in-depth introduction of the algorithmic techniques applied in bioinformatics.

The primary audiences of this book is advanced undergraduate students and graduate students who are from mathematics or computer science departments. We assume no prior knowledge of molecular biology beyond the high school level. In fact, the first chapter gives a brief introduction to molecular biology. Moreover, we do assume the reader has some training in college-level discrete mathematics and algorithms.

This book was developed from the teaching material for the course on "Combinatorial Methods in Bioinformatics", which I taught at the National University of Singapore, Singapore. The chapters in this book is classified based on the biology application domains. For each topic, an in-depth biological motivation is given and the corresponding computation problems are precisely defined. Different methods and the corresponding algorithms are also provided. Furthermore, the book gives detailed examples to illustrate each algorithm. At the end of each chapter, a set of exercises is provided.

Below, we give a brief overview of the chapters in the book.

Chapter 1 introduces the basic concepts in molecular biology. It describes the building blocks of our cells which include DNA, RNA, and protein. Then, it describes the mechanism in the cell and some basic biotechnologies. It also briefly describes the history of bioinformatics.

Chapter 2 describes methods to measure sequence similarity, which is fundamental for comparing DNA, RNA, and protein sequences. We discuss various alignment methods, including global alignment, local alignment, and semi-global alignment. We also study gap penalty and scoring function.

Chapter 3 introduces the suffix tree and gives its simple applications. We also present Farach's algorithm for constructing a suffix tree. Furthermore, we study variants of the suffix tree including suffix array and FM-index. We also discuss how to use suffix array for approximate matching.

Chapter 4 discusses methods for aligning whole genomes. We discuss MUMmer and Mutation Sensitive Alignment. Both methods apply the suffix tree and the longest common subsequence algorithm.

Chapter 5 considers the problem of searching a sequence database. Due to the advance of biotechnology, sequence data (including DNA, RNA, and protein) increases exponentially. Hence, it is important to have methods which allow efficient database searching. In this chapter, we discuss various biological database searching methods including FASTA, BLAST, BLAT, QUASAR, BWT-SW, etc.

Chapter 6 introduces methods for aligning multiple biological sequences. The chapter describes four algorithms: an exact solution based on dynamic programming, an approximation algorithm based on star alignment, and two heuristics. The two heuristics are ClustalW (a progressive alignment method) and MUSCLE (an iterative method).

Chapter 7 first describes a phylogenetic tree and its applications. Then, we discuss how to construct a phylogenetic tree given a character-based dataset or a distance-based dataset. We cover the methods: maximum parsimony, compatibility, maximum likelihood, UPGMA, and neighbor-joining. Lastly, we discuss whether the character-based methods and the distance-based methods can reconstruct the correct phylogenetic tree.

Chapter 8 covers the methods for comparing phylogenetic trees. We discuss methods for computing similarity and distance. For similarity, we consider maximum agreement subtree (MAST). For distance, we consider Robinson-Foulds distance, nearest neighbor interchange (NNI) distance, subtree transfer (STT) distance, and quartet distance. Furthermore, we discuss methods for finding consensus of a set of trees. We consider strict consensus tree, majority rule consensus tree, median tree, greedy consensus tree, and R^* consensus tree.

Chapter 9 investigates the problem of genome rearrangement. We discuss various possible genome rearrangements including reversal, transposition, etc. Since reversal can simulate other types of genome rearrangements, this chapter focuses on reversal distance. For computing the unsigned reversal distance, the problem is NP-hard. We describe a 2-approximation algorithm for this problem. For computing the signed reversal distance, we present the Hannenhalli-Pevzner theorem and Bergeron's algorithm.

Chapter 10 introduces the problem of motif finding. We discuss a number of de novo motif finding methods including Gibb Sampler, MEME, SP-star, YMF, and suffix tree-based methods like Weeder. Since multiple motif finders exist, we also discuss ensemble methods like MotifVoter, which combines information from multiple motif finders. All the above methods perform de novo motif finding with no extra information. Last, two motif finding methods which utilize additional information are described. The first method REDUCE improves motif finding by combining microarray and sequence data. The second method uses phylogenetic information to improve motif finding.

Chapter 11 discusses methods for predicting the secondary structure of RNA. When there is no pseudoknot, we discuss the Nussinov algorithm and the ZUKER algorithm. When pseudoknot is allowed, we discuss Akutsu's algorithm.

Chapter 12 covers methods for reconstructing the peptide sequence using a mass spectrometer. We discuss both de novo peptide sequencing and database search methods. For de novo peptide sequencing, we discuss Peaks and Sherenga. For database searching, we discuss SEQUEST.

Chapter 13 covers the computational problems related to population genetics. We discuss Hardy-Weinberg equilibrium and linkage disequilibrium. Then, we discuss algorithms for genotype phasing, tag SNP selection, and association study.

Supplementary material can be found at

http://www.comp.nus.edu.sg/~ksung/algo_in_bioinfo/.

I would like to thank the students who took my classes. I thank them for their efforts in writing the lecture scripts. I would also like to thank Xu Han, Caroline Lee Guat Lay, Charlie Lee, and Guoliang Li for helping me to proofread some of the chapters.

I would like to thank my PhD supervisors Tak-Wah Lam and Hing-Fung Ting and my collaborators Francis Y. L. Chin, Kwok Pui Choi, Edwin Chung, Wing Kai Hon, Jansson Jesper, Ming-Yang Kao, Hon Wai Leong, See-Kiong Ng, Franco P. Preparata, Yijun Ruan, Kunihiko Sadakane, Chialin Wei, Limsoon Wong, Siu-Ming Yiu, and Louxin Zhang. My knowledge of bioinformatics was enriched through numerous discussions with them. I would also like to thank my parents Kang Fai Sung and Siu King Wong, my three brothers Wing Hong Sung, Wing Keung Sung, and Wing Fu Sung, my wife Lily Or, and my two daughters Kelly and Kathleen for their support.

Finally, if you have any suggestions for improvement or if you identify any errors in the book, please send an email to me at ksung@comp.nus.edu.sg. I thank you in advance for your helpful comments in improving the book.

Wing-Kin Sung

Chapter 1

Introduction to Molecular Biology

1.1 DNA, RNA, and Protein

Our bodies consist of a number of organs. Each organ is composed of a number of tissues, and each tissue is a collection of similar cells that group together to perform specialized functions. The individual cell is the minimal self-reproducing unit in all living species. It performs two types of functions: (1) stores and passes the genetic information for maintaining life from generation to generation; and (2) performs chemical reactions necessary to maintain our life.

For function (1), our cells store the genetic information in the form of double-stranded DNA. For function (2), portions of the DNA called genes are transcribed into closely related molecules called RNAs. RNAs guide the synthesis of protein molecules. The resultant proteins are the main catalysts for almost all the chemical reactions in the cell. In addition to catalysis, proteins are involved in transportation, signaling, cell-membrane formation, etc.

Below, we discuss these three main molecules in our cells, namely, protein, DNA, and RNA.

1.1.1 Proteins

Proteins constitute most of a cell's dry mass. They are not only the building blocks from which cells are built, but also execute nearly all cell functions. Understanding proteins can guide us to understand how our bodies function and other biological processes.

A protein is made from a long chain of amino acids, each linking to its neighbor through a covalent peptide bond. Therefore, proteins are also known as polypeptides. There are 20 types of amino acids and each amino acid carries different chemical properties. The length of a protein is in the range of 20 to more than 5000 amino acids. On average, a protein contains around 350 amino acids.

In order to perform their chemical functions, proteins need to fold into certain 3 dimensional shapes. The folding of the proteins is caused by the weak interactions among amino acid residues. The weak interactions include hydrogen bonds, ionic bonds, van der Waals attractions, and the hydrophobic interactions. These interactions determine the shape of a protein, which is vital to its functionality.

1.1.1.1 Amino Acids

Amino acids are the building blocks of proteins. Each amino acid consists of:

1. Amino Group (-NH$_2$ group)

2. Carboxyl Group (-COOH group)

3. R Group (Side Chain), which determines the type of an amino acid

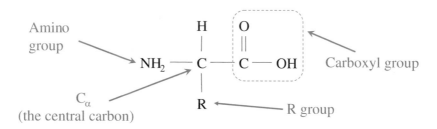

FIGURE 1.1: Structure of an amino acid.

All three groups are attached to a single carbon atom called α-carbon or C_α (see Figure 1.1). There are 20 common amino acids, characterized by different R groups. These 20 amino acids can be classified according to their mass, volume, acidity, polarity, and hydrophobicity. Figure 1.2 shows a table describing the properties of these 20 amino acids. The mass and volume are given in Daltons and in Å^3, respectively, of the R groups of the 20 amino acids. The acidity indicates if the R group of an amino acid is basic, acidic, or neutral. A basic amino acid is positively charged while an acidic amino acid is negatively charged.

The polarity indicates if the charge distribution within the R group of an amino acid is uneven or not. All acidic and basic amino acids are polar. For neutral amino acids, they are non-polar if the R groups are overall uncharged; otherwise, if they have uneven charge distribution, they are polar.

Amino acids can be hydrophilic or hydrophobic. A hydrophilic amino acid can form hydrogen bonds with water; otherwise, the amino acid is hydrophobic. The hydropathy index measures the hydrophobicity of an amino acid. A positive index indicates the amino acid is hydrophobic; a negative index indicates it is hydrophilic. As polar amino acids can form hydrogen bonds,

Amino Acid	1-Letter	3-Letter	Avg. Mass (Da)	volume (Å^3)	Side chain polarity	Side chain acidity or basicity	Hydropathy index
Alanine	A	Ala	89.09404	67	non-polar	Neutral	1.8
Cysteine	C	Cys	121.15404	86	polar	basic (strongly)	-4.5
Aspartic acid	D	Asp	133.10384	91	polar	Neutral	-3.5
Glutamic acid	E	Glu	147.13074	109	polar	acidic	-3.5
Phenylalanine	F	Phe	165.19184	135	polar	neutral	2.5
Glycine	G	Gly	75.06714	48	polar	acidic	-3.5
Histidine	H	His	155.15634	118	polar	neutral	-3.5
Isoleucine	I	Ile	131.17464	124	non-polar	neutral	-0.4
Lysine	K	Lys	146.18934	135	polar	basic (weakly)	-3.2
Leucine	L	Leu	131.17464	124	non-polar	neutral	4.5
Methionine	M	Met	149.20784	124	non-polar	neutral	3.8
Asparagine	N	Asn	132.11904	96	polar	basic	-3.9
Proline	P	Pro	115.13194	90	non-polar	neutral	1.9
Glutamine	Q	Gln	146.14594	114	non-polar	neutral	2.8
Arginine	R	Arg	174.20274	148	non-polar	neutral	-1.6
Serine	S	Ser	105.09344	73	polar	neutral	-0.8
Threonine	T	Thr	119.12034	93	polar	neutral	-0.7
Valine	V	Val	117.14784	105	non-polar	neutral	-0.9
Tryptophan	W	Trp	204.22844	163	polar	neutral	-1.3
Tyrosine	Y	Tyr	181.19124	141	non-polar	neutral	4.2

FIGURE 1.2: Table for the 20 standard amino acids and their chemical properties.

polar amino acids are usually hydrophilic while non-polar amino acids are hydrophobic. Moreover, there are exceptions since the hydrophilic and hydrophobic interactions do not have to rely only on the side chains of amino acids themselves. Usually, hydrophilic amino acids are found on the outer surface of a folded protein and hydrophobic amino acids tend to appear inside the surface of a folded protein.

In addition to the 20 common amino acids, there are non-standard amino acids. Two non-standard amino acids which appear in proteins and can be specified by genetic code (see Section 1.4.3.1) are selenocysteine (Sec, U) and pyrrolysine (Pyl, O). There are also non-standard amino acids which do not appear in proteins. Examples include lanthionine, 2-aminoisobutyric acid, and dehydroalanine. They often occur as intermediates in the metabolic pathways for standard amino acids. Some non-standard amino acids are formed through modification to the R-groups of standard amino acids. One example is hydroxyproline, which is made by a post-translational modification (see Section 1.5) of proline.

1.1.1.2 Protein Sequence and Structure

A protein or a polypeptide chain is formed by joining the amino acids together via a peptide bond. The formation and the structure of the peptide bond are shown in Figure 1.3. Note that the polypeptide is not symmetric. One end of the polypeptide is the amino group which is called the N-terminus while the other end of the polypeptide is the carboxyl group which is called the C-terminus. To write a peptide sequence, the convention is to write the sequence from the N-terminus to the C-terminus. For example, for the peptide sequence NH$_2$-QWDR-COOH, we write it as QWDR rather than RDWQ.

Although a protein is a linear chain of amino acid residues, it folds into a three-dimensional structure. We usually describe the protein structure in 4

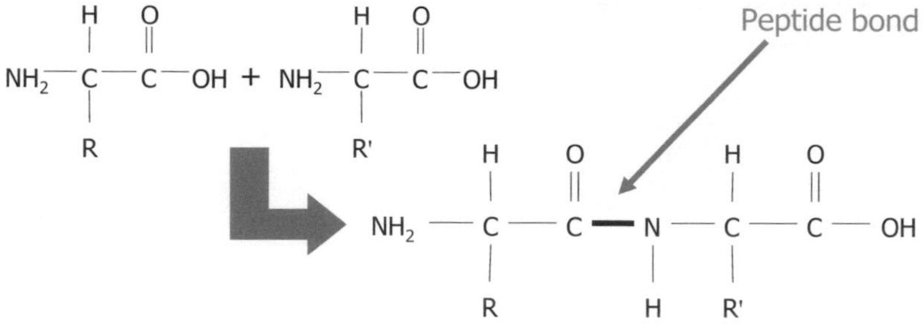

FIGURE 1.3: Formation of the peptide bond.

levels: primary, secondary, tertiary, and quaternary structures. The primary structure is the amino acid sequence itself. The secondary structure is formed through interaction between the backbone atoms due to the hydrogen bonding. They result in "local" structures such as α-helices, β-sheets, and turns. The tertiary structure is the three-dimensional structure of the protein formed through the interaction of the secondary structures due to the hydrophobic effect. The quaternary structure is formed by the packing of different proteins to form a protein complex.

1.1.2 DNA

Deoxyribonucleic acid (DNA) is the genetic material in all organisms (with certain viruses being an exception) and it stores the instructions needed by the cell to perform daily life functions.

DNA can be thought of as a large cookbook with recipes for making every protein in the cell. The information in DNA is used like a library. Library books can be read and reread many times. Similarly, the information in the genes is read, perhaps millions of times in the life of an organism, but the DNA itself is never used up.

DNA consists of two strands which are interwoven together to form a double helix. Each strand is a chain of small molecules called nucleotides.

1.1.2.1 Nucleotides

Nucleotides are the building blocks of all nucleic acid molecules. Their structure is shown in Figure 1.4. Each nucleotide consists of:

1. A pentose sugar deoxyribose

2. Phosphate (bound to the $5'$ carbon)

3. Base (bound to the $1'$ carbon) – nitrogenous base

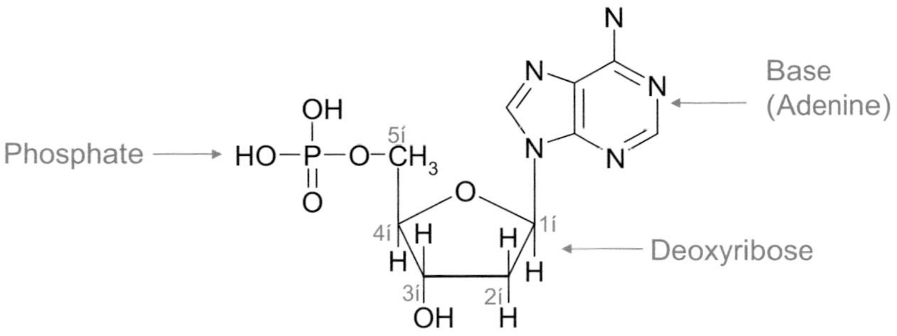

FIGURE 1.4: Structure of the nucleotide that forms DNA. The numbers $1', 2', \ldots, 5'$ label the different carbons on the ring structure.

Each nucleotide can have one, two, or three phosphates. Monophosphate nucleotides have only 1 phosphate, and are the building blocks of DNA. Diphosphate nucleotides and triphosphate nucleotides have 2 and 3 phosphate groups, respectively. They are used to transport energy in the cell. Figure 1.5 illustrates the reaction for generating energy through the conversion of triphosphate nucleotide (ATP) to diphosphate nucleotide (ADP).

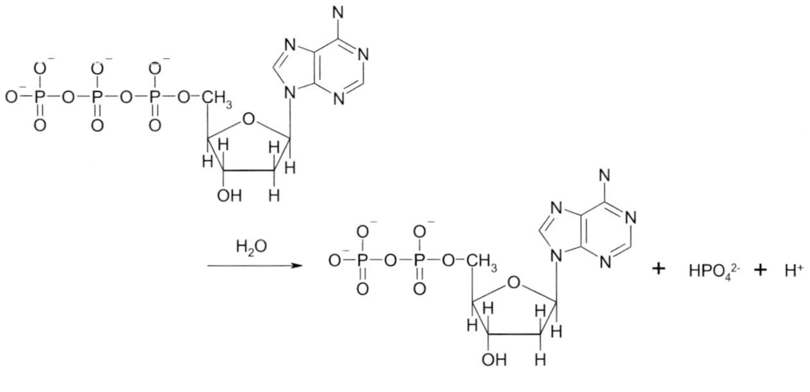

FIGURE 1.5: The conversion of adenosine triphosphate (ATP) to adenosine diphosphate (ADP), which yields about 7.3 kcal/mol energy. ATP works like a battery. Whenever power is required, ATP loses one phosphorous group to form ADP. By using the food energy, mitochondria convert ADP back to ATP.

The types of nucleotides depend on the nitrogenous bases. The five different nitrogenous bases are adenine (A), guanine (G), cytosine (C), thymine (T), and uracil (U). Their structures are shown in Figure 1.6.

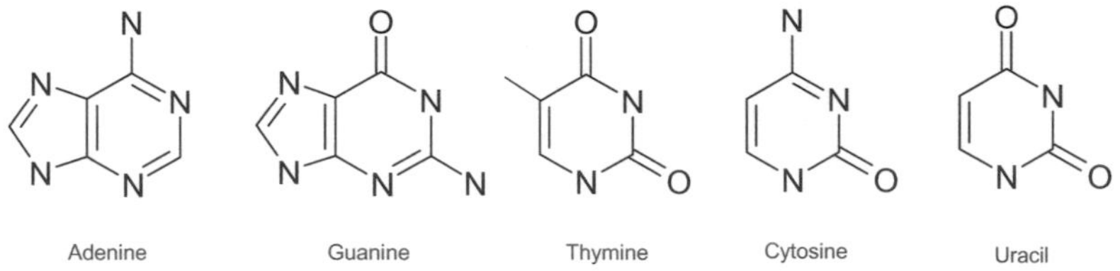

Adenine   Guanine   Thymine   Cytosine   Uracil

FIGURE 1.6:   Structures of the nitrogenous bases.

The nitrogenous bases are different, but they still show similarities. A and G are called purines and they have two ring structure. C, T, and U are pyrimidines and they have one ring structure. For DNA, only the bases A, C, G, and T are used.

Two nucleotides can be linked through the sugar-phosphate bond, which connects the phosphate (at the 5′ carbon of one nucleotide) to the 3′ carbon of another nucleotide. By repeatedly linking the nucleotides, we get a polynucleotide. Figure 1.7 shows an example of a chain of 5 nucleotides. Similar to a polypeptide, a polynucleotide is also asymmetric. Conventionally, we write the sequence from the 5′ end to the 3′ end. So, the chain in Figure 1.7 is written as ACGTA.

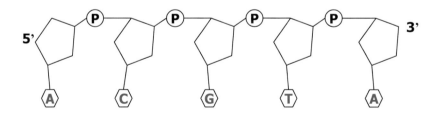

FIGURE 1.7:   An example DNA strand formed by chaining together 5 nucleotides. It consists of a phosphate-sugar backbone and 5 bases.

1.1.2.2   DNA Structure

The correct structure of DNA was first deduced by J. D. Watson and F. H. C. Crick in 1953 [304]. They deduced that DNA is double helix in structure (see Figure 1.8) based on two pieces of evidence:

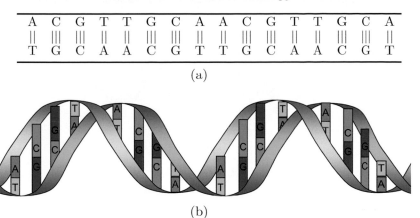

A	C	G	T	T	G	C	A	A	C	G	T	T	G	C	A
T	G	C	A	A	C	G	T	T	G	C	A	A	C	G	T

(a)

(b)

FIGURE 1.8: (a) The double-stranded DNA. The two strands show a complementary base pairing. $A = T$ and $C \equiv G$ are formed by two and three hydrogen bonds, respectively. (b) The three-dimensional structure of the same double-stranded DNA. They form double helix. (See color insert after page 270.)

1. From the analysis of E. Chargaff and colleagues, the concentration of thymine always equals the concentration of adenine and the concentration of cytosine always equals the concentration of guanine. This observation strongly suggests that A and T as well as C and G have some fixed relationship.

2. X-Ray diffraction pattern from R. Franklin, M. H. F. Wilkins, and co-workers. The data indicated that DNA has a highly ordered, multiple-stranded structure with repeating substructures spaced every 3.4Å along the axis of the molecule.

From the two observations, Watson and Crick proposed that DNA exists as a double helix in which polynucleotide chains consist of a sequence of nucleotides linked together by the sugar-phosphate bond. The two polynucleotide strands are held together by hydrogen bonds between bases in opposing strands. The base pairs are of high specificity such that: A is always paired with T and G is always paired with C. These base pairs are called the complementary base pairing (or Watson-Crick base pairing). A can form two hydrogen-bonds with T, whereas G can form three hydrogen-bonds with C. Figure 1.9 shows the two base pairings.

As shown in Figure 1.7, nucleotides are chained to form a strand of DNA. It has direction: from 5′ to 3′ (upstream). Similarly, the complementary strand are formed by chaining the complementary nucleotides. It goes from 3′ to 5′ (downstream). Two strands are held together due to the Watson-Crick base

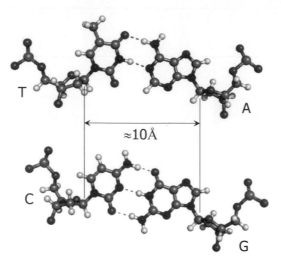

FIGURE 1.9:   Watson-Crick base pairing. Note that the distance between the complementary base pair is about 10Å. (See color insert after page 270.)

pairing. The distance between the two strands is about 10Å. Due to the weak interaction force, the two strands form double helix as shown in Figure 1.8.

1.1.2.3   Location of DNA in a Cell

There are two types of organisms, i.e., prokaryotes and eukaryotes. Prokaryotes are single-celled organisms with no nuclei (e.g., bacteria). They have no distinct nuclear compartment to house their DNA and therefore the DNA swims within the cells. Eukaryotes are organisms with single or multiple cells, for example, plant and animal. Each eukaryote's cell contains a nucleus surrounded by cytoplasm, which is contained within a plasma membrane. Most DNA is located within the nucleus. Moreover, some DNA is located in mitochondria and chloroplasts[1].

1.1.2.4   Linear or Circular Form of DNA

DNA usually exists in linear form, e.g., the nuclear DNA in human and yeast exists in linear form. However, in mitochondria, chloroplasts, viruses, and prokaryotes, DNA exists in circular form.

---

[1]Chloroplasts occur only in plants and some protists.

1.1.3 RNA

Ribonucleic acid (RNA) is the nucleic acid which is produced during the transcription process (i.e., from DNA to RNA). However, in certain organisms, such as viruses, RNA is used as genetic material instead of DNA.

1.1.3.1 Nucleotide for RNA

Figure 1.10 shows the nucleotide structure for RNA. Similar to the nucleotide of DNA, the nucleotide for RNA also has a phosphate and a base. The only difference is that the nucleotide here has a ribose sugar, instead of a deoxyribose in the DNA nucleotide. The ribose sugar has an extra OH group at 2', which is different from the H group at the same place of the deoxyribose. That's why we call these two different molecules "ribonucleic acid" and "deoxyribonucleic acid".

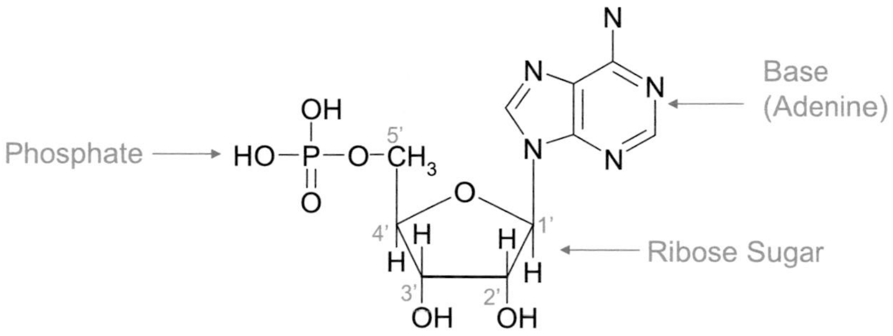

FIGURE 1.10: Diagram of nucleotide for RNA.

Besides the primitive difference of replacing a H group by a OH group, RNA has some other characteristics allowing us to differentiate it from DNA. First of all, unlike the double helix structure of DNA, RNA is single-stranded. Due to the extra OH group at 2' carbon, RNA can form more hydrogen bonds than DNA. Hence, it can form complex 3D structures to perform more functions. Second, RNA uses base U instead of base T used by DNA. Base U is chemically similar to base T. In particular, U is also complementary to A.

1.1.3.2 Functions of RNA

RNA has the properties of both DNA and proteins. First, similar to DNA, it can store and transfer information. Second, similar to protein, it can form complex 3D structures and perform functions. So, it seems that we only need RNA to accomplish all the requirements for DNA and proteins. Why do we still need DNA and proteins? The reason lies in a simple rule — when you want to do two different things at the same time, you can never do either one as perfectly as those people who only focus on one thing. For the storage of information, RNA is not as stable as DNA and that's why we still have DNA. Protein can perform more functions than RNA does, which is the reason why protein is still needed.

1.1.3.3   Different Types of RNA

In general, there are two types of RNAs: messenger RNAs (mRNAs) and non-coding RNAs (ncRNAs). Messenger RNAs carry the encoded information required to make proteins of all types. Non-coding RNAs include ribosomal RNAs (rRNAs), transfer RNAs (tRNAs), short ncRNAs, and long ncRNAs. Ribosomal RNAs form parts of ribosomes that help to translate mRNAs into proteins. Transfer RNAs serve as a molecular dictionary that translates the nucleic acid code into the amino acid sequences of proteins. Short ncRNAs include 4 different types: snoRNAs, microRNAs, siRNAs, piRNAs. They are responsible for the regulation of the process for generating proteins from genes. There are many long ncRNAs. People expected the number can be up to 30 thousands. Their functions are diverse and many of them are unknown.

1.2   Genome, Chromosome, and Gene

1.2.1   Genome

The complete set of DNA in an organism is referred to collectively as a "genome". Genomes of different organisms vary widely in size: the smallest known genome for a free-living organism (a bacterium called *Mycoplasma genitalium*) contains about 600,000 DNA base pairs, while the largest known genome is *Amoeba dubia* which contains 670 billion base pairs.

Human and mouse genomes have about 3 billion DNA base pairs. Except for mature red blood cells, sperm cells, and egg cells, all human cells contain the same complete genome, though some may have small differences due to mutations (see Section 1.3). Sperm and egg cells only contain half of the genome from father and mother, respectively. During fertilization, the sperm cell fuses with the egg cell, and forms a new cell containing the complete genome.

1.2.2   Chromosome

A genome is not one consecutive double-stranded DNA chain. For instance, the 3 billion bases of the human genome is partitioned into 23 separate pairs of DNA, called chromosomes. Each chromosome is a double-stranded DNA chain which is wrapped around histones. Humans have 23 pairs of chromosomes, where 22 pairs are autosomes (non-sex-determining chromosomes) and 1 pair is sex chromosomes (X and Y chromosomes). For the sex chromosomes, the male has one X chromosome and one Y chromosome, whereas the female carries a pair of X chromosomes. The collection of chromosomes in an individual is called his/her karyotype. For example, the typical male karyotype has 22 pairs of autosomes, one X and one Y chromosome.

1.2.3 Gene

A gene is a DNA sequence that encodes a protein or an RNA molecule. Each chromosome contains many genes, which are the basic physical and functional units of heredity. Each gene exists in a particular position of a particular chromosome. In the human genome, there are approximately 20,000 — 30,000 genes. In the prokaryotic genome, one gene corresponds to one protein, whereas in a eukaryotic genome, one gene can correspond to more than one protein because of the process called "alternative splicing" (see Section 1.4.2).

1.2.4 Complexity of the Organism versus Genome Size

Due to the huge amount of genes inside the human genome, one might argue that the complexity of one organism is somewhat related to its genome size. But in fact, it's not the truth. The genome of humans, a complex multicellular organism, is approximately 3 billion base pairs, yet the *Amoeba dubia*, a single cell organism, has up to 670 billion base pairs! Thus, genome size really has no relationship to the complexity of the organism.

1.2.5 Number of Genes versus Genome Size

Is there any relationship between the genome size and the number of genes? Before answering this question, let's take a look at the human genome and a prokaryotic genome, *E. coli*. There are about 30,000 to 35,000 genes in the human genome, which is of length 3 billion. The average length of the coding region of a gene in the human genome is around 1,000 to 2,000 base pairs. So, the total length of all coding regions is less than 70 million base pairs, which is less than 3% of the human genome. For the rest of the genome, people generally call them "junk DNA".

How about the *E. coli* genome? It has 5 million base pairs and 4,000 genes. The average length of a gene in the *E. coli* genome is 1,000 base pairs. From these figures, around 90% of the *E. coli* genome is useful, i.c., it consists of the coding regions. So, it seems that the prokaryotic organism *E. coli* has less junk DNA than human beings! The conclusion is that the genome size has little relationship to the number of genes!

1.3   Replication and Mutation of DNA

Why is DNA double-stranded? One of the main reasons is to facilitate replication. Replication is the process allowing a cell to duplicate and pass its DNA to the two daughter cells. The process first separates the two strands of DNA. Then, each DNA strand serves as a template for the synthesis of another complementary strand with the help of DNA polymerase. This process generates two identical double-stranded DNAs for the two daughter cells.

When one strand of a double-stranded DNA is damaged, the strand can be repaired with the information of another strand. The replication and repair mechanisms help to maintain the same genome for all cells in our body.

Despite replication being a near-perfect process, infrequent mistakes called "mutations" are still possible. Mutations are generally caused by modifying the way that nucleotides form hydrogen bonds. Then, replication will hybridize wrong nucleotides with the template DNA and mutations occur. One possible way to modify nucleotides is through tautomeric shift. Tautomeric shift is a rare event which changes the chemical property of the bases and affects the way the bases form hydrogen bonds. For example, tautomeric shift may enable adenine to base pair with cytosine. When a tautomeric shift occurs during replication, a wrong nucleotide will be inserted and a mutation occurs.

The mutations may affect one nucleotide or a segment of nucleotides. For instance, when one segment of DNA is reproduced, a small sub-segment of it could be lost, duplicated, or reversed. Furthermore, a new segment sometimes could be inserted into the DNA segment. In summary, we have various types of mutations: point mutation, insertion, deletion, duplication, inversion (or reversal), and translocation.

- Point mutation: The modification of a nucleotide in the genome.

- Deletion: The deletion of a segment in the genome.

- Duplication: The duplication of a segment in the genome.

- Inversion: The inversion of a segment in the genome.

- Insertion: The insertion of a DNA segment into the genome.

- Translocation: A portion of a chromosome is moved to a new location.

Mutations make the new generation of cells or organisms different from their ancestors. Occasionally, some "good" mutations make the cells or organisms survive better in the environment. The selection of the fittest individuals to survive is called "natural selection". Mutation and natural selection have resulted in the evolution of diversified organisms. On the other hand, the "bad" mutations may alter the genes, which cause fatal diseases, such as cancer.

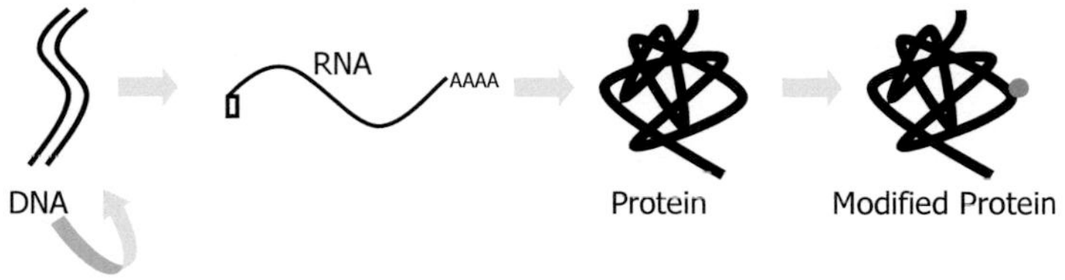

FIGURE 1.11: DNA is transcribed to mRNA, which is translated into protein (Central Dogma). Then, through post-translation modification, the protein is modified.

1.4   Central Dogma (from DNA to Protein)

Central dogma was first enunciated by Francis Crick in 1958 [68] and re-stated in a *Nature* paper published in 1970 [67]. Central dogma describes the process of transferring information from DNA to RNA to protein. It states that the information from DNA is transferred sequentially to RNA to protein and the information cannot be transferred back from protein to either protein or nucleic acid. In other words, once information gets into protein, it cannot flow back to nucleic acid.

The information transfer process consists of two steps (see Figure 1.11).

1. Transcription: DNA is transcribed to mRNA. During the transcription process, an mRNA is synthesized from a DNA template resulting in the transfer of genetic information from the DNA molecule to the mRNA.

2. Translation: mRNA is translated to protein. In the translation process, the mRNA is translated to an amino acid sequence by stitching the amino acids one by one; thus the information obtained from DNA is transferred to the protein, through mRNA.

The subsections below discuss transcription (prokaryotes), transcription (eukaryotes), and translation. Figure 1.12 summarizes the steps of information transfer from DNA to protein.

1.4.1   Transcription (Prokaryotes)

The transcription process of prokaryotes synthesizes an mRNA from the DNA gene with the help of the RNA polymerase. The process is simple and it is as follows. First, the RNA polymerase temporarily separates the double-stranded DNA. Then, it locates the transcription start site, which is a marker denoting the start of a gene. Next, the RNA polymerase synthesizes an mRNA following two rules: (1) the bases A, C, and G are copied exactly

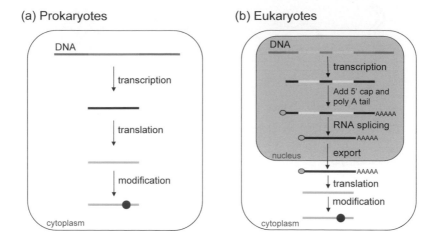

FIGURE 1.12:   Central dogma for (a) prokaryotes and (b) eukaryotes. (See color insert after page 270.)

from DNA to mRNA and (2) T is replaced by U (in RNA we only have U instead of T). Once the RNA polymerase reaches the transcription stop site (a marker denoting the end of a gene), the transcription process is stopped and an mRNA is obtained.

1.4.2   Transcription (Eukaryotes)

The transcription processes of prokaryotes and eukaryotes are different. A prokaryotic gene is completely transcribed into an mRNA by the RNA polymerase. For a eukaryotic gene, only part of it is finally transcribed into an mRNA. The segments of the eukaryotic gene which are finally transcribed into an mRNA are called exons. Introns are segments of a gene situated between exons. Each eukaryotic gene can have many introns and each intron can be very long. An extreme example is the gene associated with the disease cystic fibrosis in humans. It has 24 introns of total length approximately 1 million bases, whereas the total length of its exons is only 1 kilobase. Introns in eukaryotic genes normally satisfy the GT-AG rule, that is, the intron begins with GT and ends with AG. Spliceosomes can recognize the signal at the start and the end of an intron and splice it out.

The transcription process of eukaryotes is as follows. First, RNA polymerase produces a pre-mRNA which contains both introns and exons. Then the 5′ cap and poly-A tail are added to the pre-mRNA. After that, with the help of spliceosomes, the introns are removed and an mRNA is produced. The final mRNA is transported out of the nucleus and the translation process starts.

Finally, note that in different tissues or different conditions, the introns which will be removed are not the same. Hence, one eukaryotic gene can be transcribed into many different mRNA sequences.

1.4.3 Translation

Translation is also called protein synthesis. It synthesizes a protein from an mRNA. The translation process is handled by a molecular complex known as a ribosome which consists of both proteins and ribosomal RNA (rRNA). First, the ribosome reads mRNA from 5' to 3'. The translation starts around the start codon (translation start site). Then, with the help of transfer RNA (tRNA, a class of RNA molecules that transport amino acids to ribosomes for incorporation into a polypeptide undergoing synthesis), each codon is translated to an amino acid. Finally the translation stops once the ribosome reads the stop codon (translation stop site).

1.4.3.1 Genetic Code

In the translation process, the translation of a codon to an amino acid is specified by a translation table called the "genetic code". The genetic code is universal for all organisms.

Before we proceed, let's think of why the genetic code is not made of one or two nucleotides. Recall that there are only 4 types of nucleotides and therefore if the genetic code is only made of 1 nucleotide, it can only code for four amino acids, which is not sufficient. For 2 nucleotides, the number of amino acids that can be coded is only $4^2 = 16$ and this is also not sufficient. That's why the genetic code consists of three consecutive nucleotides.

With 3 nucleotides in a codon, there are $4^3 = 64$ different codons. As there are only 20 different amino acids, the codons are not in one-to-one correspondence to the 20 amino acids. Figure 1.13 shows the table of genetic code. We could find out that several different codons encode for the same amino acid.

The genetic code is not a random assignment of codons to amino acids. For example, amino acids that share the same biosynthetic pathway tend to have the same first base in their codons, and amino acids with similar physical properties tend to have similar codons. Also, the codon usage in the genetic code table is not uniform. Met and Trip are the only two amino acids coded by one codon. Ser is the amino acid which is coded by the maximum number of codons (6 different codons).

In the genetic code table (Figure 1.13), there are four special codons to be noted, i.e., ATG, TAA, TAG, and TGA. ATG encodes for Met. It is also a signal of the translation start site. Hence, ATG is called the start codon. On the other hand, TAA, TAG, and TGA are the stop codons. Whenever ribosomes (the protein synthesizing machines) encounter the stop codons, the ribosomes will dissociate from the mRNA and the translation process

	T	C	A	G	
T	TTT Phe [F]	TCT Ser [S]	TAT Tyr [Y]	TGT Cys [C]	T
	TTC Phe [F]	TCC Ser [S]	TAC Tyr [Y]	TGC Cys [C]	C
	TTA Leu [L]	TCA Ser [S]	TAA Ter [end]	TGA Ter [end]	A
	TTG Leu [L]	TCG Ser [S]	TAG Ter [end]	TGG Trp [W]	G
C	CTT Leu [L]	CCT Pro [P]	CAT His [H]	CGT Arg [R]	T
	CTC Leu [L]	CCC Pro [P]	CAC His [H]	CGC Arg [R]	C
	CTA Leu [L]	CCA Pro [P]	CAA Gln [Q]	CGA Arg [R]	A
	CTG Leu [L]	CCG Pro [P]	CAG Gln [Q]	CGG Arg [R]	G
A	ATT Ile [I]	ACT Thr [T]	AAT Asn [N]	AGT Ser [S]	T
	ATC Ile [I]	ACC Thr [T]	AAC Asn [N]	AGC Ser [S]	C
	ATA Ile [I]	ACA Thr [T]	AAA Lys [K]	AGA Arg [R]	A
	ATG Met [M]	ACG Thr [T]	AAG Lys [K]	AGG Arg [R]	G
G	GTT Val [V]	GCT Ala [A]	GAT Asp [D]	GGT Gly [G]	T
	GTC Val [V]	GCC Ala [A]	GAC Asp [D]	GGC Gly [G]	C
	GTA Val [V]	GCA Ala [A]	GAA Glu [E]	GGA Gly [G]	A
	GTG Val [V]	GCG Ala [A]	GAG Glu [E]	GGG Gly [G]	G

FIGURE 1.13:   Genetic code.

terminates.

Two stop codons UGA and UAG sometimes can also encode two non-standard amino acids selenocysteine (Sec, U) and pyrrolysine (Pyl, O), respectively. The UGA codon is normally a stop codon. In the presence of a SECIS element (SElenoCysteine Insertion Sequence) in the mRNA, the UGA codon encodes selenocysteine. Pyrrolysine is used by some enzymes in methanogenic archaea to produce methane. It is coded by the UAG codon in the presence of a PYLIS element.

Although most amino acids are coded by more than one codon in the genetic code table, different organisms often prefer to encode each amino acid with one particular codon. Even more interesting, some organisms use the codon usage to control the efficiency of translation. For example, in *S. pombe*, *C. elegans*, *D. melanogaster*, and many unicellular organisms, highly expressed genes like those encoding ribosomal proteins use codons whose tRNAs are the most abundant in the organism; on the other hand, low expressed genes use codons whose tRNAs are the least abundant.

1.4.3.2   More on tRNA

Recall that the translation from codon to amino acid is performed with the help of the transfer RNA, or "tRNA". Since there are 61 non-terminal codons, there are a total of 61 different tRNAs. Each tRNA has a cloverleaf-shaped structure. On one side it holds an anticodon (a sequence of three adjacent nucleotides in tRNA designating a specific amino acid that binds to a corresponding codon in mRNA during protein synthesis), and on the other side it holds the appropriate amino acid. The structure of tRNA helps to associate the codon with the corresponding amino acid.

1.4.3.3 More on Gene Structure

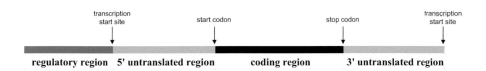

FIGURE 1.14: Gene structure.

After understanding both transcription and translation, we can have an overview of the structure of a gene (see Figure 1.14). A gene consists of three regions: the 5′ untranslated region, the coding region, and the 3′ untranslated region. The coding region contains the codons for proteins. It is also called "open reading frame". The "open reading frame" is composed of codons that occur as multiples of three nucleotides. The coding region usually begins with a start codon, and must end with a stop codon, and the rest of its codons are not stop codons. The 5′ untranslated region, coding region, and 3′ untranslated region together are also called the "mRNA transcript", because the transcription step exactly copies these three regions to form a mRNA. Finally, before the 5′ untranslated region, we have the regulatory region (also called the promoter) which regulates the transcription process. The promoter is in fact a DNA segment to which RNA polymerase binds, initiating the transcription of mRNA.

1.5 Post-Translation Modification (PTM)

Once a protein is produced, it may undergo post-translation modification (see Figure 1.11). Post-translation modification (PTM) is the chemical modification of a protein after its translation. Such modification changes or enhances the function of the protein. PTM can be classified as three types: (1) addition of functional groups (e.g., acylation, methylation, phosphorylation);

(2) addition of other peptides (e.g., ubiquitination, the covalent linkage to the protein ubiquitin); and (3) structural changes (e.g., disulfide bridges, the covalent linkage of two cysteine amino acids).

The chemical modification changes the properties of the proteins and hence their functionality. For instance, phosphorylation is a process attaching a phosphate (PO_4) to the R group of a protein. Kinases and phosphatases are enzymes which phosphorylate and dephosphorylate proteins, respectively. This process changes the conformation of proteins and causes them to become activated or deactivated. Phosphorylation can dynamically turn on or off many signaling pathways. One famous example is the phosphorylation of p53 (tumor suppressor protein) which causes apoptotic cell death.

1.6 Population Genetics

The genomes of two individuals of the same species are not exactly the same. They have some minor differences. Given the genome of two individuals of the same species, if there exists a position (called a locus) where the nucleotides between the two individuals are different, we call it a single nucleotide polymorphism (SNP). For humans, we expect SNPs are responsible for over 80% of the variation between two individuals. Hence, understanding SNPs can help us to understand the differences within a population. For example, in humans, SNPs control the color of hair, the blood type, etc., of different individuals. Also, many diseases like cancer are related to SNPs.

1.7 Basic Biotechnological Tools

A vast range of technological tools have been developed to facilitate the study of DNA in a more efficient manner. Those tools help to cut and break DNA (using restriction enzymes or sonication), to duplicate DNA fragments (using cloning or PCR), to measure the length of DNA (using gel electrophoresis), to detect the existence of certain DNA segments in a sample (using hybridization with probes), and to perform DNA sequencing. This section examines these tools.

In addition to the above technologies, Sections 12.1 and 10.2 will discuss mass spectrometry and chromatin immunoprecipitation, respectively. Other biotechnologies exist and readers are referred to the literature for details.

1.7.1 Restriction Enzymes

Restriction enzymes or restriction endonuclease are a class of bacterial enzymes. They are DNA-cutting enzymes, which recognize certain sequences, called restriction sites, in the double-stranded DNA and break the phosphodiester bonds between the nucleotides. This process is called digestion. Naturally, restriction enzymes exist in various bacterial species and are used to break foreign DNA to avoid infection or to disable the function of the foreign DNA.

Each restriction enzyme seeks out a specific DNA sequence, which is usually palindromic, 4 to 8 bp long, and precisely cuts it in one place. For instance, the enzyme shown here, EcoRI, cuts the sequence GAATTC (see Figure 1.15), cleaving between G and A. Depending on the restriction enzymes used, DNA fragments with either blunt ends or sticky ends can be produced. For EcoRI, a sticky end is formed after the digestion.

```
5'-GAATTC-3'    Digested        5'-G          AATTC-3'
3'-CTTAAG-5'    by EcoRI        3'-CTTAA  +       G-5'
```

FIGURE 1.15: Digestion by EcoRI.

Nowadays, over 3000 restriction enzymes have been studied and many of them have been isolated. The isolated restriction enzymes become a standard tool for cutting DNA at specific sites with many applications. For example, restriction enzymes are used in cloning applications (see Section 1.7.3) to assist the insertion of the target DNA segment into the plasmid vectors.

1.7.2 Sonication

Sonication is another method to cut a DNA fragment. It applies high vibration (usually ultrasound) to randomly chop a sample of the DNA fragment into small fragments. Sonication has been applied to binding site finding, genome sequencing, bacteria detection, etc.

1.7.3 Cloning

Cloning allows us to multiply the available amount of DNA in order to have enough DNA for many experiments. Precisely, given a piece of DNA X, cloning is the process of duplicating many copies of DNA X. The idea is to insert the DNA X into bacterial host cells. Through replicating the cells, DNA X is replicated. The basic steps of cloning are as follows.

1. Insert DNA X into a plasmid vector with an antibiotic-resistance gene

and obtain a recombinant DNA molecule. This step is done with the help of a restriction enzyme and a DNA ligase[2].

2. Insert the recombinant DNA into the host cell (usually *E. coli*) using the chemical transformation method, where the bacterial cells are made "competent" to take up foreign DNA by treating the cells with calcium ions. An alternative way is to insert the recombinant DNA by the electroporation method. After the recombinant DNA molecules are mixed with the bacteria cells, a brief heat shock is applied to facilitate uptake of DNA.

3. Grow the host cells in the presence of antibiotic. Note that only cells with the antibiotic-resistance gene can grow. When the host cell duplicates, X is also duplicated.

4. Select those cells that contain both the antibiotic-resistance genes and the foreign DNA X. DNA X is then extracted from these cells.

Cloning is a time-consuming process normally requiring several days, whereas we will learn a much quicker method known as PCR in the next section.

1.7.4   PCR

PCR is an acronym which stands for polymerase chain reaction. It applies DNA polymerase in a chain reaction to amplify the target DNA fragments in vitro (i.e., outside a living cell). As stated in Section 1.3, the DNA polymerase is a naturally occurring biological macromolecule that catalyzes the formation and repairs DNA. The accurate replication of all living matter depends on this activity. In the 1980s, Kary Mullis at Cetus Corporation conceived a way to start and stop a DNA polymerase's action at specific points along a single-strand of DNA. Mullis also realized that by harnessing this component of molecular reproduction technology, the target DNA could be exponentially amplified.

Inputs for PCR include:

1. The DNA fragment we hope to amplify;

2. Two oligonucleotides are synthesized, each complementary to the two ends of the DNA fragments to be amplified. They are called primers; and

3. The thermostable DNA polymerase Taq[3]. This polymerase can perform replication at a temperature up to 95°C.

---

[2]A DNA ligase is an enzyme that can catalyse the joining of two DNA fragments.

[3]Taq stands for the bacterium *Thermus aquaticus* which lives in hot springs and can be found in Yellowstone National Park. From this bacterium, the thermostable DNA polymerase was isolated.

PCR consists of repeating a cycle with three phases 25 to 30 times. Each cycle takes about 5 minutes.

- Phase 1: Denature to separate double-stranded DNA by heat;

- Phase 2: Add synthesis primers to hybridize the denatured DNA and cool down;

- Phase 3: Add DNA polymerase Taq to catalyze 5′ to 3′ DNA synthesis.

After the last cycle, Phase 3 is kept for a longer time, about 10 minutes, to ensure that DNA synthesis for all strands is complete. Then, only the flanked region by the primers has been amplified exponentially, while the other regions are not. Please refer to Figure 1.16 for a comprehensive picture of how PCR works.

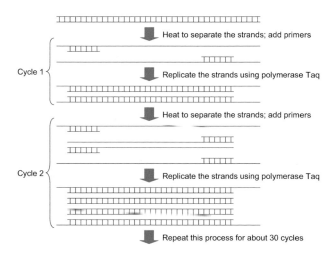

FIGURE 1.16: Cycles in the polymerase chain reaction.

The PCR method is used to amplify DNA segments to the point where it can be readily isolated for use. When scientists succeeded in making the polymerase chain reaction to perform as desired in a reliable fashion, they had a powerful technique for providing unlimited quantities of the precise genetic material for experiments. For instance, PCR has been applied to (1) clone DNA fragments from mummies and (2) detect viral infections in blood samples. For the latest developments in the PCR method, please refer to [32].

1.7.5   Gel Electrophoresis

Gel electrophoresis, developed by Frederick Sanger in 1977, is a method that separates macromolecules — either nucleic acids or proteins — on the basis of electric charge.

A gel is a colloid in a solid form. The term electrophoresis describes the migration of charged particles under the influence of an electric field. Electro refers to the energy of electricity. Phoresis, from the Greek verb *phoros*, means "to carry across". Thus, gel electrophoresis refers to the technique in which molecules are forced across a span of gel, driven by an electrical current. Activated electrodes at either end of the gel provide the driving force. A molecule's properties determine how rapidly an electric field can move the molecule through a gelatinous medium.

Gel electrophoresis can be used to separate a mixture of DNA fragments. Note that DNA is negatively charged due to the high phosphate residues in its backbone. To support the separation of the DNA fragments, a gel is prepared with holes on it. These holes will serve as a reservoir to hold the DNA fragments. DNA solutions (mixtures of different sizes of DNA fragments) are loaded in the negative end of the gel. During electrophoresis, DNA migrates from the negative electrode toward the positive electrode. The movement is retarded by the gel matrix, which acts as a sieve for DNA molecules. Large molecules have difficulty getting through the holes in the matrix. Small molecules move easily through the holes. Because of this, large fragments will lag behind small fragments as DNA migrates through the gel. As the process continues, the separation between the larger and smaller fragments increases. The mixture is separated into bands, each contains DNA molecules of the same length.

Apart from separating a mixture of DNA fragments, gel electrophoresis has been applied to reconstruct a DNA sequence of length 500–800 within a few hours. The idea is to, initially, generate all sequences that end with a particular base, e.g., A. Then we can use gel electrophoresis to separate the sequences that end with A into different bands. Such information tells us the relative positions of different bases in the sequence.

Figure 1.17 demonstrates the sequencing process. All four groups of fragments ending at A, C, G, and T are placed at the negative end of the gel. During electrophoresis, the fragments move toward the positive end. The unknown DNA sequence is reconstructed from the relative distances of the fragments.

The generation of fragments ending in a particular base can be achieved through two methods, i.e., the Maxam-Gilbert or the Sanger sequencing method. In the Maxam-Gilbert method, DNA samples are divided into four aliquots and four different chemical reactions are used to cleave the DNA at a particular base (A, C, G, or T) or base type (pyrimidine or purine). In the Sanger method, DNA chains of varying lengths are synthesized by enzymes in four different reactions. Each reaction will produce DNA ending in

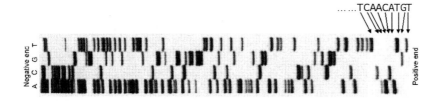

FIGURE 1.17: Autoradiograph of a dideoxy sequencing gel. We can read the DNA sequence from the gel, which is TGTACAACT.....

a particular base.

DNA sequencing can be automated by using a laser to detect the separated products in real time during gel electrophoresis. The four bases are labelled with different fluorophores and they are placed in a single lane. In this way, many DNA samples can be sequenced simultaneously. Sequences of greater than 900 bases can be obtained with $\sim 98\%$ accuracy.

1.7.6 Hybridization

Routinely, biologists need to find a DNA fragment containing a particular DNA subsequence among millions of DNA fragments. The above problem can be solved through hybridization. For example, suppose we need to find a DNA fragment which contains ACCGAT, we can perform the following steps.

1. Create an oligonucleotide, called a probe, which is inversely complementary to ACCGAT.

2. Mix the probe with the DNA fragments.

3. Due to the hybridization rule $(A = T, C \equiv G)$, DNA fragments which contain ACCGAT will hybridize with the probe and we can detect the DNA fragment.

1.7.6.1 DNA Array

The idea of hybridization leads to the evolution of DNA array technology, which enables researchers to perform experiments on a set of genes or even the whole genome. This completely changes the routine of one gene in one experiment and makes it easier for researchers to obtain the whole picture on a particular experiment.

DNA array (also called microarray or DNA chip) is an orderly arrangement of hundreds of thousands of spots, each containing many copies of the same probe (see Figure 1.18). When the array is exposed to the sample, DNA fragments in the sample and the probes on the array will match based on the hybridization rule: A matches to T while C matches to G through hydrogen

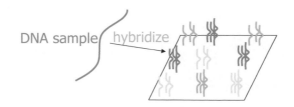

FIGURE 1.18:   This figure shows an array with 9 probes. Each probe is the reverse complement of some short DNA fragment which forms a finger-print (that is, unique identifier) of a certain target DNA sequence. When the array is exposed to a certain DNA sample, some probes will light-up if the corresponding target DNA sequences exist in the sample. (See color insert after page 270.)

bonds. This idea allows us to detect what DNA sequences appear in the sample.

1.7.6.2   Application of DNA Arrays

DNA array can help researchers to investigate the expression profile of a cell. Through spotting the complementary DNA of a short unique segment of each gene on the DNA array, we can monitor the concentrations of different genes within a cell.

DNA arrays can also be used for sequencing. The basic idea is that if the DNA array can tell us all probes that appear in the target sequence, then, the target sequence can be reconstructed from overlapping of probes by a computational method. However, due to the presence of repetitive DNA, the method is unlikely to be used in sequencing complex genomes, but it is useful for short non-repetitive DNA fragments.

There are many other applications where DNA arrays can be put to good use.

1.7.7   Next Generation DNA Sequencing

Although gel electrophoresis (like Sanger sequencing) is widely used, it has limitations. First, the cost of Sanger sequencing is high. To sequence a human genome, the cost is in millions of US dollars. Second, the throughput of Sanger sequencing is low. It takes months to generate enough sequences.

Starting in 2005, a number of technologies were developed to resolve the weakness of Sanger sequencing, including 454 life science, Solexa, and ABI SOLiD. They allow us to get tens of billions of DNA bases in a few days.

1.7.7.1 454 Pyrosequencing Technology

Pyrosequencing was first developed by Ronaghi et al.[249] in 1996. It is based on "sequencing by synthesis". Given a single-strand DNA template, the complementary strands can be synthesized one by one by sequentially adding A, C, G, and T. With the help of chemiluminescent enzyme, light is produced when the complementary nucleotide is added in the DNA template. Through detecting the light, we can reconstruct the template sequence.

454 Life Science licensed this technology and developed the first next-generation sequencing system in 2005. The technology has three steps. First, a set of single-stranded template DNA is loaded on a set of beads, which are arranged in order on a glass plate. Second, by the emulsion-based PCR method (emPCR), the DNAs on beads are amplified. Third, via pyrosequencing, we read the DNA sequences on beads in parallel. Currently, 454 technology allows us to generate 1 million reads of length 400 bases per 10 hours. The cost is about US$60 per million bases.

1.7.7.2 Illumina Solexa Technology

In November 2006, Illumina acquired Solexa and commercialized the genome analyzer, which allows us to sequence billions of DNAs in a few days. Unlike 454 technology which generates a long read, Illumina Solexa technology generates a short read of length 50 bases.

The Illumina Solexa technology has four steps. First, a set of single-stranded template DNAs is prepared. Second, the two ends of the template DNAs are randomly fixed on the surface of a flow cell. Third, through bridge PCR, the template DNAs are amplified. Fourth, using a four-color fluorescent dye and a polymerase-mediated primer extension reaction, we read the template DNAs on the flow cell in parallel. Currently, in 3 days, the Illumina Solexa technology can generate about 80 million reads, each of length 50 bases. In total, it generates about 4 billion bases per run. The cost is about US$2 per million bases.

1.7.7.3 ABI SOLiD Technology

SOLiD stands for the Supported Oligonucleotide Ligation and Detection Platform. It was introduced in 2007.

SOLiD consists of four steps. First, a set of single-stranded template DNAs is loaded on a set of beads and amplified by the emulsion-based PCR method (emPCR). Second, the beads are randomly distributed on a glass plate instead of in order as in 454. Such a change enables SOLiD to generate a higher density sequencing array. Third, via sequencing by ligation, the DNA sequences on beads are read in parallel.

Currently, SOLiD can generate about 115 million reads, each of length 35 bases in 5 days. In total, it generates about 4 billion bases per run. The cost is about US$2 per million bases.

1.8   Brief History of Bioinformatics

In 1866, Gregor Johann Mendel discovered genetics. Mendel's hybridization experiments on peas unveiled some biological elements called genes, which pass information from generation to generation. At that time, people thought genetic information was carried by some "chromosomal protein"; however, it is not.

Later, in 1869, DNA was discovered. But it was not until 1944 that Avery and McCarty demonstrated DNA is the major carrier of genetic information, not protein. Remarkable as it is, this discovery is often referred to as the start of bioinformatics. In 1953, another historic discovery enabled great advances in both biology and bioinformatics: James Watson and Francis Crick deduced the three dimensional structure of DNA, which is a double helix.

Later, in 1961, the mapping from DNA to peptide (protein), namely, the genetic code, was elucidated by Marshall Nirenberg. It is by the means of combining three nucleotides in the DNA as a codon, and mapping each of them to one amino acid in the peptide.

In 1968, the restriction enzyme was discovered and isolated from bacteria. These enzymes protect the bacteria by cutting any foreign DNA molecules at specific sites so as to restrict the ability of the foreign DNA molecules to take over the transcription and translation machinery of the bacterial cell.

Starting in the 1970s, several important biotechnology techniques were developed. First, DNA sequencing techniques like electrophoresis were developed. These enabled the identification of DNAs given just a tissue found on a human body. Moreover, in 1985, the groundbreaking technique polymerase-chain-reaction (PCR) was invented. By exploiting natural replication, DNA samples can be easily amplified using PCR, so that the amount is enough for doing experiments.

In 1986, RNA splicing in eukaryotes was discovered. This is the process of removing introns and rejoining the exons in order to produce a functional mRNA from a pre-mRNA. The splicing process is performed by a spliceosome, which consists of a group of smaller RNA-protein complexes known as snRNPs and additional proteins.

In 1998, Fire and Mello discovered a post-transcription control mechanism called RNA interference. This process prevents the translation of RNA to protein.

Starting in the 1980s, scientists began to sequence the genomes. From 1980 to 1990, complete sequencing of the genomes of various organisms, like that of the *E. coli*, was done successfully. The most remarkable event was probably the launch of the Human Genome Project (HGP) in 1990. Originally, it was planned to be completed in 15 years; however, thanks to shotgun sequencing technology, Craig Venter and Francis Collins jointly announced the publication of the first draft of the human genome in 2000. Subsequently, a more refined human genome was also published in 2003.

Starting in 2006, the second generation sequencing technology became available. We can sequence tens of billions of DNA bases within a few days.

Triggered by the Human Genome Project, many large scale international collaboration projects appeared, which study our genome in a high-throughput manner. Some projects include the Genomes to Life (GTL) project, the ENCODE project, and the HAPMAP project. The objective of the GTL project was to understand the detailed mechanism of cells, for instance, identifying the proteins involved in sustaining critical life functions and characterizing the gene regulatory networks that regulate the expression of those proteins. The ENCODE project aimed to annotate the whole human genome, that is, locate all the genes and functional elements in our human genome. The HAPMAP project would like to study the genetic variation among different individuals. They hope to understand the differences in genetic data from different populations. They also wanted to understand the relationship between our genome and diseases.

The next 10 to 20 years in bioinformatics will be challenging and fast moving.

1.9 Exercises

1. Can you visit the gene bank and check a number of protein sequences? What are the first amino acids in those protein sequences? Any explanation for your observation?

2. Within a Eukaryote's cell, how many copies of nuclear DNA and mitochondrial DNA do we have?

3. Are genes randomly distributed in our genome? For example, can you check the distribution of the Hox genes in human?

4. Different species encodes the same amino acid using different codons. For the amino acid Q, can you check the most frequent used codons in Drosophila melanogaster, human, and Saccharomyces cerevisiae? Is there any codon bias for different species?

5. For the following mRNA sequence, can you extract its 5' UTR, 3' UTR and the protein sequence?

ACTTGTCATGGTAACTCCGTCGTACCAGTAGGTCATG

Chapter 2

Sequence Similarity

2.1 Introduction

The earliest research in sequence comparison can be dated back to 1983, when Doolittle et al. [82] searched for platelet-derived growth factor (PDGF) in their database. They found that PDGF is similar to v-sis onc gene.

```
PDGF-2 01 ------SLGSLTIAEPAMIAECKTREEVFCICRRL?DR??  34
p28sis 61 LARGKRSLGSLSVAEPAMIAECKTRTEVFEISRRLIDRTN 100
```

At that time, the function of v-sis onc gene was still unknown. Based on the similarity between v-sis onc gene and PDGF, they claimed that the transforming protein of the primate sarcoma virus and the platelet-derived growth factor are derived from the same or closely related cellular genes. This was later proven by scientists. Research conducted by Riordan et al. [244] showed a similar result. They tried to understand the cystic fibrosis gene using multiple sequence alignment in 1989. They showed that similar gene sequences did imply similar functionalities.

Motivated by the above examples, there is a well-known conjecture in biology: If any two protein (or DNA, or RNA) sequences are similar, they will have similar functions or 3D structures. Bioinformaticians often compare the similarity between two biological sequences to understand their structures or functionalities. Below, we show some example applications.

- Predicting the biological function of a gene (or an RNA or a protein): If a gene is similar to some gene with known function, we can conjecture that this gene also has the same function.

- Finding the evolution distance: Species evolve through modifying their genomes. By measuring the similarity of the genomes, we could know their evolution distances.

- Helping genome assembly: Current technology can only decode short DNA sequences. Based on the overlapping information of a huge amount of short DNA pieces, the Human Genome Project reconstructs the whole human genome. The overlapping information is obtained by sequence comparison.

- Finding a common region in two genomes: If two relatively long strings from two genomes are similar or even identical, they probably represent genes that are homologous (i.e., genes evolved from the same ancestor and with common function).

- Finding repeats within a genome: The human genome is known to have many similar substrings which are known as repeats. Sequence comparison methods can help to identify them.

As a matter of fact, the reverse of the conjecture may not always be true. For example, the hemoglobin of *V. stercoraria* (bacterial) and that of *P. marinus* (eukaryotic) are similar in structure and function; their protein sequences share just 8% sequence identity.

There are many different sequence comparison problems depending on the objective functions (or the definitions of similarity). In this chapter, we will discuss the following three alignment problems: global alignment, local alignment, and semi-global alignment. We also study the alignment problems under different gap penalty functions. Finally, we discuss how similarity matrices like PAM and BLOSUM are generated.

2.2   Global Alignment Problem

Before describing the global alignment problem, we first revisit a well-known computer science problem, the string edit problem. The string edit problem computes the minimum number of operations to transform one string to another, where the operations are:

1. Replace a letter with another letter.

2. Insert a letter into a sequence.

3. Delete a letter from a sequence.

For example, given two strings $S =$ `interestingly` and $T =$ `bioinformatics`, S can be transformed to T using six replacements, three insertions, and two deletions. The minimum number of operations required to transform S to T is called the edit distance. For the above example, the edit distance is 11.

In general, we can associate a cost to every edit operation. Let Σ be the alphabet set and '$_$' be a special symbol representing a null symbol. The cost of each edit operation can be specified by a matrix σ where $\sigma(x, y)$ equals the cost of replacing x by y, for $x, y \in \Sigma \bigcup \{_\}$. Note that $\sigma(_, x)$ and $\sigma(x, _)$ denote the cost of insertion and deletion, respectively. Then, the cost of a set of edit operations E equals $\sum_{(x,y) \in E} \sigma(x, y)$. For example, if the costs of mismatch, insert, and delete are 1, 2, and 2, respectively, then the total cost to transform `interestingly` to `bioinformatics` is 16.

Unlike edit distance which measures the difference between two sequences, we sometimes want to know the similarity of two sequences. The global alignment problem is aimed at this purpose. First, let's give a formal definition of alignment.

DEFINITION 2.1 *An alignment of two sequences is formed by inserting spaces in arbitrary locations along the sequences so that they end up with the same length and there are no two spaces at the same position of the two augmented sequences.*

For example, a possible alignment of S and T is as follows:

```
-i--nterestingly
bioinformatics--
```

Two sequences are similar if their alignment contains many positions having the same symbols while minimizing the number of positions that are different. In general, we can associate a similarity score with every pair of aligned characters. Let Σ be the alphabet set and _ be a special symbol representing a null symbol. The similarity of a pair of aligned characters can be specified by a matrix δ, where $\delta(x, y)$ equals the similarity of x and y for $x, y \in \Sigma \bigcup \{\_\}$. Figure 2.1 shows an example of a similarity matrix. A more in-depth study of similarity matrices is given in Section 2.6.

The aim of the global alignment problem is to find an alignment A which maximizes $\sum_{(x,y) \in A} \delta(x, y)$. Such alignment is called the optimal alignment.

There is a one-to-one correspondence between a pair of aligned characters and an edit operation. When a pair of aligned characters are the same, it is called a *match*; otherwise it is called a *mismatch*. When a space is introduced in the first sequence, it is called an *insert* while a space in the second sequence is called a *delete*. For the above example, the alignment corresponds five *matches*, six *mismatches*, three *inserts*, and two *deletes*.

The global alignment problem and the string edit distance problem are in fact a dual problem. The lemma below shows that the solutions reported by them are in fact the same.

LEMMA 2.1

Let σ be the cost matrix of the edit distance problem and δ be the score matrix of the global alignment problem. If $\delta(x, y) = -\sigma(x, y)$ for all $x, y \in \Sigma \bigcup \{\_\}$, the solution to the edit distance problem is equivalent to the solution to the string alignment problem.

PROOF   Let (x, y), $x, y \in \Sigma \cup \{\_\}$, be an operation that transforms x to y. Recall that insertion of symbol y is $(\_, y)$ and deletion of x is $(x, \_)$. The cost of an operation (x, y) in the edit distance problem is $\sigma(x, y)$ and the score of an operation (x, y) in the string alignment problem is $\delta(x, y)$.

Given any alignment of two sequences. For any operation (x, y), let n_{xy} be the number of occurrences of the operation (x, y).

Then, the edit distance of the alignment is:

$$\sum_{x \in \Sigma \cup \{\_\}} \sum_{y \in \Sigma \cup \{\_\}} n_{x,y} \sigma(x, y)$$

The alignment score of the alignment is:

$$\sum_{x \in \Sigma \cup \{\_\}} \sum_{y \in \Sigma \cup \{\_\}} n_{x,y} \delta(x, y)$$

Because $\delta = -\sigma$, minimizing the edit distance is equivalent to maximizing the alignment score.   ☐

Although string edit and global alignment are equivalent, people in computational biology prefer to use global alignment to measure the similarity of DNA, RNA, and protein sequences.

Before we end this section, we give an example of global alignment. Consider the similarity matrix for $\{\_, A, C, G, T\}$ in Figure 2.1, where $\delta(x, y) = 2, -1, -1, -1$ for match, mismatch, delete, and insert, respectively. One possible alignment of two DNA sequences $S = ACAATCC$ and $T = AGCATGC$ is shown below.

$$S = \text{A-CAATCC}$$
$$T = \text{AGCA-TGC}$$

The above alignment has five matches, one mismatch, one insert, and one delete. Thus, the similarity score of this alignment is 7 ($2 * 5 - 1 - 1 - 1 = 7$). We can check that this alignment has the maximum score. Hence, it is an optimal alignment. Note that S and T may have more than one optimal alignment. For the example, another optimal alignment is as follows.

$$S = \text{A-CAATCC}$$
$$T = \text{AGC-ATGC}$$

2.2.1   Needleman-Wunsch Algorithm

Consider two strings $S[1..n]$ and $T[1..m]$. To compute the optimal global alignment between S and T, the brute-force method generates all possible alignments and report the alignment with the maximum alignment score. Such an approach, however, takes exponential time. This section introduces

-	-	A	C	G	T
-		-1	-1	-1	-1
A	-1	2	-1	-1	-1
C	-1	-1	2	-1	-1
G	-1	-1	-1	2	-1
T	-1	-1	-1	-1	2

FIGURE 2.1:   Example of a similarity score matrix.

the Needleman-Wunsch algorithm [218] which applies dynamic programming to find the optimal global alignment between S and T in $O(nm)$ time. Define $V(i,j)$ to be the score of the optimal alignment between $S[1..i]$ and $T[1..j]$. We devise the recursive formula for $V(i,j)$ depending on two cases: (1) either $i = 0$ or $j = 0$ and (2) both $i > 0$ and $j > 0$.

For case (1), when either $i = 0$ or $j = 0$, we will align a string with an empty string. In this case, we have either insertions or deletions. Hence, we have the following equations.

$V(0,0) = 0$

$V(0,j) = V(0,j-1) + \delta(\_, T[j])$ Insert j times

$V(i,0) = V(i-1,0) + \delta(S[i], \_)$ Delete i times

For case (2), when both $i > 0$ and $j > 0$, we observe that, in the best alignment between $S[1..i]$ and $T[1..j]$, the last pair of aligned characters should be either match/mismatch, delete, or insert. To get the optimal score, we choose the maximum value among these three cases. Thus, we get the following recurrence relation:

$$V(i,j) = \max \begin{cases} V(i-1,j-1) + \delta(S[i], T[j]) & match/mismatch \\ V(i-1,j) + \delta(S[i], \_) & delete \\ V(i,j-1) + \delta(\_, T[j]) & insert \end{cases}$$

The optimal alignment score of $S[1..n]$ and $T[1..m]$ is $V(n,m)$. This score can be computed by filling in the table $V(1..n, 1..m)$ row by row using the above recursive equations. Figure 2.2 shows the table V of the two strings $S = ACAATCC$ and $T = AGCATGC$. For example, the value in the entry $V(1,1)$ is obtained by choosing the maximum of $\{0+2, -1-1, -1-1\}$. The score of the optimal alignment is obtained at the bottom right corner of the table $V(7,7)$ which is 7.

To recover the optimal alignment, for every entry, we draw arrows to indicate the ways to get the corresponding values. We draw a diagonal arrow, a horizontal arrow, or a vertical arrow for the entry $V(i,j)$ if $V(i,j)$ equals $V(i-1,j-1) + \delta(S[i], T[j])$, $V(i,j-1) + \delta(\_, T[j])$, or $V(i-1,j) + \delta(S[i], \_)$, respectively. Figure 2.2 shows an example. In the example, since the value at $V(1,1)$ is obtained by $V(0,0) + 2$, we draw a diagonal arrow. For the entry

_	_	A	G	C	A	T	G	C
_	0	-1	-2	-3	-4	-5	-6	-7
A	-1	2	1	0	-1	-2	-3	-4
C	-2	1	1	3	2	1	0	-1
A	-3	0	0	2	5	4	3	2
A	-4	-1	-1	1	4	4	3	2
T	-5	-2	-2	0	3	6	5	4
C	-6	-3	-3	0	2	5	5	7
C	-7	-4	-4	-1	1	4	4	7

FIGURE 2.2: The dynamic programming table for running the Needleman-Wunsch algorithm for two strings $S = ACAATCC$ and $T = AGCATGC$.

$V(3, 2)$, there are two ways to get the maximum value. Hence, we draw both diagonal and vertical arrows. The optimal alignment is obtained by back-tracing the arrows from $V(7,7)$ back to $V(0,0)$. If the arrow is diagonal, two characters will be aligned. If it is horizontal or vertical, a deletion or an insertion, respectively, will be present in the alignment. In Figure 2.2, the optimal alignment score is $V(7,7) = 7$. Through back-tracing from $V(7,7)$ to $V(0,0)$, the path is $0 \leftarrow 1 \leftarrow 3 \leftarrow 5 \leftarrow 4 \leftarrow 6 \leftarrow 5 \leftarrow 7$. The corresponding optimal alignment is $A - CAATCC$ and $AGCA - TGC$. The optimal alignment is not unique. We may get another optimal alignment: $A - CAATCC$ and $AGC - ATGC$.

Finally, we analyze the time and space complexity of the Needleman-Wunsch algorithm. The Needleman-Wunsch algorithm fills in nm entries of the table $V(1..n, 1..m)$. Each entry requires $O(1)$ word space and can be computed in $O(1)$ time. Hence, the Needleman-Wunsch algorithm computes the optimal global alignment of $S[1..n]$ and $T[1..m]$ using $O(nm)$ time and $O(nm)$ space.

2.2.2   Running Time Issue

As shown in the previous section, the Needleman-Wunsch algorithm takes $O(nm)$ time to solve the global alignment problem. In other words, using an 1 GHz computer, it takes hundreds of years to align the mouse and the human genomes. Can we improve the time complexity?

The current best method runs in $O(nm/\log n)$ time. The method was proposed by Masek and Paterson [208] in 1980. It is based on the Four-Russian's paradigm. Since then, there has been no further improvement. On the other hand, two lower bound results exist.

- Aho, Hirschberg, and Ullman (1976) [4]: If we can only compare whether two symbols are equal or not, it takes $\Omega(nm)$ time to compute the global alignment.

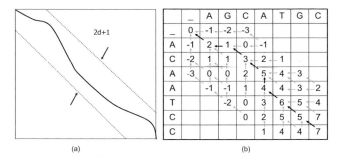

FIGURE 2.3: (a) An illustration of a $2d + 1$ band of a dynamic programming table. (b) A $2d + 1$ band example for the dynamic programming table in Figure 2.2 with $d = 3$.

- Hirschberg (1978) [148]: If symbols are ordered and can be compared, it takes $\Omega(n \log n)$ time to compute the global alignment.

These works imply that the global alignment problem is difficult to improve. Hence, people try to identify cases where the global alignment problem can be solved efficiently. One special case is that we restrict the maximum number of insertions or deletions (by shorthand, we use indel to denote an insertion or a deletion) in the alignment to be d. Obviously, $0 < d \le n + m$. Recall that, in the table V, an insertion corresponds to a horizontal arrow and a deletion corresponds to a vertical arrow. Hence, the alignment should be inside the $(2d + 1)$ band if the number of indels is at most d (see Figure 2.3). It is unnecessary to execute the Needleman Wunsch algorithm to fill in the lower and upper triangles in the table V. An algorithm which fills in only the middle band is called the banded Needleman-Wunsch alignment. For time analysis, note that the area of the $(2d + 1)$ band in the table V is $nm - (n - d)(m - d) = md + nd - d^2$. Since the time for filling in every entry inside the band is $O(1)$, the running time of the banded Needleman-Wunsch algorithm is $O((n + m)d)$.

Chapter 4 will discuss other efficient heuristic methods for aligning two long DNA sequences.

2.2.3 Space Efficiency Issue

Note that the Needleman-Wunsch algorithm requires $O(mn)$ space as the table V has nm entries. When we compare two very long sequences, memory space will be a problem. For example, if we compare the human genome with the mouse genome, the memory space is at least 9×10^{18} words, which is not feasible in real life. Can we solve the global alignment problem using less space?

Recall that, in the Needleman-Wunsch algorithm, the value of an entry $V(i, j)$ depends on three entries $V(i - 1, j - 1)$, $V(i - 1, j)$, and $V(i, j - 1)$.

Algorithm FindMid$(S[1..n], T[1..m])$

Require: Two sequences $S[1..n]$ and $T[1..m]$

Ensure: A mid-point $j \in 1..m$

1: Perform a cost-only Needleman-Wunsch algorithm to compute $AS(S[1..n/2], T[1..j])$ for all j;

2: Perform a cost-only Needleman-Wunsch algorithm to compute $AS(S[n/2 + 1..n], T[j + 1..m])$ for all j;

3: Determine the mid-point j which maximizes the sum using Equation 2.1;

FIGURE 2.4:   The algorithm FindMid.

Hence, when we fill in the i-th row in the table V, only the values in the $(i-1)$-th row are needed.

Therefore, if we just want to compute the optimal alignment score, it is unnecessary to store all values in the table V. Precisely, when we fill in the i-th row of the table V, we only keep the values in the $(i-1)$-th row. In this way, the space complexity becomes $O(m)$. This method is called the cost-only Needleman-Wunsch algorithm.

Although the space complexity is reduced from $O(nm)$ to $O(m)$, we only compute the optimal alignment score. Can we also reconstruct the optimal alignment? The answer is YES! Below, we present Hirschberg's algorithm [147] which computes the optimal alignment of two strings $S[1..n]$ and $T[1..m]$ in $O(n+m)$ space.

The main observation is that the optimal alignment score of $(S[1..n], T[1..m])$ is the sum of (1) the optimal alignment score of $(S[1..n/2], T[1..j])$ and (2) the optimal alignment score of $(S[n/2+1..n], T[j+1..m])$ for some j. Equation 2.1 illustrates this idea. Note that $AS(A, B)$ denotes the optimal alignment score between sequences A and B.

$$AS(S[1..n], T[1..m]) =$$
$$\max_{1 \leq j \leq m} \{AS(S[1..n/2], T[1..j]) + AS(S[n/2 + 1..n], T[j + 1..m])\} \quad (2.1)$$

The integer j, which maximizes the sum, is called the mid-point. The algorithm FindMid in Figure 2.4 describes how to compute the mid-point using the cost-only Needleman-Wunsch algorithm.

Figure 2.5 gives an example demonstrating how the algorithm FindMid computes the mid-point. Step 1 fills in the first half of the table to compute $AS(S[1..n/2], T[1..j])$ for all j. Then, Step 2 fills in the second half of the table in reverse to compute $AS(S[n/2 + 1..n], T[j + 1..m])$ for all j. Finally, Step 3 computes the score $AS(S[1..n/2], T[1..j]) + AS(S[n/2 + 1..n], T[j + 1..m])$, for $j = 0, 1, \ldots, m - 1$, by getting the m sums of the m diagonal pairs from the two middle rows. Then, we obtain the mid-point j which corresponds to the maximum sum. Here the maximum score is 7, which is the sum of 4 and 3 and the position $j = 4$ is determined to be the mid-point.

	_	A	G	C	A	T	G	C	_
_	0	-1	-2	-3	-4	-5	-6	-7	
A	-1	2	1	0	-1	-2	-3	-4	
C	-2	1	1	3	2	1	0	-1	
A	-3	0	0	2	5	4	3	2	
■									
T		-1	0	1	2	3	0	0	-3
C		-2	-1	1	-1	0	1	1	-2
C		-4	-3	-2	-1	0	1	2	-1
_		-7	-6	-5	-4	-3	-2	-1	0

FIGURE 2.5: Mid-point example. We split the dynamic programming table into two halves. For the top half, we fill in the table row by row in top-down order. Then, we obtain $AS(S[1..n/2], T[1..j])$ for all j. For the bottom half, we fill in the table row by row in bottom-up order. Then, we obtain $AS(S[n/2+1..n], T[j+1..m])$ for all j. Finally, the mid-point is computed by Equation 2.1.

Algorithm Alignment($S[i_1..i_2], T[j_1..j_2]$)

Require: Two sequences $S[i_1..i_2]$ and $T[j_1..j_2]$

Ensure: The optimal alignment between $S[i_1..i_2]$ and $T[j_1..j_2]$.

1: If $i_1 = i_2$, compute the alignment using the Needleman-Wunsch algorithm and report it;
2: Let $mid = (i_1 + i_2)/2$;
3: $j = \text{FindMid}(S[i_1..i_2], T[j_1..j_2])$;
4: Concatenate the alignments computed by Alignment($S[i_1..mid], T[j_1..j]$) and Alignment($S[mid+1..i_2], T[j+1..j_2]$) and report it;

FIGURE 2.6: The recursive algorithm for computing the global alignment of $S[i_1..i_2]$ and $T[j_1..j_2]$ using linear space.

If we divide the problem into two halves based on the mid-point and recursively deduce the alignments for the two halves, we can compute the optimal alignment while using $O(n+m)$ word space. The detailed algorithm is shown in Figure 2.6 and the idea is illustrated in Figure 2.7.

Below, we analyze the time and the space complexities. For algorithm FindMid, both steps 1 and 2 take $O(nm/2)$ time. Step 3 finds the maximum of m sums, which requires $O(m)$ time. In total, FindMid takes $O(nm)$ time.

For algorithm Alignment, let the running time of Alignment($S[1..n], T[1..m]$) be $Time(n, m)$. $Time(n, m) =$ time for finding the mid-point + time for solving the two subproblems $= O(nm) + Time(n/2, j) + Time(n/2, m - j)$. By solving the recursive equation, the time complexity is $Time(n, m) = O(nm)$.

For space complexity, the working memory for finding the mid-point takes $O(m)$ space. Once we find the mid-point, we can free the working memory.

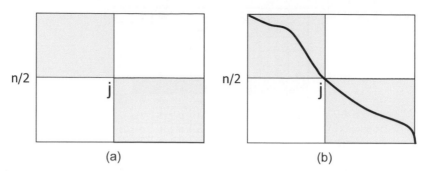

(a)   (b)

FIGURE   2.7: (a)   When   we   call   Alignment($S[1..n], T[1..m]$),
we   first   compute   the   mid-point j and   generate   two   subproblems
Alignment($S[1..n/2], T[1..j]$)   and   Alignment($S[n/2 + 1..n], T[j + 1..m]$).
(b) By recursion, we obtain the alignments for the two subproblems. By
concatenating the two alignments, we obtain the alignment of $S[1..n]$ and
$T[1..m]$.

In each recursive call, we only need to store the alignment path. Therefore
the space complexity is $O(m + n)$.

2.2.4   More on Global Alignment

There are two special cases of the global alignment problem. Both of them
can be solved using a similar dynamic programming approach with specific
score function:

1. Longest Common Subsequence (LCS): Given two sequences X and Y,
   a sequence Z is said to be a common subsequence of X and Y if Z
   is a subsequence of both X and Y. The LCS problem aims to find a
   maximum-length common subsequence of X and Y. For example, the
   LCS of "catpaplte" and "xapzpleg" is "apple," which is of length 5. The
   LCS of two strings is equivalent to the optimal global alignment with
   the following scoring function.

   - score for match $= 1$
   - score for mismatch $= -\infty$ (that is, we don't allow mismatch)
   - score for insert/delete $= 0$

   Using the same algorithm for the global alignment, LCS can be computed in $O(nm)$ time. Section 4.3.2 will discuss a case where all characters are distinct. In such a case, LCS can be computed in $O(n \log n)$
   time.

2. Hamming Distance: The Hamming distance is the number of positions
   in two strings of equal length for which the corresponding symbols are

different. For example, the Hamming distance between two strings "toned" and "roses" is 3. It is equivalent to the optimal global alignment with the following scoring function.

- score for match $= 1$
- score for mismatch $= 0$
- score for indel $= -\infty$ (that is, we don't allow any indel)

Moreover, since the number of indels allowed is $d = 0$, the Hamming distance of two length-n strings can be computed in $O((2d+1)n) = O(n)$ time and space as discussed in Section 2.2.2.

2.3 Local Alignment

Global alignment is used to align entire sequences. Sometimes we are interested in finding the most similar substring pair of the two input sequences. Such a problem is called the local alignment problem. Precisely, given two strings $S[1..n]$ and $T[1..m]$, a local alignment is a pair of substrings A of S and B of T with the highest alignment score.

By brute force, the local alignment can be computed using the following three steps.

1: Find all substrings A of S and B of T;
2: Compute the global alignment of A and B;
3: Return the substring pair (A, B) with the highest score.

Since there are $\binom{n}{2}$ choices for A and $\binom{m}{2}$ choices for B, the time complexity of this method is $O(\binom{n}{2}\binom{m}{2}nm) = O(n^3 m^3)$.

The above algorithm is too slow. In 1981, Smith and Waterman [267] proposed a better solution to the local alignment problem.

Similar to the global alignment, the Smith-Waterman algorithm computes the optimal local alignment using dynamic programming. Consider two strings $S[1..n]$ and $T[1..m]$. We define $V(i,j)$ be the maximum score of the global alignment of A and B over all substrings A of S end at i and all substrings B of T end at j, where $1 \leq i \leq n$ and $1 \leq j \leq m$. Note that we assume an empty string is also a substring of S ends at i and a substring of T ends at j. By definition, the score of the optimal local alignment is $\max_{1 \leq i \leq n, 1 \leq j \leq m} V(i,j)$.

We devise the recursive formula for $V(i,j)$ depending on two cases: (1) either $i = 0$ or $j = 0$ and (2) both $i > 0$ and $j > 0$.

For case (1), when either $i = 0$ or $j = 0$, the best alignment aligns the empty strings of S and T. Hence, we have the following equations.

$$V(i, 0) = 0 \text{ for } 0 \leq i \leq n$$
$$V(0, j) = 0 \text{ for } 0 \leq j \leq m$$

	_	C	T	C	A	T	G	C
_	0	0	0	0	0	0	0	0
A	0	0	0	0	2	1	0	0
C	0	2	1	2	1	1	0	2
A	0	1	1	1	4	3	2	1
A	0	0	0	0	3	3	2	1
T	0	0	2	1	2	5	4	3
C	0	2	1	4	3	4	4	6
G	0	1	1	3	3	3	6	5

FIGURE 2.8:   The V table for local alignment between $S = ACAATCG$ and $T = CTCATGC$.

For case (2), when both $i > 0$ and $j > 0$, there are two scenarios. The first scenario is that the best alignment aligns empty strings of S and T. In this case, $V(i, j) = 0$. For another scenario, within the best alignment between some substring of S ends at i and some substring of T ends at j, the last pair of aligned characters should be either match/mismatch, delete, or insert. To get the optimal score, we choose the maximum value among 0 and these three cases. Thus, we get the following recurrence relation:

$$V(i, j) = \max \begin{cases} 0 & \text{al} ign \text{ empty strings} \\ V(i - 1, j - 1) + \delta(S[i], T[j]) & \text{m} atch/mismatch \\ V(i - 1, j) + \delta(S[i], \_) & \text{d} elete \\ V(i, j - 1) + \delta(\_, T[j]) & \text{i} nsert \end{cases}$$

The optimal local alignment score is $\max_{i,j} V(i, j)$. Smith and Waterman proposed that this score can be computed by filling in the table V row by row using the above recursive equations. For example, consider $S = ACAATCG$ and $T = CTCATGC$. Assume a match score is $+2$ and an insert/delete score is -1. Figure 2.8 shows the table V. The maximum score in the table V is 6. Thus, the optimal local alignment score is 6. Through back-tracing from $V(7, 6)$, we can recover the optimal alignment, which corresponds to the path in Figure 2.8:

```
CAATCG
C-AT-G
```

Below, we analyze the time and space complexities of the Smith-Waterman algorithm. The algorithm fills in nm entries in the table $V(1..n, 1..m)$. Each entry can be computed in $O(1)$ time. The maximum local alignment score equals the entry with the maximum value. So, both the time and space complexity are $O(nm)$. Finally, using additional $O(n + m)$ time, we can recover the alignment by back-tracing.

Similar to the optimal global alignment, the space complexity for computing the optimal local alignment can be reduced from $O(mn)$ to $O(n+m)$ (see Exercise 12).

2.4   Semi-Global Alignment

The earlier sections discussed two kinds of sequence alignment: global alignment and local alignment. Now we introduce another type of alignment known as semi-global alignment. Semi-global alignment is similar to global alignment, in the sense that it tries to align two sequences as a whole. The difference lies in the way it scores alignments. Semi-global alignment ignores spaces at the beginning and/or the end of an alignment, while global alignment does not differentiate spaces that are sandwiched within adjacent characters and spaces that precede or succeed a sequence. In other words, semi-global alignment assigns no cost to spaces that appear before the first character and/or after the last character.

To better appreciate the motivation behind semi-global alignment, let's consider the following example. Suppose we have two sequences S and T below:

$$S = \texttt{ATCCGAACATCCAATCGAAGC}$$
$$T = \texttt{AGCATGCAAT}$$

If we compute the optimal global alignment, we get:

```
ATCCGAACATCCAATCGAAGC
A---G--CATGCAAT------
```

Since the alignment has nine matches (score = 18), one mismatch (score = -1), and eleven deletions (score = -11), the score of the alignment is 6. However, such alignment might not be desired. One might wish to disregard flanking (i.e., starting or trailing) spaces. Such a feature is desirable, for example, in aligning an exon to the original gene sequence. Spaces in front of the exon might be attributed to an untranslated region (UTR) or introns and should not be penalized. This method is also used in aligning genes on a prokaryotic genome.

Coming back to our example, the optimal semi-global alignment for S and T would be:

```
ATCCGAA-CATCCAATCGAAGC
------AGCATGCAAT------
```

The alignment has eight matches (score=16), one deletion (score=-1), and one mismatch(score=-1), which gives an alignment score of 14 instead of 6 in global alignment.

Another example of semi-global alignment is that we may ignore the starting spaces of the first sequence and the trailing spaces of the second sequence, as in the alignment below. This type of alignment finds application in sequence assembly. Depending on the goodness of the alignment, we can deduce whether the two DNA fragments are overlapping or disjoint.

```
---------ACCTCACGATCCGA
TCAACGATCACCGCA--------
```

Modifying the algorithm for global alignment to perform semi-global alignment is quite straightforward. The idea is that we give a zero score instead of a minus score to those spaces in the beginning that are not charged. To ignore spaces at the end, we choose the maximum value in the last row or the last column. Figure 2.9 summarizes the charging scheme.

Spaces that are not charged	Action
Spaces in the beginning of S	Initialize first row with zeros
Spaces in the ending of S	Look for the maximum in the last row
Spaces in the beginning of T	Initialize first column with zeros
Spaces in the end of T	Look for maximum in the last column

FIGURE 2.9:   Charging scheme of spaces in semi-global alignment.

2.5   Gap Penalty

A gap in an alignment is defined as a maximal substring of contiguous spaces in any sequence of alignment. By definition, the following alignment has 2 gaps and 8 spaces.

```
A-CAACTCGCCTCC
AGCA-------TGC
```

Previous sections assumed the penalty for indel is proportional to the length of a gap. (Such gap penalty model is known as linear gap penalty model.) This assumption may not be valid in the biology domain. For example, mutation may cause insertion/deletion of a substring instead of a single base. This kind of mutation may be as likely as the insertion/deletion of a single base. Another example is related to mRNA. Recall that mRNA misses the introns during splicing. When aligning mRNA with its gene, the penalty should not be proportional to the length of the gaps.

Hence, it is natural not to impose a penalty that is strictly proportional to the length of the gap. Such a scheme is also preferable in situations when we expect spaces to appear contiguously.

2.5.1 General Gap Penalty Model

In general, if we define the penalty of a gap of length q as $g(q)$, then we can align $S[1..n]$ and $T[1..m]$ using the following dynamic programming algorithm. Let $V(i, j)$ be the global alignment score between $S[1..i]$ and $T[1..j]$. We derive the recursive formula for V depending on two cases: (1) $i = 0$ or $j = 0$ and (2) $i > 0$ and $j > 0$.

When $i = 0$ or $j = 0$, we either insert or delete one gap. Hence, we have the following equations.

$$V(0,0) = 0$$
$$V(0,j) = -g(j)$$
$$V(i,0) = -g(i)$$

When $i > 0$ and $j > 0$, $V(i, j)$ can be computed by the following recursive equation.

$$V(i,j) = \max \begin{cases} V(i-1,j-1) + \delta(S[i], T[j]) & \text{match/mismatch} \\ \max_{0 \le k \le j-1}\{V(i,k) - g(j-k)\} & \text{Insert } T[k+1..j] \\ \max_{0 \le k \le i-1}\{V(k,j) - g(i-k)\} & \text{Delete } S[k+1..i] \end{cases}$$

To compute the optimal global alignment score under the general gap penalty, we just fill in the table V row by row. Then, the optimal alignment score is stored in $V(n, m)$. The optimal alignment can be recovered by back-tracing the dynamic programming table V.

Below, we analyze the time and space complexities. We need to fill in nm entries in the table $V(1..n, 1..m)$. Each entry can be computed in $O(n + m)$ time according to the recursive equation. The global alignment of two strings $S[1..n]$ and $T[1..m]$ can be computed in $O(nm(n+m))$ time and $O(nm)$ space.

Similarly, for computing local or semi-global alignment, we can also modify the algorithms to handle the general gap penalty.

2.5.2 Affine Gap Penalty Model

Although the general gap penalty model is flexible, computing alignment under the general gap penalty model is inefficient. Here, we study a simple gap penalty model, called the affine gap model, which is used extensively in the biology domain. Under the affine gap model, the penalty of a gap equals the sum of two parts: (1) a penalty h for initiating the gap and (2) a penalty s depending on the length of the gap. Thus, the penalty for a gap of length q is:

$$g(q) = h + qs$$

Note that, when $h = 0$, the affine gap penalty is reduced to the uniform gap penalty.

Given two strings $S[1..n]$ and $T[1..m]$, their global alignment under the affine gap penalty can be computed as efficient as that under the uniform gap penalty (that is, $h = 0$). The idea is again to use dynamic programming [120]. Moreover, we require four dynamic programming tables instead of one. The first table is table V where $V(i, j)$ is defined as the global alignment score between $S[1..i]$ and $T[1..j]$. The other three tables are defined depending on the last pair of aligned characters.

- $G(i, j)$: the global alignment score between $S[1...i]$ and $T[1...j]$ with $S[i]$ matches $T[j]$

- $F(i, j)$: the global alignment score between $S[1...i]$ and $T[1...j]$ with $S[i]$ matches a space

- $E(i, j)$: the global alignment score between $S[1...i]$ and $T[1...j]$ with $T[j]$ matches a space

The global alignment score between $S[1..n]$ and $T[1..m]$ is $V(n, m)$. Below, we devise the recursive formula for the tables $V, G, F,$ and E depending on whether (1) $i = 0$ or $j = 0$ and (2) $i \neq 0$ and $j \neq 0$.

When $i = 0$ or $j = 0$, the basis for the recurrence equations are as follows:

$$V(0, 0) = 0$$
$$V(i, 0) = F(i, 0) = -g(i) = -h - is$$
$$V(0, j) = E(0, j) = -g(j) = -h - js$$
$$E(i, 0) = F(0, j) = -\infty$$
$$G(i, 0) = G(0, j) = -\infty$$

When $i > 0$ and $j > 0$, the recurrence of $V(i, j)$, $G(i, j)$, $E(i, j)$, and $F(i, j)$ are defined as follows:

$$V(i, j) = \max\{G(i, j), F(i, j), E(i, j)\}$$
$$G(i, j) = V(i - 1, j - 1) + \delta(S[i], T[j])$$
$$F(i, j) = \max\{F(i - 1, j) - s, V(i - 1, j) - h - s\}$$
$$E(i, j) = \max\{E(i, j - 1) - s, V(i, j - 1) - h - s\}$$

The recurrences for V and G are straightforward. For $F(i, j)$, it is the score of the global alignment of $S[1..i]$ and $T[1..j]$ with $S[i]$ matches a space. There are two cases:

- $S[i - 1]$ matches with a space.
  In this case, we need to add the space penalty, that is, $F(i, j) = F(i - 1, j) - s$.

- $S[i - 1]$ matches with $T[j]$.
  In this case, we need to add the space penalty and the gap initiating penalty, that is, $F(i, j) = V(i - 1, j) - h - s$.

Algorithm Affine_gap($S[1..n], T[1..m]$)

Require: Two sequences $S[1..n]$ and $T[1..m]$

Ensure: The optimal alignment score between $S[1..n]$ and $T[1..m]$.

1: Compute $E(i,0), F(i,0), G(i,0), V(i,0)$ for $i = 1, \ldots, n$
2: Compute $E(0,j), F(0,j), G(0,j), V(0,j)$ for $j = 1, \ldots, m$
3: **for** $i = 1$ to n **do**
4:    **for** $j = 1$ to m **do**
5:       $G(i,j) = V(i-1, j-1) + \delta(S[i], T[j])$
6:       $F(i,j) = \max\{F(i-1, j) - s, V(i-1, j) - h - s\}$
7:       $E(i,j) = \max\{E(i, j-1) - s, V(i, j-1) - h - s\}$
8:       $V(i,j) = \max\{G(i,j), F(i,j), E(i,j)\}$
9:    **end for**
10: **end for**
11: Report $V(n,m)$

FIGURE 2.10: The global alignment algorithm under the affine gap penalty.

Then, by taking the maximum value of the two cases, we have $F(i,j) = \max\{F(i-1,j) - s, V(i-1,j) - h - s\}$. Similarly, we can derive the recurrence for E.

To compute the optimal alignment of $S[1..n]$ and $T[1..m]$, we need to fill in the four tables according to the algorithm in Figure 2.10. The optimal alignment score is stored in $V(n,m)$. The optimal alignment can be recovered by back-tracing on the four tables.

Below, we analyze the time complexity and the space complexity. We maintain four $n \times m$ matrices E, F, G, and V. Hence, the space complexity is $O(nm)$. Since the 4 tables have $4nm$ entries where each entry can be computed in $O(1)$ time, the time complexity is $O(nm)$.

2.5.3 Convex Gap Model

The affine gap penalty assigns a fixed cost for opening a gap and a fixed cost for extending a gap. Such a penalty function may be too rigid to truly represent the underlying biological mechanism. In particular, affine gap penalty is not in favor of long gaps in the alignment.

Because of this reason, non-affine gap penalty functions have been studied by a number of works. Both Benner et al. [22] and Gu and Li [124] proposed the logarithmic gap penalty function (which is in the form of $g(q) = a \log q + b$ where q is the number of spaces in the gap). Gotoh [121] and Mott [214] modified affine gap penalty so that long gaps are desirable in the alignment of DNA. Altschul [8] generalized affine gap penalty for the protein sequence alignment.

In 1984, Waterman [302] proposed and studied the convex gap penalty function. Under the convex gap model, the gap penalty function employed is any non-negative increasing function $g()$ such that $g(q+1) - g(q) \leq g(q) - g(q-1)$ for any $q \geq 1$. In other words, the convex gap model required that the penalty incurred by additional space in a gap decreases as the gap gets longer. Note that both the affine gap penalty function and the logarithmic gap penalty function are kind of convex gap function.

The $O(mn(n+m))$-time dynamic programming algorithm for the general gap penalty can be readily applied for the convex gap penalty. Moreover, Miller and Myers [212] and Galil and Giancarlo [113] independently proposed a practically more efficient method, which runs in $O(mn\log(mn))$ time. This section will describe the $O(mn\log(mn))$-time algorithm.

First, we restate the $O(mn(n+m))$-time dynamic programming algorithm. We introduce two substitution functions $A(i,j)$ and $B(i,j)$ that serve to ease the discussion of the upcoming proof.

Let $V(i,j)$ be the global alignment score between $S[1..i]$ and $T[1..j]$. For either $i = 0$ or $j = 0$, we have

$$V(0,0) = 0, V(0,j) = -g(j), V(i,0) = -g(i)$$

For both $i > 0$ and $j > 0$, we have

$$V(i,j) = \max \begin{cases} V(i-1,j-1) + \delta(S[i], T[j]) & match/mismatch \\ A(i,j) & \texttt{insert } T[k+1..j] \\ B(i,j) & \texttt{delete } S[k+1..i] \end{cases}$$

where

$$A(i,j) = \max_{0 \leq k \leq j-1} \{V(i,k) - g(j-k)\}$$

$$B(i,j) = \max_{0 \leq k \leq i-1} \{V(k,j) - g(i-k)\}$$

The most time consuming part for filling up the dynamic programming table for this algorithm is to compute the tables A and B. It takes $O(n)$ time and $O(m)$ time to compute each $A(i,j)$ and each $B(i,j)$, respectively. Since there are nm entries in tables A and B, computing $A(i,j)$ and $B(i,j)$ for all $1 \leq i \leq n$ and $1 \leq j \leq m$ requires $O(nm^2 + n^2m)$ time.

When the gap penalty function is a convex function, we claim that $\{A(i,1), \ldots, A(i,m)\}$ can be computed in $O(m\log m)$ time for any fixed i while $\{B(1,j), \ldots, B(n,j)\}$ can be computed in $O(n\log n)$ time for any fixed j. Hence, in total, $A(i,j)$ and $B(i,j)$ for all $1 \leq i \leq n$ and $1 \leq j \leq m$ can be computed in $O(nm\log(nm))$ time. Then, the table V can be filled in using $O(nm)$ time. In total, the optimal alignment can be found in $O(nm\log(nm))$ time.

What remains is to prove the claim that, for any fixed i, $\{A(i,1), \ldots, A(i,m)\}$ can be computed in $O(m\log m)$ time. To simplify the discussion, we define the following functions.

$$E(j) = A(i,j)$$
$$C_k(j) = V(i,k) - g(j-k)$$

Hence, the recurrence function of $A(i, j)$ can now be written as

$$E(j) = \max_{0 \leq k \leq j-1} C_k(j)$$

Regular dynamic programming can fill $E(1), \ldots, E(m)$ in $O(m^2)$ time. We show that they can be filled in $O(m \log m)$ time. The speed up is based on the property of $C_k(j)$: when any two curves $C_{k_1}()$ and $C_{k_2}()$ cross, there can be only one intersection (see Lemma 2.2). Figure 2.11 illustrates this phenomena graphically.

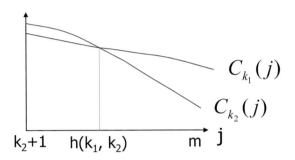

$C_{k_1}(j)$

$C_{k_2}(j)$

k_2+1 $h(k_1, k_2)$ m j

FIGURE 2.11: Graphical interpretation of Lemma 2.2. (See color insert after page 270.)

LEMMA 2.2
For any $k_1 < k_2$, let $h(k_1, k_2) = \arg\min_{k_2 < j \leq m}\{C_{k_1}(j) \geq C_{k_2}(j)\}$. We have $j < h(k_1, k_2)$ if and only if $C_{k_1}(j) < C_{k_2}(j)$.

PROOF   If $j < h(k_1, k_2)$, by the definition of $h(k_1, k_2)$, we have $C_{k_1}(j) < C_{k_2}(j)$. On the other hand, if $j \geq h(k_1, k_2)$, we show that $C_{k_1}(j) \geq C_{k_2}(j)$ by induction. When $j = h(k_1, k_2)$, by the definition of $h(k_1, k_2)$, $C_{k_1}(j) \geq C_{k_2}(j)$. Assume $C_{k_1}(j) \geq C_{k_2}(j)$ for some $j \geq h(k_1, k_2)$. Then, we have

$$
\begin{aligned}
C_{k_1}(j+1) &= C_{k_1}(j) - g(j+1-k_1) + g(j-k_1) \\
&\geq C_{k_2}(j) - g(j+1-k_1) + g(j-k_1) \quad \text{since } C_{k_1}(j) \geq C_{k_2}(j) \\
&\geq C_{k_2}(j) - g(j+1-k_2) + g(j-k_2) \quad \text{since } g(q) \text{ is convex} \\
&= C_{k_2}(j+1)
\end{aligned}
$$

Hence, by induction, the claim is proved.   ☐

LEMMA 2.3
For any $k_1 < k_2$, $h(k_1, k_2)$ can be computed in $O(\log m)$ time.

PROOF   By Lemma 2.2, $j < h(k_1, k_2)$ if and only if $C_{k_1}(j) < C_{k_2}(j)$. Hence, we can perform binary search in the interval $k_2 + 1..m$ to identify $h(k_1, k_2)$. The running time is $O(\log m)$.   ⬚

Let $Env_\ell(j) = \max_{0 \le k < \ell} C_k(j)$ for any $j \ge \ell$. Note that $E(j) = Env_j(j)$. $Env_\ell()$ is a curve formed by merging the segments of the curves C_k for $k \le \ell$. For example, $Env_5()$ in Figure 2.12 is formed by three segments $C_3(5..h(2,3))$, $C_2(h(2,3) + 1..h(0,2))$, and $C_0(h(0,2) + 1..m)$. We represent $Env_5()$ as $((3, h(2,3)), (2, h(0,2)), (0, m))$.

Formally, for any ℓ, suppose $Env_\ell()$ is formed by t's segments $C_{k_t}(\ell..z_t)$, $C_{k_{t-1}}(z_t + 1..z_{t-1}), \ldots, C_{k_1}(z_2 + 1..z_1)$ where $z_1 = m$ and $z_i = h(k_{i-1}, k_i)$ for $i = t, \ldots, 2$. We represent $Env_\ell()$ as $((k_t, z_t), \ldots, (k_1, z_1))$.

LEMMA 2.4
Let $((k_t, z_t), (k_{t-1}, z_{t-1}), \ldots, (k_1, z_1))$ be the representation of $Env_\ell()$. We have $k_1 < \ldots < k_t < \ell < z_t < \ldots < z_1 = m$.

PROOF   Since z_1 is the end position of the last segment, $z_1 = m$. By definition, $k_i < \ell$ and $z_i > \ell$ for $1 \le i \le t$.

Note that the i-th segment of $Env_\ell()$ is $C_{k_{t+1-i}}(z_{t+2-i} + 1..z_{t+1-i})$. Since the i-th segment must be on the left of the $(i+1)$-th segment, we have $z_{t+1-i} < z_{t-i}$. Since the i-th segment crosses the $i + 1$-th segment, we have $k_{t+1-i} > k_{t-i}$. The lemma follows.   ⬚

Now, we describe how to construct $Env_\ell()$ inductively. For the curve $Env_1()$, by definition, it is the same as $C_0()$. We represent $Env_1()$ as $(0, m)$.

For the curve $Env_\ell()$ where $\ell > 1$, suppose it is represented by $(k_0, z_0), \ldots, (k_t, z_t)$. Below, we describe how to generate $Env_\ell()$ from $Env_{\ell-1}()$. There are two cases: (1) $C_\ell(\ell) \le C_{k_t}(\ell)$ and (2) $C_\ell(\ell) > C_{k_t}(\ell)$.

1. Case 1: If $C_\ell(\ell) \le C_{k_t}(\ell)$, then the curve $C_\ell(j)$ can't cross $C(k_t, j)$ and will always be below $C(k_t, j)$ (by Lemma 2.2). Thus, $Env_{\ell+1}$ is still the same as Env_ℓ. Figure 2.13a illustrates this case.

2. Case 2: If $C_\ell(\ell) > C_{k_t}(\ell)$, then $C_\ell(\ell) > Env_\ell(\ell)$. We need to find position h' such that $C_\ell(h') > Env_\ell(h')$ and $C_\ell(h' + 1) < Env_\ell(h' + 1)$. We remove all pairs (k_i, z_i) such that $z_i < h'$; then, we insert the pair (ℓ, h'). Figure 2.13b describes this pictorially.

Now, we describe the algorithm to compute $E(\ell)$ for $1 \le \ell \le m$. The algorithm iteratively builds Env_ℓ for $1 \le \ell \le m$. The algorithm maintains

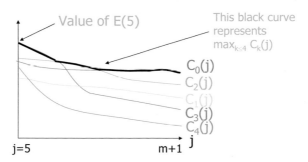

FIGURE 2.12: The figure shows 5 curves $C_k(j)$ for $0 \leq k \leq 4$. The thick black curve corresponds to $Env_4(j) = \max_{0 \leq k \leq 4} C_k(j)$. (See color insert after page 270.)

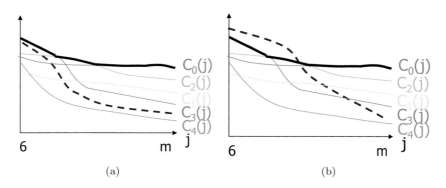

FIGURE 2.13: Possible outcomes of overlaying $C_5()$ (purple color) and $Env_5()$: (a) $C_5()$ is below $Env_5()$ and (b) $C_5()$ intersects with (or above) $Env_5()$. (See color insert after page 270.)

$Env_\ell() = ((k_t, z_t), \ldots, (k_1, z_1))$ in a stack S. Precisely, the stack S stores the t-th pairs (k_i, z_i) with (k_t, z_t) at the top of the stack. Then, it reports $E(\ell) = Env_\ell(\ell) = C_{k_t}(\ell)$. Figure 2.14 describes the algorithm.

It is easy to see that for every ℓ, we push at most one pair onto the stack S and thus, we push at most m pairs onto the stack S, which in turn limits the number of pops from stack S to m pops. Since the value $h(k, k')$ can be computed in $O(\log m)$ time by Lemma 2.3, the total time to fill $E(j)$ is $O(m \log m)$.

Algorithm Filling_E

1: Push $(0, m)$ onto stack S
2: $E(1) = C_{k_t}(1)$
3: **for** $\ell = 2$ to $m - 1$ **do**
4:     **if** $C_\ell(\ell + 1) > C_{k_t}(\ell + 1)$ **then**
5:         **while** $S \neq \emptyset$ and $C_\ell(z_t - 1) > C_{k_t}(z_t - 1)$ **do**
6:             pop S
7:         **end while**
8:         **if** $S = \emptyset$ **then**
9:             push (ℓ, m) onto S
10:        **else**
11:            push $(\ell, h(k_t, \ell))$
12:        **end if**
13:    **end if**
14:    report $E(\ell) = C_{k_t}(\ell)$
15: **end for**

FIGURE 2.14:   Algorithm for computing $E(\ell)$ for $1 \leq \ell \leq m$.

2.6   Scoring Function

To measure the similarity between two sequences, we need to define the scoring function $\delta(x, y)$ for every pair of bases x and y. This section describes how the scoring function is derived.

2.6.1   Scoring Function for DNA

For DNA, since there are only 4 nucleotides, the score function is simpler. In BLAST, it simply gives a positive score for match and a negative score for mismatch. The NCBI-BLASTN set the match score and the mismatch score to +2 and -1, respectively. The WU-BLASTN and FastA set the match score and the mismatch score to +5 and -4, respectively. The score function for NCBI-BLASTN is better for detecting homologous alignment with 95% identity. For WU-BLASTN and FastA, their scoring function is more suitable to detect homologous alignment with 65% identity.

In the evolution of real sequences, transitions are observed more often than transversions. Thus, some people use transition transversion matrix. In this model, the four nucleotides are separated into two groups: purines (A and G) and pyrimidines (C and T). It gives a mild penalty (-1) for replacing purine by purine or pyrimidine by pyrimidine. For substitution between purine and pyrimidine, it gives a bigger penalty (-5). Match score is +1.

2.6.2 Scoring Function for Protein

For protein, there are two approaches to assign a similarity score. The first approach assigns a similarity score based on the chemical or physical properties of amino acids, including hydrophobicity, charge, electronegative, and size. The basic assumption is that an amino acid is more likely to be substituted by another if they share a similar property [169]. For example, we give a higher score for substituting a non-polar amino acid by another non-polar amino acid.

Another approach assigns a similarity score purely based on statistics. The two popular scoring functions PAM and BLOSUM belong to this category. The idea is to count the observed substitution frequency and compare it with the expected substitution frequency. Two residues a and b are expected to be similar if they have a big log-odds score

$$\log \frac{O_{a,b}}{E_{a,b}}$$

where $O_{a,b}$ and $E_{a,b}$ are the observed substitution frequency and the expected substitution frequency, respectively, between amino acid residues a and b. With the assumption that the aligned residue pairs are statistically independent, the alignment score of an aligned sequence equals the sum of individual log-odds score. Below, we detail both PAM and BLOSUM methods.

2.6.2.1 Point Accepted Mutation (PAM) Score Matrix

Dayhoff developed the PAM score matrix in 1978 [75]. Actually, it is a family of score matrices which are suitable for aligning protein sequences of different levels of divergence. The divergence is measured in terms of point accepted mutation (PAM). A point mutation refers to a substitution of one residue by another. A point accepted mutation (PAM) is a point mutation which does not change the protein's function or is not fatal. Two sequences S_1 and S_2 are said to be 1 PAM diverged if S_1 can be converted to S_2 with an average of 1 accepted point mutation per 100 residues. For every i, Dayhoff constructed the PAM-i matrix which is suitable to compare sequences which are PAM-i diverged.

To obtain a PAM-1 matrix, Dayhoff collected a set of ungapped alignments which is constructed from high similarity amino acid sequences (usually > 85%). Then, a phylogenetic tree is constructed to identify the set of mutations. Figure 2.15 shows an example ungapped alignment and the corresponding phylogenetic tree.

Based on the phylogenetic tree, $O_{a,b}$ and $E_{a,b}$ can be computed for any two amino acid residues a and b as follows. Since PAM-1 assumes 1 mutation per 100 residues, $O_{a,a}$ is set to be $99/100$ for any residue a. For any residues $a \neq b$, $O_{a,b}$ is set to be $F_{a,b}/(100 \sum_{x,y \in \mathcal{A}} F_{x,y})$ where $\mathcal{A}$ is the set of amino acids and $F_{x,y}$ is the frequency of substitution between x and y in the phylogenetic tree.

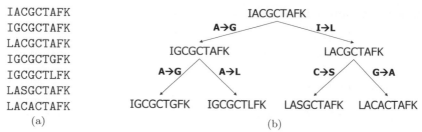

```
IACGCTAFK
IGCGCTAFK
LACGCTAFK
IGCGCTGFK
IGCGCTLFK
LASGCTAFK
LACACTAFK
   (a)
```

FIGURE 2.15: The left figure is an ungapped alignment of seven amino acid sequences and the right figure is the corresponding phylogenetic tree of those seven sequences.

$E_{a,b}$ is set to be $f_a f_b$ where f_a is the number of residue a divided by the total number of residues. Given $O_{a,b}$ and $E_{a,b}$, the similarity score $\delta(a,b)$ equals $\log(O_{a,b}/E_{a,b})$.

For example, in Figure 2.15, $F_{A,G} = 3$, $F_{A,L} = 1$, $f_A = f_G = 10/63$. Then, we obtain $O_{A,G} = 3/(100 * 2 * 6) = 0.0025$, $E_{A,G} = (10/63)(10/63) = 0.0252$, and $\delta(A, G) = \log(0.0025/0.0252) = \log(0.09925) = -1.0034$.

To obtain the PAM-i matrix, Dayhoff generated it by extrapolating the PAM-1 matrix. Let $M_{a,b}$ be the probability that a is mutated to b, which equals $O_{a,b}/f_a$. Let M^i be the matrix formed by multiplying M i's times. Then, $M^i_{a,b}$ is the probability that a is mutated to b after i steps. Therefore, the (a,b) entry of the PAM-i matrix is

$$\log \frac{f_a M^i_{a,b}}{f_a f_b} = \log \frac{M^i_{a,b}}{f_b}.$$

Which PAM matrix should we use when we align protein sequences? In general, to align closely related protein sequences, we should use the PAM-i matrix where i is small. To align distant related protein sequences, we should use the PAM-i matrix where i is large.

2.6.2.2 BLOSUM (BLOck SUbstitution Matrix)

PAM did not work well for aligning evolutionarily divergent sequences since the PAM matrix is generated by extrapolation. Henikoff and Henikoff [142] proposed BLOSUM, which is a scoring matrix constructed directly from the observed alignments (instead of extrapolation).

The BLOSUM matrix is constructed using multiple alignments of non-redundant groups of protein families. Using PROTOMAT [141], blocks of ungapped local alignments are derived. Each block represents an ungapped conserved region of a protein family. For any two amino acid residues a and b, the expected frequency $E_{a,b}$ is set to be $f_a f_b$ where f_a is the frequency of a in the blocks. The observed frequency $O_{a,b}$ is set to be the (a,b) pairs among all aligned residue pairs in the block. BLOSUM defines the similarity score as the log-odds score

$$\delta(a, b) = \frac{1}{\lambda} \ln \frac{O_{a,b}}{f_a f_b}$$

where λ is some normalization constant.

For example, in Figure 2.15, there are $7 * 9 = 63$ residues, including 9 A's and 10 G's. Hence, $f_A = 9/63$, $f_G = 10/63$. Thus, $E_{A,G} = 0.0227$. There are $9\binom{7}{2} = 189$ aligned residue pairs, including 23 (A, G) pairs. Hence, $O_{A,G} = 23/189$. Suppose $\lambda = 0.347$, we have $\delta(A, G) = 4.84$.

To reduce the contribution of closely related protein sequences to the residue frequencies and the residue pair frequencies, similar sequences are merged within blocks. The BLOSUM p matrix is created by merging sequences with no less than $p\%$ similarity. For example, consider the following set of sequences.

```
AVAAA
AVAAA
AVAAA
AVLAA
VVAAL
```

The first four sequences have at least 80% similarity. The similarity of the last sequence with the other four sequences is less than 62%. For BLOSUM 62, we merge the first four sequences and we get sequences

$$AV[A_{0.75}L_{0.25}]AA$$
$$VVAAL$$

The sequences consist of 10 residues and 5 residue pairs. We have $f_A = 5.75/10$, $O_{A,V} = 1/5$, and $O_{A,L} = (0.25 + 1)/5$.

To align distant related sequences, we use BLOSUM p for small p; otherwise, we use BLOSUM p for large p. The relationship between BLOSUM and PAM is as follows.

- BLOSUM 80 ≈ PAM 1

- BLOSUM 62 ≈ PAM 120

- BLOSUM 45 ≈ PAM 250

BLOSUM 62 is the default matrix for BLAST 2.0.

2.7 Exercises

1. Calculate the score for the following alignment.

```
--TCATAC-TCATGAACT
GGTAATCCCTC---AA--
```

    (a) Match= 1, mismatch= 0, indel= -1

    (b) Match= 1, mismatch= -1, initial gap= -2, each space= -1

    (c) Match= 0, mismatch= -1, initial gap= -2, each space= -1

2. Can you propose a dynamic programming solution to solve the longest common subsequence problem?

3. Given two sequences S and T (not necessarily the same length), let G, L, and H be the scores of an optimal global alignment, an optimal local alignment, and an optimal global alignment without counting the beginning space of S and end space of T, respectively.

    (a) Give an example of S and T so that all three scores G, L, and H are different.

    (b) Prove or disprove the statement $L \geq H \geq G$.

4. Consider two DNA sequences of the same length n and let the scoring function be defined as follows: 1 for match, -1 for mismatch, -2 for indel. Let the score of the optimal global alignment be G and that of the optimal local alignment be L.

    (a) Prove that $L \geq G$ and construct an example such that $L = G$.

    (b) What is the maximum value of $L - G$? Construct an example with this maximum value.

    (c) If we want to find a pair of non-overlapping substrings within a given sequence S with the maximum global alignment score, can we simply compute the optimal local alignment between S and itself? Explain your answer.

5. Under the affine gap penalty model, can you describe an $O(nm)$-time algorithm to compute the optimal local alignment between two sequences $S[1..n]$ and $T[1..m]$.

6. Given two DNA sequences S_1 and S_2 of length n and m, respectively, can you give an efficient algorithm which returns the number of possible optimal global alignments between S_1 and S_2? What is the time complexity of your algorithm?

7. Consider the following alignment

```
A C - G G T T A T -
- C T G G - - A T C
```

and the following score matrix.

	A	C	G	T
A	1			
C	-1	2		
G	-1	-2	1	
T	-2	-1	-2	1

(a) Suppose the gap penalty is $-5 - g$ for a gap of size g. Compute the alignment score for the above alignment using the above score matrix.

(b) Is the above alignment an optimal global alignment? If not, what should be the optimal global alignment score and the corresponding optimal global alignment?

(c) Give a score matrix and a gap penalty so that the above alignment is an optimal global alignment.

8. Given a pattern P of length m and a text T of length n. Let P^n be a length nm sequence formed by concatenating n copies of P. Assume the score for match is 1 and the score for mismatch/indel is -1. Can you give an $O(nm)$ time algorithm to compute the optimal local alignment between P^n and T?

9. Given two DNA sequences S_1 and S_2 of length n and m, respectively, we would like to compute the minimum number of operations required for transforming S_1 to S_2, where the allowed operations include (1) insertion of one base, (2) deletion of one base, (3) replacement of one base, and (4) reversal of a DNA substring. In addition, for operation (4), once it is applied on a segment of the DNA sequence, the bases in the segment cannot be further transformed using any operation. Can you give an efficient algorithm which returns the minimum number of operations to transform S_1 to S_2? What is the time complexity?

10. Consider two sequences $S[1..n]$ and $T[1..m]$. Assume match, mismatch, and indel scores 1, -1, and -2, respectively. Give an algorithm which computes the maximum score of the best alignment (S', T') of S and T, forgiving the space at the beginning of S' and the space at the end of T'.

11. Consider a genome sequence $G[1..n]$ and an mRNA sequence $S[1..m]$. Suppose S consists of at most x exons and there is no penalty for the introns. Can you give an efficient algorithm to compute the maximum alignment score between G and S? What is the time complexity? (Note: Assume the scores for match and mismatch are 1 and -1, respectively. Assume the score for indel is -1.)

12. Given two sequences $S[1..n]$ and $T[1..m]$, can you design an algorithm which computes the optimal local alignment between S and T using $O(nm)$ time and $O(n + m)$ space?

13. Given two strings $S_1[1..n]$ and $S_2[1..m]$, we would like to find two non-overlapping alignments $(S_1[i_1..i_2], S_2[j_1..j_2])$ and $(S_1[i_3..i_4], S_2[j_3..j_4])$ such that $i_2 < i_3$ and $j_2 < j_3$ and the total alignment score is maximized. Give an efficient algorithm to compute the total alignment score. What is the running time of your algorithm? (We assume the match score is 2 and the mismatch/indel score is -1.)

Chapter 3

Suffix Tree

3.1 Introduction

The suffix tree of a string is a fundamental data structure for pattern matching [305]. It has many biological applications. The rest of this book will discuss some of its applications, including

- Biological database searching (Chapter 5)

- Whole genome alignment (Chapter 4)

- Motif finding (Chapter 10)

In this chapter, we define a suffix tree and present simple applications of a suffix tree. Then, we discuss a linear suffix tree construction algorithm proposed by Farach. Finally, we discuss the variants of a suffix tree like suffix array and FM-index. We also study the application of suffix tree related data structures on approximate matching problems.

3.2 Suffix Tree

This section defines a suffix tree. First, we introduce a suffix trie. A trie (derived from the word re*trie*val) is a rooted tree where every edge is labeled by a character. It represents a set of strings formed by concatenating the characters on the unique paths from the root to the leaves of the trie. A suffix trie is simply a trie storing all possible suffixes of a string S. Figure 3.1(a) shows all possible suffixes of a string $S[1..7] = acacag\$$, where $\$$ is a special symbol that indicates the end of S. The corresponding suffix trie is shown in Figure 3.1(b). In a suffix trie, every possible suffix of S is represented as a path from the root to a leaf.

A suffix tree is a rooted tree formed by contracting all internal nodes in the suffix trie with single child and single parent. Figure 3.1(c) shows the suffix tree made from the suffix trie in Figure 3.1(b). For each edge in the suffix tree, the edge label is defined as the concatenation of the characters on the

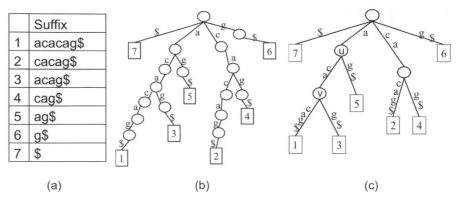

	Suffix
1	acacag$
2	cacag$
3	acag$
4	cag$
5	ag$
6	g$
7	$

(a)   (b)   (c)

FIGURE 3.1:   For $S = acacag\$$, (a) shows all suffixes of S, (b) shows the suffix trie of S, and (c) shows the suffix tree of S.

edges of the suffix trie which are merged. The path label of a node is defined as the concatenation of the edge labels from the root to the node. We define the string depth of a node to be the length of the path label of the node. For example, in Figure 3.1(c), the edge label of the edge (u, v) is ca, the path label of the node v is aca, and the string depth of the node v is 3.

Below, we analyze the space complexity of a suffix tree. Unlike in other chapters, the space complexity is analyzed in terms of bits to visualize the actual size in memory. Consider a string S of length n over the alphabet $\mathcal{A}$. Let σ be the size of $\mathcal{A}$ ($\sigma = 4$ for DNA and $\sigma = 20$ for protein). The suffix tree T of S has exactly n leaves and at most $2n$ edges. Since each character can be represented in $\log \sigma$ bits and every edge in T can be as long as n, the space required to store all edges can be as large as $O(n^2 \log \sigma)$ bits. Observe that every edge label is some substring of S. To avoid storing the edge label explicitly, the label of each edge can be represented as a pair of indices, (i, j), if the edge label is $S[i..j]$. For example, the edge "a" in Figure 3.1(c) could be represented by $(1, 1)$, the edge "ca" is represented by $(2, 3)$ and the edge "$cag\$$" is represented by $(4, 7)$. Each index is an integer between 1 and n and hence it can be represented by $\log n$ bits. Thus, each edge label can be stored in $O(\log n)$ bits and the suffix tree T can be stored in $O(n \log n)$ bits.

In general, we can store two or more strings using a generalized suffix tree. For example, the generalized suffix tree of $S_1 = acgat\#$ and $S_2 = cgt\$$ is shown in Figure 3.2. Note that different terminating symbols are used to represent the ends of different strings.

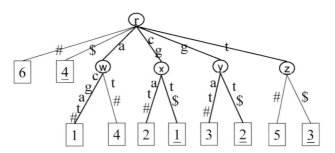

FIGURE 3.2: The generalized suffix tree of $S_1 = acgat\#$ and $S_2 = cgt\$$. The leaves with underlined integers represent the suffixes of S_2, and the rest of the leaves represent the suffixes of S_1.

3.3 Simple Applications of a Suffix Tree

Given a string S of length n, its suffix tree can be constructed in $O(n)$ time as shown in Section 3.4. Before we detail the construction algorithm, this section first discusses some basic applications of a suffix tree.

3.3.1 Exact String Matching Problem

Given a string S of length n, the exact string matching problem asks if a query pattern Q of length m exists in S. If yes, we would like to report all occurrences of Q in S.

This problem can be solved easily in $O(n+m)$ time using the Knuth-Morris-Pratt (KMP) algorithm. However, when n is big, the running time is long. (In genomic application, if S is human/mouse genome, n equals 3 billions.) If we can preprocess the string S, can we solve the exact string matching problem in a time independent of n? Here, we give a definite answer. We show that given the suffix tree for S, the occurrences of any pattern Q of length m can be found in $O(m + occ)$ time where occ is the number of occurrences of the pattern Q in S. The algorithm has two steps.

1: Starting from the root of the suffix tree, we traverse down to find a node x such that the path label of x matches Q. If such a path is found, Q exists in S; otherwise, Q does not occur in S.
2: If Q exists in S, all leaves in the subtree rooted at x are the occurrences of Q. By traversing the subtree rooted at x, we can list all occurrences of Q using $O(occ)$ time .

For example, consider the string $S = acacag\$$ whose suffix tree is shown in Figure 3.1(b). To find the occurrences of $Q = aca$, we traverse down the suffix tree of S along the path aca. Since such a path exists, we confirm Q

occurs in S. The positions of occurrences are 1 and 3, which are the leaves of the corresponding subtree.

Consider the case when $Q = acc$. We can find a path with label ac when we traverse down the suffix tree. However, from this point, we cannot extend the path for the character c; thus we report that the pattern acc does not exist in S.

Next, we give the time analysis. Consider the suffix tree T for S. For any query Q of length m, we can identify if there exists a path label in T matches Q in $O(m)$ time. Then, traversing the subtree to report all occurrences takes $O(occ)$ time where occ is the number of occurrences of Q in S. In total, all occurrences of Q in S can be found in $O(m + occ)$ time.

3.3.2   Longest Repeated Substring Problem

Two identical substrings in a sequence S are called repeated substrings. The longest repeated substring is the repeated substring which is the longest. In the past, without the use of a suffix tree, people solved this problem using $O(n^2)$ time. Moreover, with a suffix tree, we know that this problem can be solved in $O(n)$ time.

The basic observation is that if a substring is repeated at positions i and j in S, the substring is the common prefix of suffix i and suffix j. Hence, the longest repeated substring corresponds to the longest path label of some internal node in the suffix tree of S. Thus, the longest repeated substring can be found as follows.

1: Build a suffix tree T for the string S using $O(n)$ time.
2: Traverse T to find the deepest internal node. Since T has $O(n)$ nodes, this step can be done in $O(n)$ time. The length of the longest repeat is the length of the path label of the deepest internal node.

Thus, a suffix tree can be used to solve the longest repeated substring problem in $O(n)$ time. For example, consider $S = acacag\$$. As shown in Figure 3.1(c), the deepest internal node in the suffix tree of S is the node with the path label aca. Indeed, aca is the longest repeated substring of S.

3.3.3   Longest Common Substring Problem

Scientists are interested in finding similar regions between two sequences. This problem can be modeled as finding the longest common substring between those sequences. Given two strings P_1 and P_2 of total length n, the longest common substring problem asks for the longest common substring between P_1 and P_2. In 1970, Don Knuth conjectured that the longest common substring problem is impossible to solve in linear time. Below, using a generalized suffix tree, we show that the problem can be solved in linear time. The idea is to find a common prefix between the suffixes of P_1 and P_2. Below is the detail of the algorithm.

1: Build a generalized suffix tree for P_1 and P_2. This takes $O(n)$ time.
2: Mark each internal node with leaves representing suffixes of P_1 and P_2. This takes $O(n)$ time. This will ensure that the path labels of the marked internal nodes are common substrings of P_1 and P_2.
3: Report the deepest marked node. This step can be done in $O(n)$ time.

For example, consider $P_1 = acgat$ and $P_2 = cgt$, the longest common substring between P_1 and P_2 is cg. See Figure 3.3.

Note that this solution can be generalized to find the longest common substring for more than 2 sequences (see Exercise 13). By the way, the longest common substring is not the same as the longest common subsequence. The longest common subsequence is a sequence of characters that are not necessary contiguous, whereas the longest common substring is a contiguous substring! Besides, the longest common substring problem can be solved in $O(n)$ time using a suffix tree, whereas the longest common subsequence takes $O(n^2)$ time for a general alphabet (see Section 2.2.4).

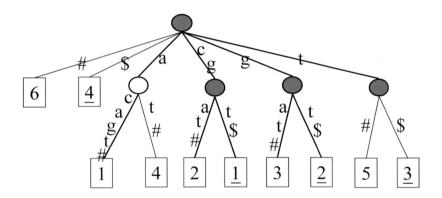

FIGURE 3.3: This is the generalized suffix tree for $P_1 = acgat\#$ and $P_2 = cgt\$$. All square nodes are leaves representing suffixes. For suffixes of P_2, the corresponding integers are underlined. The gray colored internal nodes are nodes whose subtrees contain suffixes of both P_1 and P_2. The path label of the deepest gray colored node is of depth 2, whose path label is cg, which is the longest common substring of P_1 and P_2.

3.3.4 Longest Common Prefix (LCP)

Given a string S of length n, for any $1 \le i, j \le n$, we denote $LCP(i, j)$ as the length of the longest common prefix of suffixes i and j of S. For example, for $S = acacag\$$, the longest common prefix of suffix 1 and suffix 3 is aca; hence, $LCP(1, 3) = 3$. This section describes an $O(n)$-time preprocessing to create a data-structure such that, for any i, j, $LCP(i, j)$ can be reported in $O(1)$ time.

Let T be the suffix tree for S. The key observation is that $LCP(i, j)$ equals the length of the path label of the lowest common ancestor of the two leaves representing suffixes i and j (this lowest common ancestor is denoted as $LCA(i, j)$).

The problem of finding the lowest common ancestor (LCA) of two nodes in a tree is well studied. Harel and Tarjan [137] showed that for a tree of size n, after an $O(n)$-time preprocessing, the LCA for any two nodes in the tree can be computed in $O(1)$ time. The solution was then simplified by Schieber and Vishkin [259] and Bender and Farach-Colton [21]. Based on the $O(n)$-time preprocessing for the lowest common ancestor, the longest common prefix of any two suffixes can be computed in $O(1)$ time.

The LCP data-structure finds many applications. Section 3.3.5 demonstrates one such application.

3.3.5   Finding a Palindrome

In DNA, a complemented palindrome is a sequence of base pairs that reads the same backward and forward across the double strand. As described in Section 1.7.1, those sequences may be specific sites which are cut by restriction enzymes. This section describes an algorithm to locate all palindromes in a string.

First, we formally define the term palindrome and complemented palindrome. Given a string S, a palindrome is a substring u of S such that $u = u^r$ where u^r denotes the reverse string of u. For example, $acagaca$ is a palindrome since $acagaca = (acagaca)^r$. A palindrome $u = S[i..i + |u| - 1]$ is called a maximal palindrome in S if $S[i - 1..i + |u|]$ is not a palindrome. A complemented palindrome is a substring u of S, such that $u = \bar{u}^r$ where $\bar{u}$ is a complement of u. For example, $acaugu$ is a complemented palindrome. A complemented palindrome $u = S[i..i + |u| - 1]$ is said to be maximal if $S[i - 1..i + |u|]$ is not a complemented palindrome.

With the definition of maximal palindrome, every palindrome is contained in a maximal palindrome. In other words, a maximal palindrome is a compact way to represent all palindromes. Similarly, a maximal complemented palindrome is a compact way to represent all complemented palindromes. Below, we give a solution to find all maximal palindromes of a string in linear time based on a suffix tree. Note that maximal complemented palindromes can be found in a similar manner.

The basic observation of the solution is that every palindrome can be divided into two halves where the first half equals the reversal of the second half. Precisely, $u = S[i - (k - 1)..i + (k - 1)]$ is a length $(2k - 1)$ palindrome if $S[i + 1..i + (k - 1)] = (S[i - (k - 1)..i - 1])^r = S^r[n - i + 2..n - i + k]$. Similarly, $u = S[i - k..i + k - 1]$ is a length $2k$ palindrome if $S[i..i + k - 1] =$

$(S[i - k..i - 1])^r = S^r[n - i + 2..n - i + k + 1]$. Utilizing this observation, we can find all maximal palindromes in four steps.

1: Build a generalized suffix tree for S and S^r.
2: Enhance the suffix tree so that we can answer the longest common prefix query in $O(1)$ time (see Section 3.3.4).
3: For $i = 1, \ldots, n$, find the longest common prefix for (S_i, S^r_{n-i+1}) in $O(1)$ time. If the length of the longest prefix is k, we have found an odd length maximal palindrome $S[i - k + 1..i + k - 1]$.
4: For $i = 1, \ldots, n$, find the longest common prefix for (S_i, S^r_{n-i+2}) in $O(1)$ time. If the length of the longest prefix is k, we have found an even length maximal palindrome $S[i - k..i + k - 1]$.

Finally, we analyze the time complexity. The generalized suffix tree and the longest common prefix data structure can be constructed in $O(n)$ time. Since there are $O(n)$ longest common prefix queries and each such query can be answered in $O(1)$ time, we can find all the maximal palindromes in $O(n)$ time.

3.3.6 Extracting the Embedded Suffix Tree of a String from the Generalized Suffix Tree

Given a generalized suffix tree T for K strings $S_1, \ldots, S_K$, we aim to extract the suffix tree T_k for one particular string S_k.

Note that, for every $1 \leq k \leq K$, the suffix tree T_k for S_k is actually embedded in the generalized suffix tree T. Precisely, such an embedded suffix tree T_k consists of the leaves corresponding to S_k and their lowest common ancestors. The edges of T_k can be inferred from the ancestor-descendant relationship among the nodes. For example, in Figure 3.2, the embedded suffix tree for S_1 consists of r, w, and six leaves whose integers are not underlined. The edges are (r, w), $(r, 6)$, $(r, 2)$, $(r, 3)$, $(r, 5)$, $(w, 1)$, and $(w, 4)$. The embedded suffix tree for S_2 consists of r and four leaves whose integers are underlined. It has four edges, each attaches r to each of the four leaves.

To construct the embedded suffix tree T_k from T, we assume we have (1) the generalized suffix tree T for $S_1, \ldots, S_K$, (2) the lexicographical ordering of the leaves in T, and (3) the lowest common ancestor data structure [21, 137, 259]. Then, the embedded suffix trees T_k can be constructed in $O(|S_k|)$ time as follows. Figure 3.4 demonstrates an example.

1: Let the set of leaves be $a_1, \ldots, a_{|S_k|}$ ordered in lexicographical order;
2: We generate a tree R with the leaf a_1 and the root;
3: **for** $i = 2$ to $|S_k|$ **do**
4:     We insert the internal node $LCA(a_i, a_{i+1})$ into R if it does not exist and attach a_i to this node;
5: **end for**

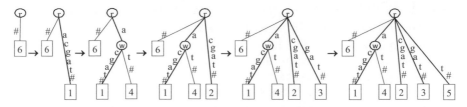

FIGURE 3.4: Constructing the embedded suffix tree of S_1 from the generalized suffix tree in Figure 3.2 for $S_1 = acgat\#$ and $S_2 = cgt\$$.

3.3.7 Common Substring of 2 or More Strings

Consider K strings $S_1, S_2, \ldots, S_K$ whose total length is n. Let $\ell(k)$ be the length of the longest substring common to at least k of these strings, for every $2 \leq k \leq K$.

For example, consider a set of 5 strings {sandollar, sandlot, handler, grand, pantry}. The longest substring appearing in all five strings is "an"; hence, $\ell(5) = 2$. $\ell(4) = 3$ and the corresponding substring is "and", which appears in {sandollar, handlot, handler, grand}. Using the same method, we can check that $\ell(3) = 3$ and $\ell(2) = 4$. The corresponding substrings are "sand" and "and", respectively.

This section aims to compute $\ell(k)$ for all $2 \leq k \leq K$. We solve the problem using a generalized suffix tree. Let T be the generalized suffix tree for all K strings. Then, for every internal node v in T, we let (1) $d(v)$ be the length of its path label and (2) $C(v) = |\mathcal{C}(v)|$ where $\mathcal{C}(v)$ is the set of distinct terminating symbols in the subtree rooted at v. Note that $\ell(k) = \max_{j=k}^{K} \max_{v:C(v)=j} d(v)$.

We observe that, for any node u in T, $d(u) = d(v) + len(u, v)$ where v is the parent of u and $len(u, v)$ is the length of the edge label of (u, v). Also, $\mathcal{C}(u) = \cup_{v \text{ is a child of } u} \mathcal{C}(v)$. Below, we detail the steps for computing $\ell(k)$ for $2 \leq k \leq K$.

1: Build a generalized suffix tree T for all K strings, each string having a distinct terminating symbol. This can be done in $O(n)$ time.
2: For every internal node u in T traversing in pre-order, we compute $d(u) = d(v) + len(u, v)$ where v is the parent of u. This step takes $O(n)$ time.
3: For every internal node u in T traversing in post-order, we compute $\mathcal{C}(u) = \cup\{\mathcal{C}(v) \mid v \text{ is a child of } u\}$ and set $C(u) = |\mathcal{C}(u)|$. This step takes $O(Kn)$ time.
4: This step computes, for $2 \leq k \leq K$, $V(k) = \max_{v:C(v)=k} d(v)$. The detail is as follows. We initialize $V(k) = 0$ for $2 \leq k \leq K$. For every internal node v in T, if $V(C(v)) < d(v)$, set $V(C(v)) = d(v)$. This step takes $O(n)$ time.
5: Set $\ell(K) = V(K)$. For $k = K - 1$ down to 2, set $\ell(k) = \max\{\ell(k + 1), V(k)\}$.

For time analysis, Step 3 runs in $O(Kn)$ time and all other steps take $O(n)$

time. Hence, the overall time complexity of the above algorithm is $O(Kn)$.

Can we further speed up the algorithm? Below, we give a definite answer by showing a way to reduce the running time of Step 3 to $O(n)$. Thus, we can compute $\ell(k)$ for $k = 2, \ldots, K$ using $O(n)$ time.

Prior to describe a faster solution for Step 3, we need some definitions. For every internal node v in T, let $N(v)$ be the number of leaves in the subtree of v. Let $n_k(v)$ be the number of leaves in the subtree of v representing suffixes of S_k. Intuitively, $n_k(v) - 1$ is the number of duplicated suffixes of S_k in the subtree of v. Hence $U(v) = \sum_{k:n_k(v)>0}(n_k(v) - 1)$ is the number of duplicate suffixes in the subtree of T rooted at v. Therefore, $C(v) = N(v) - U(v)$.

The difficulty is in computing $U(v)$ efficiently for all internal nodes v of T. Let $deg_k(v)$ be the degree of v in T_k where T_k is the embedded suffix tree of S_k in T. We observe that, for every internal node v in T, $n_k(v) - 1$ equals $\sum\{deg_k(u) - 1 \mid u$ is a descendant of v in T and u is in $T_k\}$. Based on this observation, we define $h(u)$ to be $\sum_{k=1..K, u \in T_k}(deg_k(u) - 1)$. Then $U(v) = \sum_{k:n_k(v)>0}(n_k(v) - 1) = h(v) + \sum\{U(u) \mid u$ is a child of $v\}$.

Note that, for any node v in T, $N(v) = \sum\{N(u) \mid u$ is a child of $v\}$. Therefore, $C(v)$ for all $v \in T$ can be computed as follows using $O(n)$ time.

1: Compute all the embedded suffix trees T_k in T for $k = 1, \ldots, K$ using the method in Section 3.3.6;
2: **for** every internal node v in T visiting in post-order **do**
3:　　compute $h(v) = \sum_{k=1..K, v \in T_k}(deg_k(v) - 1)$;
4:　　compute $U(v) = h(v) + \sum\{U(u) \mid u$ is a child of $v\}$;
5:　　compute $N(v) = \sum\{N(u) \mid u$ is a child of $v\}$;
6:　　compute $C(v) = N(v) - U(v)$;
7: **end for**

3.4　Construction of a Suffix Tree

Given a string $S = s_1 s_2 \ldots s_n$, where $s_n = \$$. We denote S_i be the i-th suffix of S, that is, $s_i s_{i+1} \ldots s_n$. A straightforward approach to build the suffix tree T for S is as follows. We first initialize a tree T with only a root. Then, the suffixes S_i are inserted into T one by one for $i = n$ downto 1. To insert the suffix S_i into T, we identify the maximal path in T which matches the prefix of S_i. Then, a new leaf edge is created whose label is the remaining part of S_i. It will take $O(n)$ time for adding suffix S_i. Since there are n suffixes, the straightforward algorithm will take $O(n^2)$ time. Figure 3.5 shows an example for constructing the suffix tree for $S = acca\$$ and Figure 3.6 shows an example for constructing the generalized suffix tree for $S = acca\$$ and $S' = c\#$.

The straightforward method for constructing the suffix tree is not efficient.

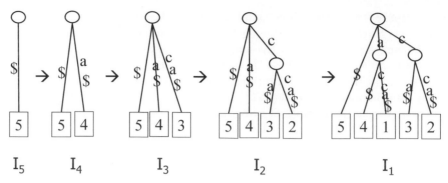

FIGURE 3.5: Constructing the suffix tree for $S = acca\$$.

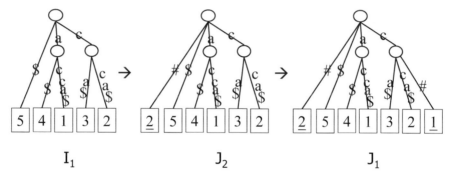

FIGURE 3.6: Constructing the generalized suffix tree for $S = acca\$$ and $S' = c\#$.

Many works tried to address this problem and we can now construct a suffix tree using $O(n)$ time. As early as 1973, Weiner [305] introduced the suffix tree data structure and he gave the first linear time suffix tree construction algorithm when the alphabet size σ is constant. Weiner's algorithm, however, requires a lot of memory. McGreight [211] improved the space complexity to $O(n^2)$ in 1976. Later, in 1995, Ukkonen [295] presented a simplified on-line variant of the algorithm. In 1997, Farach [100] showed that, even for a general alphabet of unbounded size, a suffix tree can be constructed in linear time. The rest of this section will discuss the Farach suffix tree construction algorithm.

Before presenting Farach's algorithm, we first introduce the concept of suffix link. Let T be the suffix tree of S. For any internal node u in T whose path label is ap where a is a character and p is a string, a suffix link is a pointer from u to another node whose path label is p. Based on Lemma 3.1, every internal node has a suffix link. Figure 3.7 shows the suffix links for all internal nodes of the suffix tree of the string $acacag\$$.

LEMMA 3.1

Let T be a suffix tree. If the path label of an internal node v in T is ap, then there must be another internal node w in T whose path label is p. [305]

PROOF   Since the path label of v is ap, there exist two suffixes S_i and S_j such that the longest common prefix between them is ap. Then, the longest common prefix between S_{i+1} and S_{j+1} is p. Hence, there exists a node w in T whose path label is p. ▯

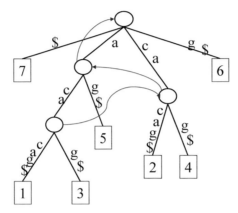

FIGURE 3.7:   The arrows indicate the suffix links of the suffix tree for $S = acacag\$$.

Farach's algorithm performs 3 steps to construct the suffix tree T of S. First, the algorithm constructs the suffix tree for all the odd suffixes of S, denoted as odd suffix tree T_o. Second, the algorithm constructs the suffix tree for all the even suffixes of S, denoted as even suffix tree T_e. Last, the algorithm merge T_o and T_e to construct the suffix tree T of S. The algorithm is summarized as follows.

Algorithm **Construct_Suffix_Tree**(S)

1: Rank $s_{2i-1}s_{2i}$ for $i = 1, 2, \ldots, n/2$. Let $S'[1..n/2]$ be a string such that $S'[i]$ = rank of $s_{2i-1}s_{2i}$. By recursively call Construct_Suffix_Tree(S'), we compute the suffix tree T' of S'. Then, we refine T' to generate the odd suffix tree T_o.
2: From T_o, compute the even suffix tree T_e.
3: Merge T_o and T_e to form the suffix tree T of S.

Below, we detail all three steps.

3.4.1 Step 1: Construct the Odd Suffix Tree

Step 1 pairs up the characters of S and forms a string $PS[1 \cdots n/2]$, in which $PS[i] = s_{2i-1}s_{2i}$. By bucket sort on PS (see Exercise 10), the rank of each character pair can be computed. Then a new string $S' = s'_1 \ldots s'_{n/2}$ is generated, where s'_i equals the rank of $PS[i]$ in the sorted list. Now S' is a string of length $n/2$ over integers $[1 \cdots n/2]$. By calling Construct_Suffix_Tree(S'), the suffix tree T' for S' can be constructed recursively.

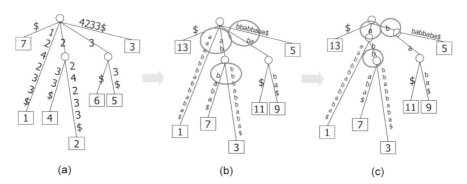

(a)   (b)   (c)

FIGURE 3.8: Given the string $S = aaabbbabbaba\$$, by replacing every character pair by its rank, we obtain $S' = 123233\$$. (a) The suffix tree T' of S'. By (1) replacing every leaf j by $2j - 1$ and (2) replacing the ranks on each edge by the corresponding character pairs, we obtain the tree in (b). Note that this is not a suffix tree since some edges attached to the same internal node start with the same character (as shown in the circle). After resolving the problems in the edges, we obtain the odd suffix tree T_o for S in (c).

For example, for the string $S = aaabbbabbaba\$$, its PS is (aa, ab, bb, ab, ba, ba). By bucket sort, these character pairs could be sorted to $aa < ab < ba < bb$ and the rank of each pair is $\{Rank(aa) = 1, Rank(ab) = 2, Rank(ba) = 3, Rank(bb) = 4\}$. Now we can form a new string $S' = 124233\$$ and get the suffix tree of S' as shown in Figure 3.8(a) by recursion.

The last job of Step 1 is to refine T' into the odd suffix tree T_o. To do this, the characters s'_i on every edge of T' should be replaced by the corresponding character pairs and the leaf labels j in T' should be replaced by the leaf labels $2j - 1$. For example, after replacement, the suffix tree of S' in Figure 3.8(a) becomes a tree in Figure 3.8(b). However, the tree after replacement may not be a suffix tree. For example, in Figure 3.8(b), the root node has two edges whose labels are $aaabbbabbaba\$$ and ab and both labels start with a. There are two other faults in the same tree. To correct the tree into a suffix tree, for each internal node u, if some of its child edges whose labels start with the same character, say x, a new node v is introduced between u and these child edges and the edge (u, v) is labeled by x. Because the edges coming from any node are lexicographically sorted, we can identify and correct all those faults by checking the first characters on the adjacent edges. Hence, the correction takes linear time. Figure 3.8(c) shows the odd suffix tree of $S = aaabbbabbaba\$$ generated by Step 1.

For time analysis, the bucket sorting and the refinement of T' take $O(n)$ time in Step 1. Suppose $Time(n)$ is the time to build the suffix tree for a string of length n, then Step 1 takes $Time(n/2) + O(n)$ time in total.

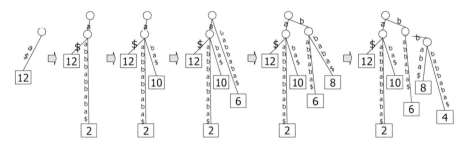

FIGURE 3.9: Process of constructing the even suffix tree T_e of $S = aaabbbabbaba\$$.

3.4.2   Step 2: Construct the Even Suffix Tree

Step 2 constructs the even suffix tree T_e. Interestingly, Farach observed that T_e can be constructed from T_o. The process consists of three substeps: (i) Find the lexicographical ordering of the even suffixes, (ii) for adjacent even suffixes, we compute their longest common prefixes, and (iii) construct T_e using information from (i) and (ii).

For (i), we notice that $S_{2i} = s_{2i}S_{2i+1}$. Given the lexicographical order of all the odd suffixes, we can generate the lexicographical order of all even suffixes by bucket sorting on the pairs (s_{2i}, S_{2i+1}) in linear time. For instance, the even suffixes $S_2, S_4, \cdots, S_{12}$ of $S = aaabbbabbaba\$$ could be represented as $(a, S_3), (b, S_5), (b, S_7), (b, S_9), (a, S_{11}), (a, S_{13})$, respectively. Note that the lex-ordering of the suffixes in T_o is $S_{13} < S_1 < S_7 < S_3 < S_{11} < S_9 < S_5$; so after bucket sorting, the lex-ordering of the pairs is $(a, S_{13}), (a, S_3), (a, S_{11}), (b, S_7), (b, S_9), (b, S_5)$, which can be generated in $O(n)$ time. Hence, the lex-ordering of the leaves of T_e is $S_{12} < S_2 < S_{10} < S_6 < S_8 < S_4$.

For (ii), we compute $LCP(2i, 2j)$ for every adjacent suffixes S_{2i} and S_{2j}. Observe that for any two adjacent suffixes S_{2i} and S_{2j} in T_e, if $S_{2i} = S_{2j}$, then $LCP(2i, 2j) = LCP(2i+1, 2j+1)+1$; otherwise $LCP(2i, 2j) = 0$. Hence, we can calculate the LCP for a pair of adjacent suffixes S_{2i} and S_{2j} in linear time. In our example, the LCP for the adjacent leaves of T_e is $\{LCP(12, 2) = 1,$

$LCP(2, 10) = 1$, $LCP(10, 6) = 0$, $LCP(6, 8) = 1$, $LCP(8, 4) = 2$}.

For (iii), given the order of the leaves and the LCP information, we can construct T_e incrementally from left to right in linear time using an approach similar to Section 3.3.6. Figure 3.9 shows the process of constructing the even suffix tree for S. In summary, Step 2 takes $O(n)$ time to construct the even tree T_e for S.

3.4.3   Step 3: Merge the Odd and the Even Suffix Trees

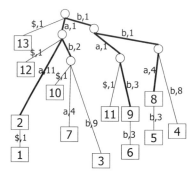

FIGURE 3.10:   The over-merged tree of the odd and even suffix trees of $S = aaabbbabbaba\$$. Every edge is labeled by the first character of the edge and the length of the edge. All the thick edges are formed by merging between the odd and even suffixes.

After the first two steps, we get the odd suffix tree T_o and the even suffix tree T_e of S. In the final step, T_o and T_e will be merged to generate the suffix tree T of S.

Consider the uncompressed version of both T_o and T_e so that every edge is labeled by one character (that is, they are treated as suffix tries). We can merge T_o and T_e using depth first search (DFS). Precisely, we start by the two roots of both trees. Then, we simultaneously take the edges in both trees whose labels are the same and recursively merge the two subtrees. If only one tree has an edge labeled a, then we do nothing since there is nothing to merge. Merging trees in this way may take $O(n^2)$ time since there may be $O(n^2)$ characters in the suffix tree.

Merging the uncompacted version of T_o and T_e is time consuming. Farach introduced a method which merges the original T_o and T_e directly in $O(n)$ time. The method is also based on DFS. Moreover, two edges are merged as long as they start with the same character. The merge is ended when the label of one edge is longer than that of the other. Figure 3.10 shows an example of an over-merged tree M, which merges our example odd and even suffix trees T_o and T_e of $S = aaabbbabbaba\$$. The thick edges in Figure 3.10 are the merged edges and they may be over-merged. What remains is to unmerge these incorrectly merged parts of the tree. Before explaining the unmerging process, let's first introduce two definitions d and L.

For any node v created by merging, its incoming edge may be over-merged. We let $L(v)$ be the correct depth of v.

Since v is a node created by merging, there must exist one even suffix S_{2i} and one odd suffix S_{2j-1} such that $v = LCA(2i, 2j-1)$ in M. The suffix link $d(v)$ is the node $LCA(2i+1, 2j)$ if v is not a root; otherwise $d(v)$ is undefined. The dotted edges in Figure 3.11 show the d relationship of the vertices in M. Since every vertex has at most one out-going dotted edge, the dotted edges form a tree, which is denoted as d tree. By construction, $L(v) = 1 + L(d(v))$ if v is not the root; otherwise $L(v) = 0$. Hence, $L(v)$ equals the depth of v in the d tree.

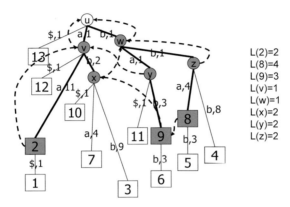

FIGURE 3.11: The dotted edges form the d tree for the over-merged tree M in Figure 3.10. We also show the L values for the nodes on the d tree, which are the depths of the nodes with respect to the d tree.

The depth of every node in the d tree can be determined in linear time by DFS. Then, we identify all the real over-merged nodes. In Figure 3.11, the real over-merged nodes are 2, 8, 9, and x, whose L values are smaller than the depth in M. For any over-merged node v, we adjust its depth to $L(v)$. Then, all the children of v are appended under v' in the lexicographical order. The final tree computed is just the suffix tree T for S. Figure 3.12 shows the suffix tree for $S = aaabbbabbaba\$$.

When these three steps are completed, we get the suffix tree T for S.

Finally, we analyze the time complexity of Farach's algorithm. Let $Time(n)$ be the time required to construct a suffix tree for a sequence of length n. Then, Step 1 takes $Time(n/2) + O(n)$ time. Steps 2 and 3 take $O(n)$ time. In total,

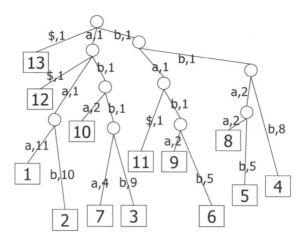

FIGURE 3.12:   The suffix tree for $S = aaabbbabbaba\$$.

the running time of the whole algorithm is $Time(n) = Time(n/2) + O(n)$. By solving this equation, we have $Time(n) = O(n)$. In conclusion, given a string $S = s_1 s_2 \cdots s_n$, Farach's algorithm can construct the suffix tree for S using $O(n)$ time and space.

3.5   Suffix Array

Although a suffix tree is a very useful data structure, it has limited usage in practice. This is mainly due to its large space requirement. For a length-n sequence over an alphabet $\mathcal{A}$, the space requirement is $O(n|\mathcal{A}|\log n)$ bits. (Note that the size of $|\mathcal{A}|$ is 4 for DNA and 20 for protein.)

To solve this problem, Manber and Myers [204] proposed a new data structure called a suffix array, in 1993, which has a similar functionality as a suffix tree but only requires $n \log n$ bits space.

Let $S[1..n]$ be a string of length n over an alphabet $\mathcal{A}$. We assume that $S[n] = \$$ is a unique terminator which is alphabetically smaller than all other characters. The suffix array $(SA[1..n])$ stores the suffixes of S in a lexico-graphically increasing order. Formally, $SA[1..n]$ is an array of integers such that $S[SA[i]..n]$ is lexicographically the i-th smallest suffix of S. For example, consider $S = acacag\$$. Its suffix array is shown in Figure 3.13(b).

For the space complexity, as each integer in the suffix array is less than n and can be stored using $\log n$ bits, the whole suffix array of size n can be stored in $n \log n$ bits.

Suffix	Position
$acacag\$$	1
$cacag\$$	2
$acag\$$	3
$cag\$$	4
$ag\$$	5
$g\$$	6
$\$$	7

i	$SA[i]$	Suffix
1	7	$\$$
2	1	$acacag\$$
3	3	$acag\$$
4	5	$ag\$$
5	2	$cacag\$$
6	4	$cag\$$
7	6	$g\$$

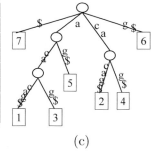

(a)   (b)   (c)

FIGURE 3.13: (a) The set of suffixes of $S = acacag\$$, (b) the corresponding suffix array, and (c) the corresponding suffix tree. Note that the lexicographical order of the leaves in the suffix tree equals the order of the suffixes in the suffix array.

3.5.1 Construction of a Suffix Array

Observe that when the leaves of a suffix tree are traversed in lexicographical depth-first search order, they form the suffix array of that string. This is shown in Figure 3.13(b,c). Thus the suffix array of $S[1..n]$ can be constructed in $O(n)$ time by first constructing the suffix tree T and then the suffix array is generated by traversing T using lexicographical depth-first traversal.

However, this naïve approach requires a large working space, since it needs to build the suffix tree itself which requires $O(n|\mathcal{A}|\log n)$ bits space. Thus this defeats the purpose of using a suffix array.

To date, if we only have $O(n)$ bits of working memory available, the best known technique for constructing a suffix array takes $O(n)$ time. Please refer to [152] for more details.

3.5.2 Exact String Matching Using a Suffix Array

Most applications using a suffix tree can be solved using a suffix array with some overhead. This section demonstrates how to use a suffix array to solve the exact string matching problem. Assume the suffix array of a length-n string S is given. For any query Q of length m, our aim is to check if Q exists in S.

We first state a property. Let Q be a string. Suppose suffix $SA[i]$ and suffix $SA[j]$ are lexicographically the smallest and largest suffixes, respectively, having Q as their prefix. This implies that Q occurs at positions $SA[i], SA[i+1], \ldots, SA[j]$ in S. The interval $[i..j]$ is called the SA range of Q. Notationally, we denote $[i..j]$ as $range(S, Q)$. For example, considering the text $S = acacag\$$, ca occurs in $SA[5]$ and $SA[6]$. In other words, $range(S, ca) = [5..6]$.

The idea is to perform a binary search on the suffix array of S. The algorithm is shown in Figure 3.14. Initially, Q is expected to be in the SA range

$L..R = 1..n$. Let $M = (L + R)/2$. If Q matches $SA[M]$, report Q exists. If Q is smaller than suffix $SA[M]$, Q is in the SA range $L..M$; otherwise, Q is in the SA range $M..R$. Using binary search on the suffix array, we will perform at most $\log n$ comparisons. Each comparison between the query Q and a suffix takes at most $O(m)$ time. Therefore, in the worst case, the algorithm takes $O(m \log n)$ time.

Comparison between the query Q and the suffix $SA[M]$ is time consuming. We have the following observation which reduces the amount of comparisons. Suppose l is the length of the longest common prefix between Q and the suffix $SA[L]$ and r is the length of the longest common prefix between Q and the suffix $SA[R]$. Then, the length of the longest common prefix between Q and the suffix $SA[M]$ is at least $mlr = \min\{l, r\}$.

The above observation allows us to reduce the redundant comparisons and we obtain the algorithm shown in Figure 3.15. Figure 3.16 demonstrates how to find the occurrence of $Q = acag$ in the text $S = acacag\$$ by the algorithm. First, as shown in Figure 3.16(a), we initialize L and R to the smallest index 1 and the biggest index 7, respectively. Also we initialize l as the length of the longest common prefix between Q and suffix $SA[L]$ and r as that between Q and suffix $SA[R]$. Second, as shown in Figure 3.16(b), we compute the mid-point $M = (L + R)/2 = 4$. Note that Q and suffix $SA[M]$ must match at least the first $mlr = \min\{l, r\} = 0$ character. We check the rest of the strings to compute the length of the longest common prefix between Q and $SA[M]$, which equals $m = 1$. Since the $(m + 1)$-th character of suffix $SA[M]$ is g and the $(m + 1)$-th character of Q is c, we have suffix $SA[M] > Q$. Hence, we can halve the range (L, R) by setting $R = M = 4$ and $r = m = 1$. Third, as shown in Figure 3.16(c), we compute $M = (L + R)/2 = 2$. Note that Q and suffix $SA[M]$ must match the first $mlr = \min\{l, r\} = 0$ characters. We check the rest of the strings to compute the length of the longest common prefix between Q and $SA[M]$, which equals $m = 3$. Since the $(m + 1)$-th character of suffix $SA[M]$ is c and the $(m+1)$-th character of Q is g, we have suffix $SA[M] < Q$. Hence, we can halve the range (L, R) by setting $L = M = 2$ and $l = m = 3$. Fourth, as shown in Figure 3.16(d), we compute $M = (L + R)/2 = 3$. Note that Q and suffix $SA[M]$ must match at least the first $mlr = \min\{l, r\} = 1$ characters. We check the rest of the strings to compute the length of the longest common prefix between Q and $SA[M]$, which equals $m = 4$. Hence, Q is found at $SA[2] = 3$.

Using binary search on a suffix array of size n, we will perform at most $\log n$ comparisons. Each comparison takes at most $O(m)$ time. Therefore, in the worst case, the algorithm takes $O(m \log n + occ)$ time. Moreover, since the number of comparisons is reduced, the practice running time is reported to be $O(m + \log n + occ)$ for most of the cases [204]. When space is limited, solving the exact string matching problem using a suffix array is a good alternative to a suffix tree.

Furthermore, if we can afford to store the information $\{LCP(1, 2), LCP(2, 3), \ldots, LCP(n - 1, n)\}$ using an additional $n \log n$-bit space, the exact string

---

SA_binary_search(Q)

1: Let $L = 1$ and $R = n$;
2: **while** $L \leq R$ **do**
3:     Let $M = (L + R)/2$.
4:     Find the length m of the longest common prefix of Q and suffix $SA[M]$;
5:     **if** $m = |Q|$ **then**
6:         Report suffixes $SA[M]$ contain the occurrence of Q;
7:     **else if** suffix $SA[M] > Q$ **then**
8:         set $R = M$;
9:     **else**
10:        set $L = M$;
11:     **end if**
12: **end while**
13: **if** $L > R$ **then**
14:     Report Q does not exist in S;
15: **end if**

FIGURE 3.14: Checking the existence of a query Q through binary search on the suffix array of S.

SA_binary_search_with_lcp(Q)

1: Let $L = 1$ and $R = n$;
2: Let l be the length of the longest common prefix of Q and suffix$_{SA[1]}$;
3: Let r be the length of the longest common prefix of Q and suffix$_{SA[n]}$;
4: **while** $L \leq R$ **do**
5:     Let $M = (L + R)/2$.
6:     $mlr = \min(l, r)$;
7:     Starting from the position mlr of Q, find the length m of the longest common prefix of Q and suffix $SA[M]$;
8:     **if** $m = |Q|$ **then**
9:         Report suffixes $SA[M]$ contain the occurrence of Q;
10:     **else if** suffix$_{SA[M]} > Q$ **then**
11:        set $R = M$ and $r = m$;
12:     **else**
13:        set $L = M$ and $l = m$;
14:     **end if**
15: **end while**
16: **if** $L > R$ **then**
17:     Report Q does not exist in S;
18: **end if**

FIGURE 3.15: Checking the existence of a query Q through binary search on the suffix array of S. In this algorithm, we speed up the computation using the longest common prefix.

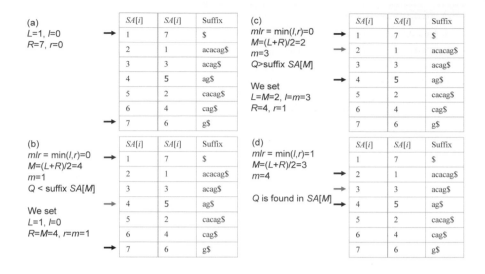

FIGURE 3.16:   This example demonstrates how to check the existence of $Q = acag$ in the text $S = acacag\$$ by performing binary search on the suffix array of S.

matching of a length-n pattern can be found in $O(m + \log n + occ)$ worst case time (see Exercise 5).

3.6   FM-Index

Although the space requirement of a suffix array is less than that of a suffix tree, it is still unacceptable in many real life situations. For example, to store the human genome, with its 3 billion base pairs, it takes up to 40 gigabytes using a suffix tree and 13 gigabytes using a suffix array. This is clearly beyond the capacity of a normal personal computer. A more space-efficient solution is required for practical solutions. Two such alternatives were proposed. They are the compressed suffix array by Grossi and Vitter [123] and the FM-index by Ferragine and Manzini [106]. Both data structures require only $O(n)$ bits where n is the length of the string. For instance, we can store the FM-index of the whole human genome using less than 1.5 gigabytes. We will introduce the FM-index data structure below.

3.6.1 Definition

The FM-index is a combination of the Burrows-Wheeler (BW) text [45] and an auxiliary data structure. Let $S[1..n]$ be a string of length n, and $SA[1..n]$ be its suffix array. The FM-index for S stores the following three data structures:

1. The BW text is defined as a string of characters $BW[1..n]$ where

$$BW[i] = \begin{cases} S[SA[i] - 1] & \text{if } SA[i] \neq 1 \\ S[n] & \text{if } SA[i] = 1 \end{cases}$$

   In other words, the BW text is an array of preceding characters of the sorted suffixes. For example, for $S = acacag\$$, its BW text is $g\$ccaaa$ as shown in Figure 3.17.

2. For every $x \in \mathcal{A}$, $C[x]$ stores the total number of occurrences of characters which are lexicographically less than x. For example, for $S = acacag\$$, we have $C[a] = 1$, $C[c] = 4$, $C[g] = 6$, and $C[t] = 7$.

3. A data structure which supports $O(1)$ computation of $occ(x, i)$, where $occ(x, i)$ is the number of occurrences of $x \in \mathcal{A}$ in $BW[1..i]$. For example, for the BW text $g\$ccaaa$, we have $occ(a, 4) = 1$ and $occ(c, 3) = 2$. Section 3.6.2 will detail a succinct way to store the occ data structure using $O(\frac{n \log \log n}{\log n})$ bits.

We will now analyze the space complexity of the FM-index. In the case of the DNA sequence, structure one can be stored in $2n$ bits because we are storing n characters, each character having four possible choices ($\log 4 = 2$). Structure 2 can be stored in $O(\log n)$ bits and structure 3 can be stored in $O(\frac{n \log \log n}{\log n})$ bits. Therefore, in total the size of the FM-index is $O(n)$ bits.

i	$SA[i]$	Suffix	$S[SA[i] - 1]$
1	7	$\$$	g
2	1	$acacag\$$	$\$$
3	3	$acag\$$	c
4	5	$ag\$$	c
5	2	$cacag\$$	a
6	4	$cag\$$	a
7	6	$g\$$	a

FIGURE 3.17: Consider the string $S = acacag\$$. The second column shows the suffix array $7, 1, 3, 5, 2, 4, 6$. The last column is the BW text, which is the list of preceding characters of the sorted suffixes $g\$ccaaa$.

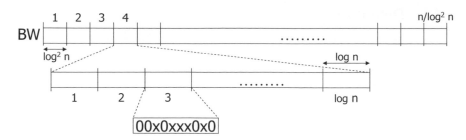

FIGURE 3.18:   The data structure for answering the $occ(x, i)$ query.

3.6.2   The occ Data Structure

To describe the occ data structure, we need to conceptually divide the text $BW[1..n]$ into $\frac{n}{\log^2 n}$ buckets, each of size log^2n. Each bucket is further subdivided into $\log n$ sub-buckets of size $\log n$. Figure 3.18 demonstrates how the BW text is partitioned into buckets and sub-buckets.

The occ data structure stores an integer for each bucket and each sub-bucket. For each bucket $i = 1, \cdots, \frac{n}{\log^2 n}$, we store $P[i] =$ number of x's in $BW[1..i \log^2 n]$. For each sub-bucket j of the bucket i, we store $Q[i][j] =$ number of x's in the first j sub-buckets of the bucket i. In addition, the occ data structure also requires a lookup table $rank(b, k)$ for every string b of length $\log n/2$ and $1 \leq k \leq \log n/2$. Each entry $rank(b, k)$ stores the number of x's occurring in the first k characters of b.

In total, the occ data structure needs to store two arrays and one lookup table. The array $P[1..n/\log^2 n]$ stores $\frac{n}{\log^2 n}$'s $(\log n)$-bit integers, which uses $O(\frac{n}{\log n})$-bit space. The array $Q[1..n/\log^2 n][1..\log n]$ stores $\frac{n}{\log n}$'s $(\log \log n)$-bit integers, which uses $O(\frac{n \log \log n}{\log n})$-bit space. The table $rank(b, k)$ has $2^{\frac{1}{2} \log n} \frac{\log n}{2} = \sqrt{n} \log n/2$ entries of $\log \log n$-bit integers, which uses $o(n)$-bit space. Thus, the occ data structure uses $O(\frac{n \log \log n}{\log n})$ bits.

How can we compute $occ(x, i)$ using the occ data structure? Observe that, for any i,

$$i = i_1 \log^2 n + i_2 \log n + i_3$$

where $i_1 = \lfloor \frac{i}{\log^2 n} \rfloor$, $i_2 = \lfloor \frac{i \mod \log^2 n}{\log n} \rfloor$, and $i_3 = i \mod \log n$.

Based on the occ data structure, the number of x's in $BW[1..i - i_3]$ is $P[i_1]+Q[i_1+1][i_2]$. The number of x's in $BW[i-i_3+1..i]$ equals $rank(BW[i-i_3+1..i], x_3)$ if $i_3 < \log n/2$; otherwise, it equals $rank(BW[i - i_3 + 1..i - i_3 + \log n/2], \log n/2) + rank(BW[i - i_3 + \log n/2 + 1..i], i_3 - \log n/2)$. Hence, we can compute $occ(x, i)$ in $O(1)$ time. For example, suppose $\log n = 10$. As shown in Figure 3.18, to compute $occ(x, 327)$, the result is $P[3] + Q[4][2] + rank(00x0x, 5) + rank(xx0x0, 2)$.

Algorithm BW_search($Q[1..m]$)

1: $x = Q[m]$; $st = C[x] + 1$; $ed = C[x + 1]$;
2: $i = m - 1$;
3: **while** $st \leq ed$ and $i \geq 1$ **do**
4:     $x = Q[i]$;
5:     $st = C[x] + occ(x, st - 1) + 1$;
6:     $ed = C[x] + occ(x, ed)$;
7:     $i = i - 1$;
8: **end while**
9: if $st > ed$, then pattern not found else report $[st..ed]$.

FIGURE 3.19: Given the FM-index of S, the backward search algorithm finds $range(S, Q)$.

3.6.3 Exact String Matching Using the FM-Index

Here, we revisit the exact string matching again and we describe the application of the FM-index to the exact string matching problem. Consider the FM-index of a string $S[1..n]$. For any query string Q, our task is to determine if Q exists in the string S. This section gives the backward search algorithm to perform the task. First, we state a simple property of the FM-index.

LEMMA 3.2

The number of suffixes which are lexicographically smaller than or equal to $xT[SA[i]..n]$ equals $C[x] + occ(x, i)$, where $x \in \mathcal{A}$.

PROOF A suffix $S[j..n]$ which is smaller than $xS[SA[i]..n]$ has two types: (1) $S[j] < x$ and (2) $S[j] = x$ and $S[j + 1..n] < S[SA[i]..n]$. The number of suffixes of type 1 equals $C[x]$. The number of suffixes of type 2 equals $occ(x, i)$. The lemma follows. ◻

As an example, for the FM-index of $S = acacag\$$, the number of suffixes smaller than $cS[SA[5]..n] = ccag\$$ is equal to $C[c] + occ(c, 5) = 4 + 2 = 6$. Based on Lemma 3.2, we can derive the key lemma for backward search.

LEMMA 3.3

Consider a string $S[1..n]$ and a query Q. Suppose $range(S, Q)$ is $[st..ed]$. Then, we have $range(S, xQ) = [p..q]$ where $p = C[x] + occ(x, st - 1) + 1$ and $q = C[x] + occ(x, ed)$.

PROOF By definition, $p - 1$ equals the number of suffixes strictly smaller than xQ. By Lemma 3.2, we have $p - 1 = C[x] + occ(x, st - 1)$. For q, it equals the number of suffixes smaller than or equal to $xS[SA[st-1]..n]$. By Lemma 3.2, we have $q = C[x] + occ(x, ed)$. ▯

Now, we describe the backward search algorithm. It first determines *range* $(T, Q[m])$ which is $[C[x] + 1..C[x+1]]$ where $x = Q[m]$. Then, we compute $range(T, Q[i..m])$ through applying Lemma 3.3 iteratively for $i = m - 1$ to 1. Figure 3.19 gives the detail of the backward search algorithm. Figure 3.20 gives an example to demonstrate the execution of the algorithm.

We now analyze the time complexity of backward search. To find a pattern $Q[1..m]$, we need to process the characters in Q one by one iteratively. Each iteration takes $O(1)$ time. Therefore, the whole backward search algorithm takes $O(m)$ time.

(a) First iteration (initial values),
$Q[3..3] = a$
$range(S, Q[3..3]) = [sp..ep]$
$sp = C[a] + 1 = 1 + 1 = 2$
$ep = C[c] = 4$

SA[i]	Suffix	BW[i]
7	$	g
1	acacag$	$
3	acag$	c
5	ag$	c
2	cacag$	a
4	cag$	a
6	g$	a

(b) Second iteration,
$Q[2..3] = ca$
$range(S, Q[2..3]) = [sp'..ep']$
$sp' = C[c] + occ(c, sp - 1) + 1 = 4 + 0 + 1 = 5$
$ep' = C[c] + occ(c, ep) = 4 + 2 = 6$

SA[i]	Suffix	BW[i]
7	$	g
1	acacag$	$
3	acag$	c
5	ag$	c
2	cacag$	a
4	cag$	a
6	g$	a

(c) Third iteration,
$Q[1..3] = aca$
$range(S, Q[1..3]) = [sp''..ep'']$
$sp'' = C[a] + occ(a, sp' - 1) + 1 = 1 + 0 + 1 = 2$
$ep'' = C[a] + occ(a, ep') = 1 + 2 = 3$

SA[i]	Suffix	BW[i]
7	$	g
1	acacag$	$
3	acag$	c
5	ag$	c
2	cacag$	a
4	cag$	a
6	g$	a

FIGURE 3.20: Given the FM-index for the text $S = acacag\$$. This figure shows the three iterations for searching pattern $Q = aca$ using backward search. (a) $range(S, a) = [2..4]$, (b) $range(S, ca) = [5..6]$, and (c) $range(S, aca) = [2..3]$.

i	$SA[i]$	$SA^{-1}[i]$	Suffix
1	7	2	$
2	1	5	$acacag$
3	3	3	$cacag$
4	5	6	$acag$
5	2	4	cag
6	4	7	ag
7	6	1	g

FIGURE 3.21: The suffix array and inverse suffix array of $S = acacag$.

3.7 Approximate Searching Problem

Although the suffix tree, the suffix array, and their variants are very useful, they are usually used to solve problems related to exact string matching. In bioinformatics, many problems involve approximate matching. This section describes how to apply suffix tree related data structures to solve an approximate matching problem. In particular, we describe the 1-mismatch problem.

Given a pattern $Q[1..m]$, the 1-mismatch problem finds all occurrences of Q in the text $S[1..n]$ that has hamming distance at most 1. For example, for $Q = acgt$ and $S = aacgtggccaacttgga$, the underlined substrings of S are the 1-mismatch occurrences of Q.

A naïve solution for solving the 1-mismatch problem is to construct the suffix tree for T and find the occurrences of every 1-mismatch pattern of Q in the suffix tree. Note that the number of possible 1-mismatch patterns of Q is $(|\mathcal{A}| - 1)m$ and it takes $O(m)$ time to find occurrences of each 1-mismatch pattern. Hence the naïve algorithm takes $O(|\mathcal{A}|m^2 + occ)$ time in total, where occ is the total number of occurrences.

Many sophisticated solutions for solving the indexed 1-mismatch problem have been proposed in the literature. Below, we present the algorithm by Trinh et al. [158].

Consider a text $S[1..n]$. The data structures used in [158] are the suffix array (SA) of S and the inverse suffix array (SA^{-1}) of S, where $SA[SA^{-1}[i]] = i$. SA can be constructed as shown in Section 3.5.1. Given SA, obtaining SA^{-1} in $O(n)$ time is trivial. Both SA and SA^{-1} can be stored in $O(n \log n)$ bits space. See Figure 3.21 for an example of the suffix array and the inverse suffix array of $S = acacag$.

Trinh et al. showed that, given the SA and SA^{-1}, we can compute all 1-mismatch occurrences of a query $Q[1..m]$ in $S[1..n]$ using $O(|\mathcal{A}|m \log n + occ)$ time. Precisely, the algorithm tries to compute the SA range $range(S, Q) = [st..ed]$.

The key trick used in the algorithm is that, for any strings P_1 and P_2, we can compute $range(S, P_1P_2)$ in $O(\log n)$ time given $range(S, P_1)$ and $range(S, P_2)$. The lemma below shows the correctness of the trick.

LEMMA 3.4
Suppose $[st_1..ed_1] = range(S, P_1)$ and $[st_2..ed_2] = range(S, P_2)$ then we can compute $[st..ed] = range(S, P_1P_2)$ in $O(\log n)$ time.

PROOF   Let the length of P_1 be k. Observe that $S[SA[st_1]..n], S[SA[st_1 + 1]..n], \ldots, S[SA[ed_1]..n]$ are lexicographically increasing. Since they share the common prefix P_1, $S[SA[st_1]+k..n], S[SA[st_1+1]+k..n], \ldots, S[SA[ed_1]+k..n]$ are also lexicographically increasing. Thus, we have

$$SA^{-1}[SA[st_1] + k] < SA^{-1}[SA[st_1 + 1] + k] < \ldots < SA^{-1}[SA[ed_1] + k]$$

Therefore, to find st and ed, we need to find

- the smallest st such that $st_2 < SA^{-1}[SA[st] + k] < ed_2$ and

- the largest ed such that $st_2 < SA^{-1}[SA[ed] + k] < ed_2$

This can be done using binary search.   ☐

Suppose we want to find the occurrence of the pattern formed by replacing the character at position j of the query pattern $Q[1..m]$ by x, we can compute $range(S, Q[1..j-1]xQ[j+1..m])$ from $range(S, Q[1..j-1])$, $range(S, x)$, and $range(S, Q[j+1..m])$ by applying Lemma 3.4 twice. Applying this trick on all m positions, we can identify all 1-mismatch occurrences of Q. The detail of the algorithm is shown in Figure 3.22.

Below, we analyze the running time of Trinh et al.'s algorithm. Steps 1 to 3 compute $range(S, x)$ for all $x \in \mathcal{A}$, which takes $O(|\mathcal{A}|)$ time. Steps 4 to 6 compute $range(S, Q[j..m])$ for $1 \le j \le m$, which takes $O(m \log n)$ time. Steps 7 to 9 compute $range(S, Q[1..j])$ for $1 \le j \le m$, which takes $O(m \log n)$ time. For Steps 11 to 18, we enumerate all possible $(|\mathcal{A}| - 1)m$ mismatches, and for each mismatch it takes $O(\log n)$ time to find the corresponding 1-mismatch SA range. Reporting all occurrences takes occ time. Hence the time complexity of the algorithm is $O(|\mathcal{A}|m \log n + occ)$. Note that this algorithm can be generalized to handle k mismatches in $O(|\mathcal{A}|^k m^k \log n + occ)$ time.

3.8   Exercises

1. Given three DNA sequences S_1, S_2, and S_3 of total length n.

Algorithm 1-mismatch_search($Q[1..m]$)

1: **for** each character $x \subset \mathcal{A}$ **do**
2:     Set $range(S, x) = [C[x] + 1..C[x + 1]]$;
3: **end for**
4: **for** $j = m$ to 1 **do**
5:     Compute $range(S, Q[j..m])$ from $range(S, Q[j])$ and $range(S, Q[j + 1..m])$ using Lemma 3.4;
6: **end for**
7: **for** $j = 1$ to m **do**
8:     Compute $range(S, Q[1..j])$ from $range(S, Q[1..j - 1])$ and $range(S, Q[j])$ using Lemma 3.4;
9: **end for**
10: Report all occurrences in $range(S, Q[1..m])$;
11: **for** $j = 1$ to m **do**
12:     Let $P_1 = Q[1..j - 1]$ and $P_2 = Q[j + 1..m]$;
13:     **for** each character $x \in \mathcal{A} - \{Q[j]\}$ **do**
14:         By Lemma 3.4, find $range(S, P_1x)$;
15:         Compute $range(S, P_1x)$ from $range(S, P_1)$ and $range(S, x)$ using Lemma 3.4;
16:         Compute $range(S, P_1xP_2)$ from $range(S, P_1x)$ and $range(S, P_2)$ using Lemma 3.4;
17:         Report all occurrences in $range(S, P_1xP_2)$;
18:     **end for**
19: **end for**

FIGURE 3.22: 1-mismatch search.

(a) Suppose $S_1 = acgatca$, $S_2 = gattact$, $S_3 = aagatgt$. What is the longest common substring of S_1, S_2, and S_3?

(b) Describe an efficient algorithm that computes the length of the longest common substring. What is the time complexity?

(c) It is possible that the longest common substring for S_1, S_2, and S_3 is not unique. Can you describe an efficient algorithm to report the number of possible longest common substrings of S_1, S_2, and S_3? What is the time complexity?

2. Please give the generalized suffix tree for $S_1 = ACGT\$$ and $S_2 = TGCA\#$.

3. Given a DNA sequence S and a pattern P, can you describe an $O(|P|^2 + |S|)$ time algorithm to find all occurrences of P in S with hamming distance ≤ 1?

4. Consider the string $S = ACGTACGT\$$.

(a) What is the suffix array for S?

(b) Report the values of (1) $LCP(k, k+1)$ for $k = 1, 2, \ldots, 8$ and (2) $LCP(k, k+4)$ for $k = 1, 2, 3, 4$.

5. Consider a string S which ends at $\$$. Suppose we are given the suffix array $SA[1..|S|]$ of S and all $LCP(i, j)$ values. Can you propose an algorithm for searching pattern P which takes $O(|P| + log|S|)$ time?

6. Given a DNA sequence S of length n.

(a) Given another DNA sequence W of length smaller than n, describe an efficient algorithm that reports the number of occurrences of W in S. What is the time complexity?

(b) Describe an efficient algorithm that computes the length of the longest substring which appears exactly k times in S. What is the time complexity?

(c) Consider a DNA sequence $X=$ "acacagactacac". What is the number of occurrences of "aca" in X? What is the length of the longest substring which appears exactly 3 times in X?

7. Given a string $S = S[1..n]$ and a number k, we want to find the smallest substring of S that occurs in S exactly k times, if it exists. Show how to solve this problem in $O(n)$ time.

8. For the suffix tree for a DNA sequence S and a query Q of length m, describe an algorithm to determine whether there exists a substring x of S such that the hamming distance between Q and x is smaller than or equal to 1. What is the time complexity?

9. For the suffix tree for S, describe a linear time algorithm for searching maximal complementary palindromes.

10. Consider a set Z of n pairs $\{(a_i, b_i) \mid i = 1, 2, \ldots, n\}$, where each symbol a_i or b_i can be represented as a positive integer smaller than or equal to n. Can you describe an $O(n)$-time algorithm to sort the pairs into a sequence $(a_{(1)}, b_{(1)}), \ldots, (a_{(n)}, b_{(n)})$?

11. Given only the FM-index without the text, can you recover the text from the FM-index? If yes, can you describe the detail of the algorithm? What is the time complexity?

12. Given the FM-index, can you generate the suffix array in $O(n)$ time using $O(n)$ bits working space? (For counting the working space, we do not include the space for outputting the suffix array. Also, we are allowed to write the output once. Based on this model, the output can be written directly to the secondary storage, without occupying the main memory.)

13. Given a set of k strings of total length n, give an $O(n)$-time algorithm which computes the longest common substring of all k sequences.

14. Given a set of k strings, we would like to compute the longest common substring of each of the $\binom{k}{2}$ pairs of strings.

    (a) Assume each string is of length n. Show how to find all the longest common substrings in $O(k^2 n)$ time.

    (b) Assume the string lengths are different but sum to m. Show how to find all the longest common substrings in $O(km)$ time.

15. Given a text of length n with constant alphabet, propose an $O(n \log n)$-bit space index and a query algorithm such that, for any pattern P of length m, the 2-mismatch query can be answered in $O(m^2 \log n + occ)$ time.

16. Let T be a DNA sequence of length n. Suppose we preprocess T and get its suffix array SA, its inverse suffix array SA^{-1}, and its FM-index. Consider a pattern P of length m whose alphabet is $\{a, c, g, t, *\}$, where $*$ is a wild card (which represents any DNA base). Assuming P contains x wild cards, can you propose an $O(m + 4^x \log n + occ)$-time algorithm to find all the occurrences of P? (occ is the number of occurrences of P in T.)

17. Consider two DNA sequences S and T of total length n. Describe an $O(n)$ time algorithm which detects whether there exists a substring T' of T such that the score of the global alignment between S and T' is bigger than or equal to -1. (Suppose the score for a match is 0 and the score for insert, delete, or mismatch is -1.)

18. For a sequence S, give an efficient algorithm to find the length of the longest pattern which has at least two non-overlapping occurrences in S. What is the running time?

19. Prove or disprove each of the following.

    (a) Let T be a suffix tree for some string. Let the string α be the label of an edge in T and β be a proper prefix of α. It is not possible to have an internal node in T with path label β.

    (b) Let u be an internal node with path label $xy\alpha$ (where x and y are single characters and α is a string) in a suffix tree. An internal node with path label α always exists.

20. Given a text $T = acgtcga\$$. Can you create a suffix array for T? Can you create $BW[1..8]$? Can you create $C[a]$, $C[c]$, $C[g]$, and $C[t]$? Also, can you demonstrate the steps for finding the pattern $P = cg$ using backward search?

Chapter 4

Genome Alignment

4.1 Introduction

Due to the advance in sequencing technologies, complete genomes for many organisms are available. Biologists start to ask if we can compare two or more related organisms to extract their similarities and differences. Such a question can be answered by solving the whole genome alignment problem, which tries to align the two closely related genomes.

In the literature, a lot of methods are available to align two DNA sequences, including the Smith-Waterman algorithm, the Needleman-Wunsch algorithm, etc. (See Chapter 2 for details.) Those methods are effective when we compare two genes or two proteins which are short. When we compare two genomes, those methods become ineffective due to the high time and space complexity, which is $O(nm)$ where n and m are the lengths of the two genomes.

Hence, special tools for comparing large scale genomes were developed, including: MUMmer [76], Mutation Sensitive Alignment [56], SSAHA [220], AVID [34], MGA [149], BLASTZ [261], and LAGAN [39].

In general, all genome alignment methods assume the two genomes should share some conserved regions. Then, they align the conserved regions first; afterward, the alignment is extended to cover the non-conserved regions. More precisely, genome alignment methods often run in three phases.

- Phase 1 identifies potential anchors. Anchors are short regions between two genomes which are highly similar. Those regions are possible conserved regions between two genomes. Different methods define anchors differently. One simple way is to define anchors as shared k-mers (i.e., the length-k substrings shared by the two genomes).

- One single anchor may not correspond to a conserved region. Moreover, a set of co-linear anchors is likely to be a conserved region. Phase 2 aims to identify a set of co-linear non-overlapping anchors. Such co-linear anchors form the basis of the alignment.

- Phase 3 closes the gaps between the anchors to obtain the final alignment.

$$S = \text{acga } \underline{ctc} \text{ a } \underline{gctac} \text{ t } \underline{ggtcagctatt} \; \underline{acttaccgc} \; \#$$
$$T = \text{actt } \underline{ctc} \text{ t } \underline{gctac} \; \underline{ggtcagctatt} \text{ c } \underline{acttaccgc} \; \$$$

FIGURE 4.1: A pair of conserved sequences S and T. The underlined regions are MUMs.

To illustrate the idea of the whole genome alignment, this chapter discusses two methods, MUMmer and Mutation Sensitive Method. The organization of this chapter is as follows. We first study the concept Maximal Unique Match (MUM), which is a type of anchor. Then, we will detail the algorithms of MUMmer (Versions 1 to 3) and Mutation Sensitive Alignment algorithm. Last, we discuss dot plot, a way to visualize the alignment of two genomes.

4.2   Maximum Unique Match (MUM)

Although conserved regions shared by two genomes rarely contain exactly the same sequence, the conserved regions of two genomes usually share some short common substrings which are unique in the two genomes. For example, conserved genes usually share some short common substrings which are unique since they need to preserve functionality.

Hence, Maximal Unique Match (MUM) was proposed to be an anchor for defining candidate conserved regions. Precisely, MUM is defined as follows.

DEFINITION 4.1   *Given two genomes A and B, Maximal Unique Match (MUM) substring is a common substring of A and B of length longer than a specific minimum length d (by default, d = 20) such that*

- *it is maximal, that is, it cannot be extended on either end without incurring a mismatch; and*

- *it is unique in both sequences.*

For example, consider the pair of conserved sequences S and T in Figure 4.1. Assume $d = 3$. Sequences S and T contain four MUMs: *ctc*, *gctac*, *ggtcagctatt*, *acttaccgc*. Substring *ac* is not a MUM because its length is smaller than $d = 3$ and it is not unique in both sequences. Substring *ggt* is not a MUM since *ggt* is not maximal.

As another example, consider $S = acgat\#$ and $T = cgta\$$ (see Figure 4.2). Assuming $d = 1$, there are two MUMs: *cg* and *t*. However, *a* is not a MUM because it occurs twice in S.

$$S = a \text{ } \underline{cg} \text{ } a \text{ } \underline{t} \text{ } \#$$
$$T = \underline{cg} \text{ } \underline{t} \text{ } a\$$$

FIGURE 4.2: A pair of conserved sequences S and T. The underlined regions are MUMs.

Algorithm Brute-force-MUM

Require: Two genome sequences $S[1..m_1]$ and $T[1..m_2]$, a threshold d
Ensure: The set of MUMs of S and T
  1: **for** every position i in S **do**
  2:   **for** every position j in T **do**
  3:     Find the longest common prefix P of $S[i..m_1]$ and $T[j..m_2]$
  4:     Check whether $|P| \geq d$ and whether P is unique in both genomes.
         If yes, report it as a MUM.
  5:   **end for**
  6: **end for**

FIGURE 4.3: Algorithm for computing the MUMs of two genomes S and T using the brute-force method.

4.2.1 How to Find MUMs

Consider two genomes $S[1..m_1]$ and $T[1..m_2]$. How can we find the MUMs? A simple solution is to compute all MUMs in S and T by brute-force. Precisely, we first find all common substrings of the two sequences. Then, for each substring, we check whether it is longer than d and unique in both sequences. The algorithm is shown in Figure 4.3. Since we need to check every pair of positions in S and T, the brute-force solution requires at least $O(m_1m_2)$ time. Such a solution is too slow as the size of the genomes is in the order of millions.

We can in fact compute all MUMs in linear time by utilizing the generalized suffix tree for genomes S and T. The algorithm consists of three steps. Figure 4.4 details the algorithm.

First, we build the generalized suffix tree for genomes S and T. Then, we mark internal nodes v with exactly one leaf from S and one leaf from T. The path labels of those marked nodes v are shared by both sequences and at the same time unique to both sequences. Third, to make sure that the path labels of those nodes are of maximal length, we check whether $S[i - 1] = T[j - 1]$. If $S[i - 1] = T[j - 1]$, it means that the substring is not maximal as we can extend it at least by one character by adding $S[i - 1]$ before its first character; otherwise, the path label is a MUM. Finally, we report all MUMs of length at least d.

Figure 4.5 demonstrates how the algorithm constructs the MUMs for $S = acgat\#$ and $T = cgta\$$ when $d = 1$ (see Figure 4.2). In Step 1, the generalized

---

Algorithm Suffix-tree-MUM

Require: Two genome sequences $S[1..m_1]$ and $T[1..m_2]$, a threshold d

Ensure: The set of MUMs of S and T

1: Build a generalized suffix tree for S and T.

2: Mark all the internal nodes that have exactly two leaf children, which represent both suffixes of S and T.

3: For each marked node of depth at least d, suppose it represents the i-th suffix S_i of S and the j-th suffix T_j of T. We check whether $S[i-1] = T[j-1]$. If not, the path label of this marked node is a MUM.

FIGURE 4.4: A suffix tree-based algorithm for computing the MUMs of two genomes S and T.

suffix tree for S and T is constructed. Step 2 marks internal nodes with path labels cg, g, and t since they have exactly two leaf children, which represent both suffixes of S and T. In Step 3, for the node with the path label g, it represents the third suffix $gat\#$ of S and the second suffix $gta\$$ of T. Since both $S[3-1]$ and $T[2-1]$ equal c, the path label g of this marked node is not a MUM. So the set of MUMs is $\{cg, t\}$.

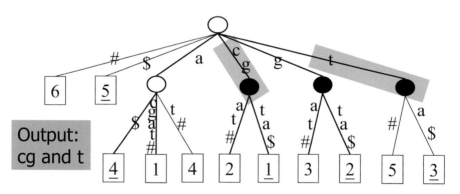

FIGURE 4.5: Consider two genomes $S = acgat\#$ and $T = cgta\$$. Step 1 constructs the generalized suffix tree for genomes S and T. Step 2 identifies internal nodes with exactly one descendant leaf representing S and exactly one descendant leaf representing T. Those nodes are filled with solid color. Step 3 identifies MUMs, which correspond to those highlighted paths.

Below, we analyze the time complexity. Step 1 takes $O(m_1 + m_2)$ time to build the generalized suffix tree for S and T (see Section 3.4). Since the generalized suffix tree has at most $m_1 + m_2$ internal nodes, Step 2 takes $O(m_1 + m_2)$ time to mark the internal nodes. In Step 3, comparing $S[i-1]$ and $T[j-1]$ for each marked node takes $O(1)$ time. Since there are at most

$O(m_1 + m_2)$ marked nodes, the running time of Step 3 is also $O(m_1 + m_2)$. In total, the algorithm takes $O(m_1 + m_2)$ time to find all MUMs of the two genomes S and T. For space complexity, the method stores the generalized suffix tree, which takes $O((m_1 + m_2) \log(m_1 + m_2))$ bits space.

To verify if MUMs can identify the conserved regions, we performed an experiment on a set of chromosome pairs of humans and mice. Figure 4.6 shows the number of conserved gene pairs on those chromosome pairs. If MUMs can identify the conserved regions between humans and mice, MUMs should cover the majority of those conserved gene pairs. In fact, our experiment showed that MUMs can cover 100% of the known conserved gene pairs. However, Figure 4.6 also shows the number of MUMs in each pair of chromosomes is much larger than the number of conserved gene pairs. This may imply that most of the MUMs are noise. We need some methods to filter those noisy MUMs. The sections below propose some methods.

Mouse Chr.	Human Chr.	# of Published Gene Pairs	# of MUMs
2	15	51	96,473
7	19	192	52,394
14	3	23	58,708
14	8	38	38,818
15	12	80	88,305
15	22	72	71,613
16	16	31	66,536
16	21	64	51,009
16	22	30	61,200
17	6	150	94,095
17	16	46	29,001
17	19	30	56,536
18	5	64	131,850
19	9	22	62,296
19	11	93	29,814

FIGURE 4.6: Mice and humans share a lot of gene pairs. The set of known conserved genes was obtained from *GenBank* [217]. The number of MUMs was computed using the method in Section 4.2.1.

4.3   MUMmer1: LCS

It has been observed that two closely related species preserve the ordering of most of the conserved genes. For example, if we compare mouse chromosome 16 and human chromosome 16, we find that the ordering of 30 conserved gene pairs (out of 31 gene pairs) is preserved. Refer to Figure 4.7 for the visualization.

Mouse Chromosome 16

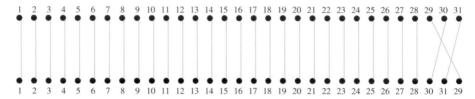

Human Chromosome 16

FIGURE 4.7:   There are 31 conserved gene pairs found between mouse chromosome 16 and human chromosome 16. The genes are labeled according to their positions in the chromosomes. The figure shows that the ordering of 30 conserved gene pairs (out of 31 gene pairs) is preserved. (See color insert after page 270.)

Based on this observation, Delcher et al. [76] suggested that MUMs which preserve order in both S and T are likely to cover the conserved regions of S and T. They proposed MUMmer1 to identify those order preserving MUMs.

Consider two closely related genomes S and T. Suppose S and T contain n MUMs. Let $A = (a_1, a_2, \cdots, a_n)$ and $B = (b_1, b_2, \cdots, b_n)$ be the ordering of the n MUMs in S and T, respectively. A sequence $C = (c_1, c_2, \cdots, c_m)$ is a subsequence of A (or B) if C can be formed by removing $n - m$ elements from A (or B).

The aim of MUMmer1 is to compute the longest common subsequence (LCS) of A and B, that is, the longest possible common subsequence of both A and B. For example, for the set of MUMs of two genomes S and T in Figure 4.8, $(2, 4, 5, 6, 7)$ is a longest common subsequence.

The naïve way to solve the longest common subsequence problem is to use dynamic programming. Suppose there are n MUMs. Section 4.3.1 shows a dynamic programming algorithm which solves the LCS problem using $O(n^2)$ time and space. This approach quickly becomes infeasible as the number of MUMs increases.

Moreover, since all MUMs are unique (by definition), we can represent each

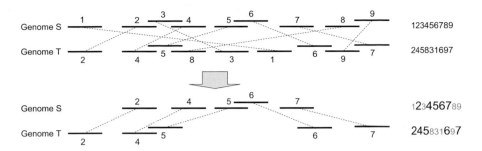

FIGURE 4.8: Example of an LCS.

MUM by a different character. For this special case, we can solve the LCS problem in $O(n \log n)$ time. Below, we first briefly describe the $O(n^2)$-time algorithm; then, we detail the $O(n \log n)$-time algorithms.

4.3.1 Dynamic Programming Algorithm in $O(n^2)$ Time

The crux of the dynamic programming algorithm lies in the fact that the LCS of $A[1..i]$ and $B[1..j]$ is related to the LCS of $A[1..(i-1)]$ and $B[1..(j-1)]$. Let us define $V_i[j]$ to be the length of the longest common subsequence of $A[1..i]$ and $B[1..j]$. Also, we define $\delta(i)$ to be the index in A such that $A[i] = B[\delta(i)]$. There are two possibilities in the longest common subsequence of $A[1..i]$ and $B[1..j]$: (1) $A[i]$ is not involved in the longest common subsequence and (2) $A[i]$ is involved in the longest common subsequence. Hence, we have the following recursive formula.

$$V_i[j] = \max \begin{cases} V_{i-1}[j], & j < \delta(i) \\ 1 + V_{i-1}[\delta(i) - 1] & j \geq \delta(i) \end{cases} \tag{4.1}$$

For example, consider $A[1..9] = 123456789$ and $B[1..9] = 245831697$. The corresponding dynamic programming table V is shown in Figure 4.9. The length of the LCS of A and B is $V_9[9] = 5$.

For time analysis, we need to fill in n^2 entries of $V_i[j]$ where each entry can be computed in constant time. Hence, the whole V table can be computed in $O(n^2)$ time. Then, the length of the LCS of A and B is $V_n[n]$. We can recover the LCS of A and B by back-tracing using additional $O(n)$ time.

4.3.2 An $O(n \log n)$-Time Algorithm

The dynamic programming method in Section 4.3.1 runs in $O(n^2)$ time, which is too slow when n is big. This section gives an $O(n \log n)$-time algorithm. The idea is to sparsify the dynamic programming method in Section 4.3.1.

V	0	1	2	3	4	5	6	7	8	9
0	0	0	0	0	0	0	0	0	0	0
1	0	0	0	0	0	0	1	1	1	1
2	0	1	1	1	1	1	1	1	1	1
3	0	1	1	1	1	2	2	2	2	2
4	0	1	2	2	2	2	2	2	2	2
5	0	1	2	3	3	3	3	3	3	3
6	0	1	2	3	3	3	3	4	4	4
7	0	1	2	3	3	3	3	4	4	5
8	0	1	2	3	4	4	4	4	4	5
9	0	1	2	3	4	4	4	4	5	5

FIGURE 4.9: The dynamic programming table V when $A[1..9] = 123456789$ and $B[1..9] = 245831697$.

First, we observe that

$$V_i[0] \leq V_i[1] \leq V_i[2] \leq \ldots \leq V_i[n]$$

Instead of storing $V_i[0..n]$ explicitly, we can store only the boundaries at which the values change. We will store $(j, V_i[j])$ for all j such that $V_i[j] > V_i[j - 1]$. For example, consider the row 7 of the table V in Figure 4.9, which is $V_7[0..9] = 0123333445$. We can represent the row using a list of five tuples $(1, 1), (2, 2), (3, 3), (7, 4)$, and $(9, 5)$.

Given the row V_{i-1}, the lemma below states how to compute the row V_i.

LEMMA 4.1
Let j' be the smallest integer greater than $\delta(i)$ such that $V_{i-1}[\delta(i) - 1] + 1 < V_{i-1}[j']$. We have, for $j = 1, 2, \ldots, n$,

$$V_i[j] = \begin{cases} V_{i-1}[\delta(i) - 1] + 1 & \text{if } \delta(i) \leq j \leq j' - 1 \\ V_{i-1}[j] & \text{otherwise} \end{cases}$$

PROOF   By Equation 4.1, we have the following 3 cases:

- When $j < \delta(i)$, $V_i[j] = V_{i-1}[j]$.

- When $\delta(i) \leq j < j'$, we have $V_{i-1}[j] \leq V_{i-1}[\delta(i) - 1] + 1$. Hence, $V_i[j] = \max\{V_{i-1}[j], V_{i-1}[\delta(i) - 1] + 1\} = V_{i-1}[\delta(i) - 1] + 1$.

Algorithm Suffix-tree-MUM

Require: $A[1..n]$ and $B[1..n]$

Ensure: The LCS between A and B

1: Construct a binary search tree T_1 which contains one tuple $(\delta(1), 1)$.

2: **for** $i = 2$ to n **do**

3:      $T_i = T_{i-1}$

4:      Delete all tuples $(j, V_{i-1}[j])$ from T_i where $j \geq \delta(i)$ and $V_{i-1}[j] \leq V_{i-1}[\delta(i) - 1] + 1$;

5:      Insert $(\delta(i), V_i[\delta(i)])$ into T_i.

6: **end for**

FIGURE 4.10: An $O(n \log n)$-time algorithm for computing the length of the longest common subsequence (LCS) between A and B.

- When $j \geq j'$, we have $V_{i-1}[j] \geq V_{i-1}[\delta(i) - 1] + 1$. Hence, $V_i[j] = \max\{V_{i-1}[j], V_{i-1}[\delta(i) - 1] + 1\} = V_{i-1}[j]$.

The lemma follows. $\square$

By the above lemma, we can construct the row V_i from the row V_{i-1} as follows.

1. We delete all tuples $(j, V_{i-1}[j])$ in V_{i-1} where $j \geq \delta(i)$ and $V_{i-1}[j] \leq V_{i-1}[\delta(i) - 1] + 1$.

2. We insert $(\delta(i), V_{i-1}[\delta(i) - 1] + 1)$.

For example, consider $A = 123456789$ and $B = 245831697$ in Figure 4.9. $V_7[0..9] = 0123333445$, which is represented by a list of five tuples $(1, 1)$, $(2, 2)$, $(3, 3)$, $(7, 4)$, and $(9, 5)$. $V_8[0..9] = 0123444445$ is represented by a list of five tuples $(1, 1)$, $(2, 2)$, $(3, 3)$, $(4, 4)$, and $(9, 5)$. In fact, V_8 can be constructed from V_7 as follows. Note that $\delta(8) = 4$ and $V_7[\delta(8) - 1] + 1 = 4$. First, we need to remove all tuples $(j, V_7[j])$ in V_7 where $j \geq \delta(8)$ and $V_7[j] \leq V_7[\delta(8)-1]+1$. In other words, we need to remove the tuple $(7, 4)$. Then, V_8 can be constructed by inserting $(\delta(8), V_7[\delta(8) - 1] + 1) = (4, 4)$.

In the implementation, the tuples of each row V_i are maintained using a binary search tree T_i. This allows us to search, insert, or delete any tuple in $O(\log n)$ time. Figure 4.10 states the algorithm.

Note that, for each iteration $i = 2, \ldots, n$, we delete α_i tuples and insert one tuple. Thus, we can construct V_n using $O((\alpha_1 + \alpha_2 + .. + \alpha_n + n) \log n)$ time. Since we insert at most n tuples and every tuple can be deleted at most once, we have $\alpha_1 + \alpha_2 + .. + \alpha_n \leq n$. Hence, V_n can be constructed in $O(n \log n)$ time.

In summary, MUMmer1 performs three steps to construct the alignment between two genomes $S[1..m_1]$ and $T[1..m_2]$.

- Step 1: According to Section 4.2.1, for two sequences of length m_1 and m_2 respectively, all MUMs can be computed in $O(m_1 + m_2)$ time.

- Step 2: After finding all MUMs, we sort them according to their position in genome S, and employ the LCS algorithm to find the largest set of MUMs which has the same orderings in both genomes. This step requires $O(n \log n)$ time, where n is the number of MUMs. In general, n is much smaller than m_1 and m_2.

- Step 3: Once a global alignment is found, we can deploy several algorithms for closing the local gaps and completing the alignment. A gap is defined as an interruption in the MUM alignment which falls into one of the following four classes: (i) an SNP, (ii) an insertion, (iii) a highly polymorphic region, or (iv) a repeat. Time and space complexity are dependent on the algorithm deployed.

MUMmer1 is one of the first methods which allows us to align two sequences at the genome level. The method is fast and produces fewer false positives. However, MUMmer1 assumes that there exists a single long alignment, which may not always be true, as there may be a huge chain of reversals/transpositions between two genomes. See Figure 4.11 for an example.

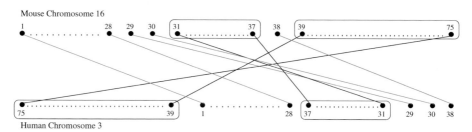

FIGURE 4.11: There are 31 conserved gene pairs found in mouse chromosome 16 and human chromosome 3. Note that the relative ordering of genes 31 to 37 in the human chromosome is exactly the reverse of that in the mouse chromosome. The same occurs in genes 39 to 75. (See color insert after page 270.)

4.4   MUMmer2 and MUMmer3

There are four improvements in MUMmer2 and MUMmer3 [181], which are listed as follows. The first two improvements enable MUMmer to align even bigger genomes. The last two improvements help to improve the coverage of the alignment.

4.4.1 Reducing Memory Usage

By employing techniques described by Kurtz [180], the amount of memory used to store the suffix tree in MUMmer2 is reduced to at most 20 bytes/base. The maximum memory usage occurs when each internal node in the suffix tree has only two children. In practice, however, many nodes have more than two children (particularly in the case of polypeptide sequences), which reduces the actual memory requirement. In MUMmer3, the memory usage is further reduced by using the compact suffix tree representation. The amount of memory used is reduced to at most 16 bytes/base.

4.4.2 Employing a New Alternative Algorithm for Finding MUMs

The MUMmer1 [76] employs an algorithm that builds a generalized suffix tree for two input sequences to find all MUMs. Although this algorithm is still available in the MUMmer2 system, a new alternative algorithm was introduced as the default algorithm to find all MUMs. Instead of storing the generalized suffix tree for the two input sequences, this new algorithm stores only one input sequence in the suffix tree. This sequence is called the reference sequence. The other sequence, which is called the query, is then "streamed" against the suffix tree, exactly as if it was being added without actually adding it. This technique was introduced by Chang and Lawler [57] and is fully described in [128].

By using this new algorithm, all maximal matches between the query sequence and any unique substring of the reference sequence can be identified in a time proportional to the length of the query sequence.

The advantage of this method is that only one of the two sequences is stored in the suffix tree, reducing the memory requirement by at least half. Furthermore, because of the streaming nature of the algorithm, once the suffix tree has been built for an arbitrarily long reference sequence, multiple queries can be streamed against it. Therefore, we don't have to re-build the suffix tree if one of the input sequences is changed. In fact, Delcher et al. [77] have used these programs to compare two assemblies of the entire human genome (each approximately 2.7 billion characters long), using each chromosome as a reference and then streaming the other entire genome pass it.

4.4.3 Clustering Matches

Biologists observed that a pair of conserved genes is likely to correspond to a sequence of MUMs that are consecutive and close in both genomes. At the same time, this sequence of MUMs generally has a sufficient length. The set of such MUMs is called a cluster.

For example, as shown in Figure 4.12, for the given two sequences, there are two clusters: 123 and 567.

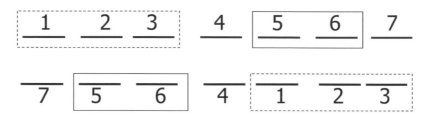

FIGURE 4.12:   Clusters of MUMs.

MUMmer1 presumed that the two input genomes have no major rearrangement between them. Hence, it computed a single longest alignment between the sequences. In order to facilitate comparisons involving unfinished assemblies and genomes with significant rearrangements, a module has been added in MUMmer2. This module first groups all the close consecutive MUMs into clusters and then finds consistent paths within each cluster. The clustering is performed by finding pairs of matches that are sufficiently close and on sufficiently similar diagonals in an alignment matrix (using thresholds set by the user), and then computing the connected components for those pairs. Within each component, a longest common subsequence is computed to yield the most consistent sequence of matches in the cluster.

As the result of the additional module, the system outputs a series of separate, independent alignment regions. This improvement enables MUMmer2 to achieve better coverage.

4.4.4   Extension of the Definition of MUM

In MUMmer3, the concept of MUM is slightly relaxed so that it needs not to be unique in both input sequences. We can opt to find all non-unique maximal matches, all matches that are unique only in the reference sequence, or all matches that are unique in both sequences. This feature was added as it was observed that the uniqueness constraint would prevent MUMmer from finding all matches for a repetitive substring.

4.5   Mutation Sensitive Alignment

The Mutation Sensitive Alignment (MSA) algorithm [56] observes that when two genomes are closely related, it is likely that they can be transformed from each other using a few transposition/reversal operations. For example, it can be seen in Figure 4.11 that mouse chromosome 16 can be transformed into human chromosome 3 by two reverse transposition operations (on sequences 31–37 and 39–75).

The MSA algorithm makes use of this observation. It looks for subsequences in the two input genome strings that differ by at most k transposition/reversal operations ($k = 4$, by default). Then, these subsequences are reported as possible conserved regions. The rest of this section describes the details of the MSA algorithm.

4.5.1   Concepts and Definitions

This section describes the similar subsequence problem (or k-similar subsequence problem), which is the basis of the MSA algorithm.

Assume A and B are the ordering of the n MUMs on the two genomes. Each MUM is assigned a weight. Normally, the weight of a MUM is set to be its length. The weight of a common subsequence of A and B is the total weight of all the MUMs that make up the subsequence. The maximum weight common subsequence (MWCS)[1] is the common subsequence of A and B with maximum weight. The k-similar subsequence is defined as follows.

DEFINITION 4.2   *Let k be a non-negative integer. Consider a subsequence X of A and a subsequence Y of B. (X,Y) is a pair of k-similar subsequences if X can be transformed into Y by performing k transposition/reversal operations on k disjoint subsequences in X.*

Intuitively, if (X,Y) is a pair of k-similar subsequences, X and Y consist of a backbone, which is a common subsequence of X and Y, which breaks X (resp., Y) into k disjoint blocks such that each block in X is either a block in Y or a reversed block of Y. For example, Figure 4.13 presents a 2-similar subsequence. In this example, the two blocks are $(2,3)$ and $(4,5,6)$, whereas the backbone is $(1,7,8)$.

The k-similar subsequence problem aims to find a k-similar subsequence of A and B which maximizes the total weight. This problem has been shown to be NP-complete [56], by a reduction from the MAX-2SAT problem [115]. Us-

---

[1]MWCS is a generalization of LCS (see Section 4.3.2). If the weight of every MWM is 1, MWCS is the same as LCS.

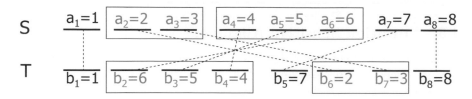

FIGURE 4.13:   An example pair of 2-similar subsequences.

ing dynamic programming, this k-similar subsequence problem can be solved in $O(n^{2k+1} \log n)$ time, where n is the number of MUMs (see Exercise 10). Unfortunately, this is too slow for practical purposes. Below, we discuss a heuristic algorithm which solves the problem in $O(n^2(\log n + k))$ time [56].

4.5.2   The Idea of the Heuristic Algorithm

A heuristic algorithm is proposed in [56] which assumes the backbone of the k-similar subsequence of A and B looks similar to the MWCS of A and B. The algorithm first generates the MWCS of A and B as the backbone; then, it refines the alignment by introducing k similar subsequences into the backbone. The detailed steps are shown below.

1. Define the backbone as the MWCS of A and B;

2. For every interval $A[i..j]$, compute the score $Score(i, j)$ of inserting such a subsequence into the backbone;

3. Find k intervals $A[i_1..j_1], \ldots, A[i_k..j_k]$ which maximize the total score;

4. Refine the k intervals so that $B[\delta(i_1)..\delta(j_1)], \ldots, B[\delta(i_k)..\delta(j_k)]$ are also disjoint.

The first step finds the maximum weight common subsequence (MWCS) of A and B. The MWCS can be computed in $O(n \log n)$ time using an algorithm similar to that described in Section 4.3.2 (see Exercise 7). We will consider the MWCS as the backbone.

For the second step, we compute $Score(i, j)$ for every interval $A[i..j]$, where $Score(i, j)$ is defined to be the difference of the following two terms.

- $MWCS(A[i..j], B[\delta(i)..\delta(j)])$;

- The total weight of characters in the backbone that fall into $A[i..j]$ or $B[\delta(i)..\delta(j)]$.

The running time of the second step is dominated by the computation of $MWCS(A[i..j], B[\delta(i)..\delta(j)])$ for all $1 < i, j < n$. By brute-force, we need to compute $\binom{n}{2}$ value where each value can be computed in $O(n \log n)$ time.

In total, the running time is $O(n^3 \log n)$. Below, we show a better algorithm which takes $O(n^2 \log n)$ time. The basic observation is that after computing the MWCS for $A[1..n]$ and $B[1..n]$ using the $O(n \log n)$-time algorithm, we obtain a set of binary search trees $T_1, \ldots, T_n$. Then, the MWCS for $A[1..i]$ and $B[1..j]$ can be retrieved from the binary search tree T_i in $O(\log n)$ time. Hence, we have the following $O(n^2 \log n)$-time algorithm.

1: **for** $i = 1$ to n **do**
2:     Find the MWCS for $A[i..n]$ and $B[\delta(i)..n]$ in $O(n \log n)$ time.
3:     Find the MWCS for $A[i..n]$ and reverse of $B[1..\delta(i)]$ in $O(n \log n)$ time.
4:     Retrieve $\mathrm{MWCS}(A[i..j], B[\delta(i)..\delta(j)])$ in $O(\log n)$ time for every $j \geq i$
5: **end for**

The third step of the heuristic algorithm finds k intervals $i_1..j_1, i_2..j_2, \ldots,$ and $i_k..j_k$ among all intervals such that they are mutually disjoint and maximize the sum of their scores. This step can be solved by dynamic programming. For $0 \leq c \leq k$ and $j \leq n$, we define $Opt(c, j)$ to be the maximum sum of the scores of c intervals in $A[1..j]$.

$$Opt(c, j) = \max \begin{cases} Opt(c, j - 1) \\ max_{1 \leq i \leq j}[Opt(c - 1, i - 1) + Score(i, j)] \end{cases}$$

Based on the recursive formula, we have the algorithm in Figure 4.14. For time complexity, since the table Opt has kn entries and each entry can be computed in $O(n)$ time, we can compute $Opt(k, n)$ in $O(kn^2)$ time.

The last step of the heuristic algorithm is to refine the k intervals $i..j$ so that $B[\delta(i)..\delta(j)]$ are disjoint. While there exist $\delta(i)..\delta(j)$ and $\delta(i')..\delta(j')$ that are overlapping, we examine all possible ways to shrink the intervals $i..j$ and $i'..j'$ so that the score is maximized and $\delta(i)..\delta(j)$ and $\delta(i')..\delta(j')$ do not overlap. This step can be done in $O(k^2)$ time. Hence, the heuristic algorithm takes $O(n^2(\log n + k))$ time.

In summary, given two genomes $S[1..m_1]$ and $T[1..m_2]$ and a non-negative parameter k as its input, the MSA algorithm computes the maximum weight k-similar subsequence as output. The MSA algorithm consists of 2 steps as follows:

1. Find all MUMs of S and T (as shown in Section 3.2.2, it can be done in linear time) and let A and B be the ordering of those MUMs on the two genomes.

2. Run the heuristics algorithm to compute the k-similar subsequences of A and B.

The total running time of the algorithm is $O(m_1 + m_2 + n^2(\log n + k))$ where n is the total number of MUMs.

Algorithm Compute-Opt

Require: A set of n MUMs, a parameter k, and a score function $Score()$

Ensure: A set of at most k disjoint intervals $\{(i_1, j_1), \ldots, (i_k, j_k)\}$ which
   maximizes $\sum_{\ell=1}^{k} Score(i_\ell, j_\ell)$

1: Set $Opt(c, 0) = 0$ for all $c = 0$ to k
2: Set $Opt(0, j) = 0$ for all $j = 0$ to n
3: **for** $c = 1$ to k **do**
4:     **for** $j = 1$ to n **do**
5:         Set $Opt(c, j) = Opt(c, j - 1)$
6:         **for** $i = 1$ to j **do**
7:             Set $Opt(c, j) = \max\{Opt(c, j), Opt(c - 1, i - 1) + Score(i, j)\}$
8:         **end for**
9:     **end for**
10: **end for**
11: Report $Opt(k, n)$

FIGURE 4.14:   An $O(kn^2)$-time algorithm for computing a set of k disjoint intervals $\{(i_1, j_1), \ldots, (i_k, j_k)\}$ which maximizes $\sum_{\ell=1}^{k} Score(i_\ell, j_\ell)$.

4.5.3   Experimental Results

This section describes two genome alignment experiments.

The first experiment applies MSA and MUMmer3 on 15 pairs of mouse and human chromosomes (described in Figure 4.6). In the experiment, the parameter k for MSA was set to be 4, while the parameter gap for MUMmer3 was set at 2000. For each of the chromosome pairs, both techniques were rated on their *coverage* and *preciseness*. The coverage indicates the percentage of published genes discovered, whereas the preciseness indicates the percentage of results returned that matches some published gene pairs. The results are given in Figure 4.15. They showed that both MUMmer3 and MSA have a good coverage. However, the preciseness is only about 30%. This means that either (1) human and mouse share a lot of non-gene conserved regions or (2) there are still many unknown conserved genes.

In the next experiment, we apply MUMmer3 and MSA to 15 pairs of Baculoviridae genomes that are in the same genus (see Figure 4.16 for details). Since viruses have a higher mutation rate, their DNA sequences show a much lower degree of similarity. We define MUMs using the translated protein sequences of the pairs of viruses instead. MUM pairs of length at least three amino acids are extracted. The parameter k for MSA was set to be 20 since we expect viruses to have more rearrangements. The results on coverage and preciseness are shown in Figure 4.17. They show that both MUMmer3 and MSA have good coverage and good preciseness.

Exp. No.	Coverage		Preciseness	
	MUMmer	MSA	MUMmer	MSA
1	76.50%	92.20%	21.70%	22.70%
2	71.40%	91.70%	21.30%	25.10%
3	87.00%	100.00%	24.80%	25.50%
4	76.30%	94.70%	27.40%	26.70%
5	92.50%	96.30%	32.50%	32.00%
6	72.20%	95.80%	31.20%	32.90%
7	67.70%	87.10%	13.50%	17.80%
8	78.10%	90.60%	37.20%	36.70%
9	80.00%	86.70%	40.70%	49.70%
10	82.00%	92.00%	30.90%	32.10%
11	65.20%	89.10%	30.50%	36.00%
12	60.00%	80.00%	27.50%	41.90%
13	89.10%	95.30%	18.20%	18.40%
14	72.70%	86.40%	10.40%	12.60%
15	78.50%	91.40%	30.00%	29.70%
average	76.60%	91.30%	26.50%	29.30%

FIGURE 4.15: Performance comparison between MUMmer3 and MSA.

Exp. No.	Virus	Virus	Length	# of MUMs	# of Conserved Genes
1	AcMNPV	BmNPV	133K*128K	35,166	134
2	AcMNPV	HaSNPV	133K*131K	64,291	98
3	AcMNPV	LdMNPV	133K*161K	65,227	95
4	AcMNPV	OpMNPV	133K*131K	59,949	126
5	AcMNPV	SeMNPV	133K*135K	66,898	100
6	BmNPV	HaSNPV	128K*131K	63,939	98
7	BmNPV	LdMNPV	128K*161K	63,086	93
8	BmNPV	OpMNPV	128K*131K	58,657	122
9	BmNPV	SeMNPV	128K*135K	66,448	99
10	HaSNPV	LdMNPV	131K*161K	57,618	92
11	HaSNPV	OpMNPV	131K*131K	59,125	95
12	HaSNPV	SeMNPV	131K*135K	64,980	101
13	LdMNPV	OpMNPV	161K*131K	75,906	98
14	LdMNPV	SeMNPV	161K*135K	62,545	102
15	OpMNPV	SeMNPV	131K*135K	63,261	101
16	CpGV	PxGV	123K*100K	59,733	97
17	CpGV	XcGV	123K*178K	63,258	107
18	PxGV	XcGV	100K*178K	81,020	99

FIGURE 4.16: Experiment datasets on Baculoviridae.

4.6 Dot Plot for Visualizing the Alignment

To understand the alignment between two genomes, we need some visualization tools. Dot plot is a simple and powerful tool to visualize the similar

Exp. No.	Coverage		Preciseness	
	MUMmer	MSA	MUMmer	MSA
1	100%	100%	44%	91%
2	58%	80%	80%	85%
3	58%	69%	64%	80%
4	83%	95%	91%	94%
5	61%	68%	70%	85%
6	59%	73%	78%	87%
7	58%	69%	47%	79%
8	83%	94%	86%	94%
9	63%	69%	65%	86%
10	75%	85%	75%	85%
11	53%	68%	77%	83%
12	75%	87%	71%	93%
13	60%	67%	52%	89%
14	74%	75%	65%	90%
15	52%	63%	66%	82%
16	61%	85%	83%	89%
17	58%	74%	83%	84%
18	61%	76%	76%	86%
average	66%	78%	71%	87%

FIGURE 4.17:   Experimental result on Baculoviridae.

regions between two sequences. It is commonly used to identify repeat regions of a sequence and to extract the similar regions in two sequences.

For any base x, y, let $\delta(x, y)$ be the similarity score. Given two sequences $A[1..n]$ and $B[1..m]$, a dot plot is a two-dimensional matrix M of size $n \times m$. There are two parameters for generating the dot plot, which are the window size w and the similarity limit d. For any $1 \leq i \leq n - w + 1$ and $1 \leq j \leq m - w + 1$, $M[i, j] = 1$ if $\sum_{p=0}^{w-1} \delta(A[i + p], B[j + p]) \geq d$; and 0 otherwise. Figure 4.18 shows example dot plots between $AGCTCA$ and $ACCTCA$. The left dot plot allows no mismatches and $w = 3$ while the right dot plot allows 1 mismatch per 3 nucleotides. Figure 4.19 shows two dot plots between a HIV type 1 subtype C (AB023804) sequence and itself, using window size 9. The left figure allows no mismatches while the right figure allows 1 mismatch. The main diagonal always appears in the two dot plots since we compare a sequence with itself. The two diagonals on the top right corner and the bottom left corner represent the repeat appears at the beginning and the end of the sequence. The short lines in the figures are noise. In general, when the window size w is longer and the similarity limit d is higher, we get less random noise but also increase the chance of missing similar regions.

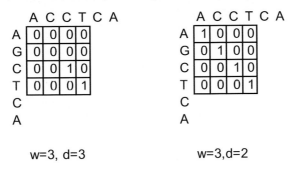

FIGURE 4.18: The two figures show the dot plot for two sequences *AGCTCA* and *ACCTCA*. Both of them use window size $w = 3$. The left figure sets $d = 3$ while the right figure sets $d = 2$.

4.7   Further Reading

This chapter details two genome alignment methods: MUMmer and Mutation Sensitive Alignment. For other alignment methods, please refer to [54] for a review.

In addition to aligning two genomes, we are sometimes interested in aligning multiple genomes. Example methods include Mauve [71] and VISTA browser [85].

4.8   Exercises

1. Suppose the minimum length of MUM is 3. What are the set of MUMs for $S_1 = $ AAAACGTCGGGATCG and $S_2 = $ GGGCGTAAAGCTCT.

2. Refer to the above question. If we further require that the MUMs are also unique in the sequence and its reverse complement, what is the set of MUMs?

3. Consider two sequences $S[1..n]$ and $T[1..m]$. We would like to find a set of strings P appearing in both S and T such that P is maximal and P is unique in S. Can you propose an efficient algorithm to find all such matches? What is the running time?

4. Consider two sequences $S[1..n]$ and $T[1..m]$. Suppose we aim to find a set of strings P appearing in both S and T such that P is maximal.

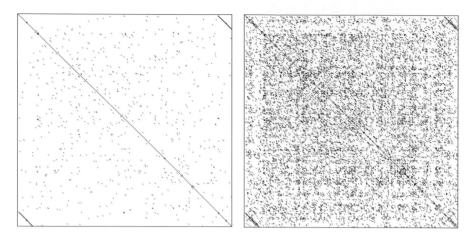

FIGURE 4.19:   The dot plot between a HIV type 1 subtype C (AB023804) and itself. The dot plot on the left was generated using window size 9 and allowing no mismatches. The dot plot on the right was generated using window size 9 and allowing at most 1 mismatch.

Can you propose an efficient algorithm to find all such matches? What is the running time?

5. Consider $S_1 =$ CATGTAT and $S_2 =$ TATGTAC. Suppose we want to select amino acid MUM of length at least 1 codon. What is the set of MUMs for S_1 and S_2?

6. Let $A = 1, 2, 3, 4, 5, 6, 7, 8, 9, 10$ and $B = 9, 1, 6, 5, 4, 7, 2, 3, 8, 10$.

   (a) Assume every MUM has the same weight. What is the maximum weight common subsequence of A and B?

   (b) Is there any 2-similar subsequence of A and B whose backbone is a maximum weight common subsequence?

   (c) Is there any 2-similar subsequence of A and B whose backbone is not a maximum weight common subsequence?

7. Let A and B be two permutations of $\{1, 2, \ldots, n\}$ where each i has a weight $w(i)$. Can you describe an $O(n \log n)$-time algorithm to compute the maximum weight common subsequence (MWCS) for A and B?

8. Consider two sequences $S[1..n]$ and $T[1..m]$. Assume, in the alignment between S and T, there are at most K non-overlapping inversions. We would like to find the alignment between S and T which maximizes the number of matches and report the number of matches. (For example, for $K = 2$, the number of matches in the optimal alignment of $ACGTACGT$

and $ATCCAGCT$ is 7.) Can you describe an algorithm to compute the number of matches in the optimal alignment? What is the time complexity?

9. Consider a set T of n intervals $i..j$ where $i \leq j$. For any interval $i..j \in T$, let $Score(i..j)$ be the score of the interval. We would like to find k non-overlapping intervals $I_1, I_2, \ldots, I_k$ in T such that $\sum_{t=1}^{k} Score(I_t)$ is maximized. Can you give an $O(nk + n \log n)$-time algorithm to compute such a maximum score?

10. Give the $O(n^{2k+1} \log n)$-time algorithm for solving the k-similar subsequences problem, where n is the number of MUMs.

11. Consider two sequences $A[1..n]$ and $B[1..m]$ where every character appears at most twice in each sequence. Give an $O(n \log n)$-time algorithm to compute the longest common subsequence of A and B.

12. Consider two genomes $S[1..n]$ and $T[1..m]$. Give an efficient algorithm to find all maximal substrings A such that A occurs at most twice in both S and T.

13. Consider two sequences $S[1..n]$ and $T[1..m]$. We aim to find the set of all strings P such that (1) P appears exactly once in S, (2) P appears exactly once in T and its reverse complement, and (3) P is maximal. Can you give a detailed algorithm to find the set of P? What is the running time?

Chapter 5

Database Search

5.1 Introduction

5.1.1 Biological Database

The number of biological sequences (protein, DNA, and RNA sequences) has grown exponentially (doubling every 18 months) in the last few years. Often these sequences are submitted to the International Nucleotide Sequence Database Collaboration (INSDC), which consists of three databases DDBJ (DNA Data Bank of Japan, Japan), EMBL (European Molecular Biology Laboratory, Germany), and GenBank (National Center for Biotechnology Information, USA). In 2008, the database had approximately 110 million sequences and 200 billion base pairs for more than 240,000 named organisms.

Searching the database has proven to be useful. For example, in November 2005, an unknown virus from a child in an Amish community was suspected to be an intestinal virus. Through database searching, the unknown sequence was found to be a polio virus. This is the first polio case in the U.S. since 1999 and it demonstrates the power of database search.

In fact, in molecular biology, it is common to search query sequences on various databases since sequence similarity often leads to functional similarity. For instance, in 2002, GenBank alone served 100,000 search queries daily which had risen to over 140,000 queries daily by 2004 (See [210]).

Considering the size of the biological databases and the large amount of daily queries, the need for time-efficient database searching algorithms is self-evident.

5.1.2 Database Searching

The database searching problem can be formulated as follows: Given a database D of genomic (or protein) sequences and a query string Q, the goal is to find the string(s) S in D which is/are the closest match(es) to Q using a given scoring function.

The scoring function can be either:

Semi-Global Alignment Score: The best possible alignment score between a substring A of S and Q; and

Local Alignment Score: The best possible alignment score between a substring A of S and a substring B of Q.

The main measures used to evaluate the effectiveness of a searching algorithm are sensitivity and efficiency which are described below.

Sensitivity is the ratio of the number of true positives (substrings in the database matching the query sequence) found by the algorithm to the actual number of true matches. It measures an algorithm's ability to find all true positives. An algorithm with 100% sensitivity finds all true positives.

Sensitivity can be viewed as the probability of finding a particular true positive. The higher/lower the sensitivity of the algorithm, the more/less likely that a given true positive will be found.

Efficiency measures the running time of the method. We prefer an algorithm which is efficient.

As such, a good search algorithm should be sensitive and at the same time efficient.

5.1.3   Types of Algorithms

This chapter studies different search algorithms which find the best alignment between the query and the database sequences.

Exhaustive Search Algorithms: Exhaustive methods enumerate all possible solutions to find the best solution. Thus, exhaustive search algorithms are the most sensitive class of algorithms. But in practice, the number of possible solutions is very large. Exhaustive algorithms may be a bit slow.

Examples of exhaustive search algorithms are: the Smith-Waterman algorithm (Section 5.2) and the BWT-SW (Section 5.8).

Heuristic Search Algorithms: Heuristic algorithms trade sensitivity for speed. Instead of performing an exhaustive search, they use heuristics to reduce the search space. However, there is no guarantee of the quality of the results. A heuristic which works well for a given problem instance could fail horribly in another situation. In practice, heuristics give acceptable results quickly and as such, many practical algorithms for database searching are heuristics.

Examples of such algorithms are: FastA (Section 5.3), BLAST (Section 5.4), BLAT (Section 5.5.2), and PatternHunter (Section 5.5.3).

Filter based algorithms: These algorithms apply filters to select candidate positions in the database where the query sequence possibly occurs with a high level of similarity.

Examples of such algorithms are: QUASAR (Section 5.6) and LSH (Section 5.7). Note that unlike heuristic methods, QUASAR does not miss any solution while LSH is within a known margin of error (i.e., the false negative rate is within a known error threshold).

Note that the above list is non-exhaustive and it describes some of the more commonly used approaches. There are many other database search methods.

5.2 Smith-Waterman Algorithm

The Smith-Waterman algorithm [267] is the de facto standard for database searching. By applying the Smith-Waterman dynamic programming algorithm discussed in Chapter 2, the query sequence is compared with all sequences in the database D. Since the query sequence is compared with all sequences in D exhaustively, it is the most sensitive method.

However, it is also the slowest and the most space consuming algorithm. The time and space complexity are both $O(nm)$ where n is the sum of the length of all sequences in the database and m is the length of the query sequence. For example, using a 3.0 GHz PC with 4G RAM, it takes more than 15 hours to align a sequence of 1000 nucleotides with the whole human genome [182].

5.3 FastA

FastA [191] is a heuristic search algorithm which was heavily used before the advent of BLAST. Given a database and a query, FastA aligns the query sequence with all sequences in the database and returns some good alignments, based on the assumption that a good alignment should have some exact match substrings. This algorithm is much faster but less sensitive than the Smith-Waterman algorithm.

The development of FastA can be dated back to 1980's. In 1983, Wilbur and Lipman [308] proposed a diagonal method to align protein and DNA sequences. Their algorithm, known as the Wilbur-Lipman algorithm, achieved a balance between sensitivity and specificity on one hand and speed on the other. In 1985, Lipman and Pearson improved the Wilbur-Lipman algorithm and proposed the FastP [190] algorithm. FastP assumes insertion and deletion (indels) are infrequent and it tries to identify good ungapped alignments. Then, in 1988, an improved version of FastP, called FastA [191], was proposed. The major modification is to allow indels during the identification of good alignments. It increased the sensitivity with a small loss of specificity over FastP and a negligible decrease in speed.

5.3.1   FastP Algorithm

Lipman and Pearson [190] observed that amino acid replacements occur far more than indels. Hence, they targeted the identification of high scoring alignments without indels between the query sequence and the database. The approach used a lookup table to identify short exact matches known as hotspots between the query and the database. Then, those hotspots are clustered to identify high scoring alignments without indels, which are known as diagonal runs. Precisely, there are three steps in the FastP algorithm.

Step 1: Identification of hotspots. Hotspots are k-tuples (length-k substrings) of the query which match exactly with the k-tuples of the database sequences. For example, Figure 5.1 shows the hotspots between the query and the database when $k = 2$. The parameter k controls the number of hotspots. A bigger k reduces the number of hotspots, which increases searching speed but reduces sensitivity. By default, $k = 4, 5,$ or 6 for DNA sequences, and $k = 1$ or 2 for protein sequences.

The hotspots can be identified efficiently through table lookup. A lookup table consists of all possible k-tuples (the table size is 4^k for DNA and 20^k for protein). For every k-tuple in the query, we flag the corresponding entry in the lookup table. Then, we scan every k-tuple in the database. A hotspot is identified if the k-tuple in the database is flagged in the lookup table. The running time of this step is $O(n+m)$ where n is the total length of all sequences in the database and m is the length of the query.

FIGURE 5.1: Assuming each hotspot is a 2-tuple, the squared boxes are the hotspots between the query sequence and the database sequences. The ellipse boxes are the diagonal runs.

Step 2: Locating the diagonal runs. A diagonal run[1] is a set S of nearby hotspots such that, for any two hotspots in S, the distance between the two hotspots in the database sequence is the same as that in the query sequence. In other words, a diagonal run is in fact an ungapped alignment between the query and the database sequence. For example, Figure 5.1 shows two diagonal runs AA − TT − CC and AA − − − −CC which appear on sequences Seq2 and Seq3, respectively. The score of a diagonal run is the sum of scores for hotspots and penalties for interspot spaces where each hotspot is assigned a positive score (analogous to a match score) and each interspot space is given a negative score (analogous to a mismatch score). Step 2 tries to identify the 10 highest scoring diagonal runs for each database sequence.

Step 3: Re-scoring of the 10 best diagonal runs. The 10 best diagonal runs are re-scored with the appropriate substitution score matrix (such as the PAM 120 matrix for amino acids or the BLAST matrix for nucleotides (see Section 2.6)). For each diagonal run, a subregion with the highest score is identified. This subregion is called the "initial region". The best score among the 10 initial regions is called the *init1* score for that sequence.

Each database sequence gets an *init1* score based on the three steps. Finally, all database sequences are ranked according to their *init1* scores. At this point the FastP algorithm terminates.

5.3.2 FastA Algorithm

The FastP algorithm only tries to find ungapped alignments. The FastA algorithm improves the FastP algorithm and attempts to find alignments with gaps (caused by indels). It uses the first three steps of the FastP algorithm; then, it executes additional steps, namely, Steps 4 and 5. Figure 5.2 shows the flow of the FastA algorithm.

For each database sequence, the first three steps identify the 10 best initial regions (see Figure 5.2(a)). Step 4 checks whether several initial regions could be joined together. First, initial regions with scores smaller than a given threshold are discarded. Then, we attempt to join the remaining initial regions by allowing indels. Given the locations of the initial regions, their respective scores, and a "joining" penalty (analogous to a gap penalty), we calculate an optimal alignment of initial regions as a combination of compatible regions with maximal score. An example is shown in Figure 5.2(b). The score of the combined regions is the sum of the scores of those compatible regions minus the penalty for gaps. Then the resulting score is used to rank the database sequences. The best score at this stage is called the *initn* score for this sequence.

---

[1]It is named as diagonal run since those hotspots lie on the same diagonal when visualizing on a dot plot (see Section 4.6).

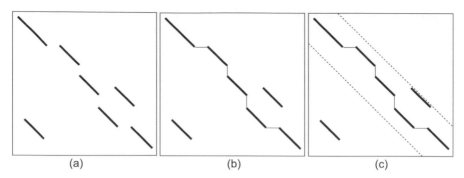

FIGURE 5.2:   The FastA algorithm: For each database sequence, it executes the three steps of FastP to identify the 10 best initial regions (see (a)). Then, Step 4 joins some of these initial regions by allowing indels (see (b)). Finally, Step 5 performs bounded Smith-Waterman dynamic programming to compute the optimal alignment score between the query and the database sequence (see (c)).

In Step 5, if the *initn* score of the database sequence is smaller than a (user-defined) threshold, it is discarded. Otherwise, the banded Smith-Waterman dynamic programming algorithm   (see Section 2.2.2) is applied to get the optimum alignment and score, as shown in Figure 5.2(c). Finally, all database sequences are ranked based on their optimum scores.

5.4   BLAST

BLAST stands for Basic Local Alignment Search Tool [114]. It is designed for speed and is faster than FastA. Although there is a corresponding reduction in the sensitivity of BLAST compared to FastA, the results returned are still good enough. As such, BLAST is now the de facto standard and is used by most public repositories like GenBank.

The first version of BLAST, namely BLAST1 [9], was proposed in 1990. It was fast and was dedicated to search for regions of local similarity without gaps. BLAST2 [10] was created as an extension of BLAST1, by allowing the insertion of gaps (similar to the difference in capability between FastP and FastA). Two versions of BLAST2 were independently developed, namely, WU-BLAST2 by Washington University in 1996 and NCBI-BLAST2 by the National Center for Biotechnology Information in 1997. NCBI-BLAST2 is the version which is more commonly used today.

5.4.1 BLAST1

BLAST1 is a heuristic method, which searches for local similarity without gaps. To state exactly what BLAST1 aims to find, we need some definitions. Given a query sequence S_1 and a database sequence S_2, a pair of equal-length substrings of S_1 and S_2 is called a segment pair. The score of a segment pair is the sum of the similarity score of every pair of aligned bases. For amino acids, we can use the PAM score or the BLOSUM score; for DNA, we can use the BLAST score, which scores match $+2$ and mismatch -1. A segment pair is called a high scoring segment pair (HSP) if we cannot extend it to give a higher score. Among all high scoring segment pairs of S_1 and S_2, the HSP with the maximum score is called the maximal segment pair (MSP). Precisely, given a query sequence Q and a database D, BLAST1 aims to report all database sequences S in D such that the maximal segment pair (MSP) score of S and Q is bigger than some threshold α. BLAST1 is a heuristic which tries to achieve the aim. There are 3 steps in the BLAST1 algorithm:

1: query preprocessing
2: database scanning
3: hit extension

Step 1: Query preprocessing. Step 1 tries to identify substrings in D which look similar to the query sequence Q. Precisely, it scans every w-tuple (length-w substring) W of Q and generates the set of (w, T)-neighbors of W, i.e., all length-w substrings W' such that the similarity between W and W' is more than a threshold T, that is, $\sum_{i=1}^{w} \delta(W[i], W'[i]) \geq T$.

For example, consider the 3-tuple NII. Using a BLOSUM 62 matrix, the set of $(3, 13)$-neighbors of NII is $\{NII, NIV, NVI\}$.

In the implementation, a lookup table for all w-tuples is constructed. For each w-tuple W' in the lookup table, we store the list of length-w substrings W of the query sequence whose (w, T)-neighbor is W'.

By default, BLAST1 uses a word size $w = 3$ and $T = 13$ for amino acid sequences (protein) and $w = 11$ and exact match for nucleic acid sequences (DNA and RNA).

Step 2: Database scanning. If a database sequence contains some neighbor of some w-tuple of a query sequence, the database sequence has the potential to match the query. Step 2 aims to scan the database D to locate those matches. Such a match is called a hit. A hit is characterized by the position pairs (x, y) where x and y are positions in the query sequence and the database sequence, respectively. All hits can be found by comparing each w-tuple in D with the lookup table. Hence, all hits can be found in $O(|D|)$ time.

Step 3: Hit extension. Every hit found in Step 2 corresponds to an ungapped alignment of length w. Step 3 extends the hit in both directions to generate a longer alignment without gap as shown in Figure 5.3. To speed up this step, any extension is truncated as soon as the score decreases by more

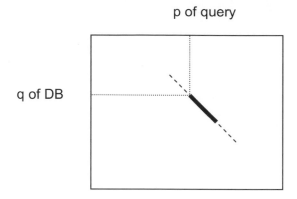

FIGURE 5.3:   Extension of a hit.

than X (X is a parameter of the algorithm) from the highest score found so far (see Exercise 4). Then, a HSP is computed. If the score of the HSP is bigger than α, it will be returned to the user. Furthermore, for every sequence in the database, the maximum scoring HSP is called the MSP.

As a matter of fact, since BLAST1 is a heuristic which speeds up the database search by applying the similarity threshold T and the extension truncation threshold X, it cannot guaranteed to find all high scoring segment pairs whose score is at least α.

5.4.2   BLAST2

BLAST1 cannot report local alignments with gaps. To address this problem, BLAST2 was developed. BLAST2 performs four steps.

1: query preprocessing
2: database scanning
3: hit extension
4: gapped extension

The first two steps in BLAST2 are the same as those in BLAST1 and they generate the list of hits. Step 3 performs the hit extension. For DNA, this step is the same as that in BLAST1. For protein, an additional requirement, known as a two-hit requirement, is needed before a hit is extended. A hit (x, y) is said to satisfy the two-hit requirement if there exists another hit (x', y') such that $x' - x = y' - y$ and $A > x' - x > w$ (by default, $A = 40$). This process is illustrated in Figure 5.4. Since only a small fraction of hits can meet this requirement, the computational time of BLAST2 is significantly reduced.

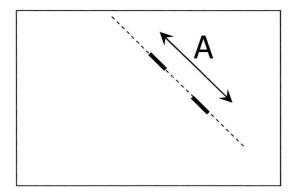

FIGURE 5.4: "Two-hits" requirement.

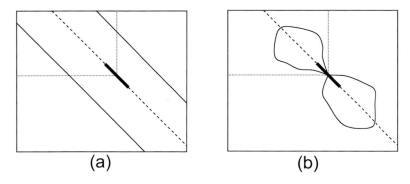

FIGURE 5.5: (a) Banded dynamic programming and (b) score-limited dynamic programming. Observe that banded dynamic programming usually needs to fill in more entries.

After BLAST2 performs the ungapped hit extension, a set of HSPs is generated. For those HSPs with scores above some threshold, Step 4 performs the gapped extension using a modified Smith-Waterman algorithm.

In the modified Smith-Waterman algorithm, the dynamic programming matrix is explored in both directions starting from the mid-point of the hit. When the alignment score drops off by more than X_g (a user-defined parameter), the extension will be truncated. (You are asked to give the detail of this algorithm in Exercise 7.) This approach only computes the scores in the area denoted by the "bubbles" in Figure 5.5(b). Hence, it reduces the running time. Note that this method saves more processing power when compared with the banded Smith-Waterman algorithm used by FastA (see Figure 5.5(a)).

Using BLAST, we can compare a protein sequence with a protein database and a DNA sequence with a DNA database. In addition, by translating the DNA sequence into a protein sequence using six different reading frames,

Program	Query Sequence	Database	Type of Alignment
Blastp	Protein	Protein	The protein query sequence is searched in the protein database.
Blastn	DNA	DNA	The DNA query sequence is searched in the DNA database.
Blastx	DNA	Protein	The DNA query sequence is converted into protein sequences in all six reading frames. Then, those translated proteins are searched in the protein database.
Tblastn	Protein	DNA	The protein query sequence is searched against the protein sequences generated from the six reading frames of the DNA sequences in the database.
Tblastx	DNA	DNA	The DNA query sequence is translated into protein sequences in all six reading frames. Then, the translated proteins are searched against the protein sequences generated from the six reading frames of the DNA sequences in the database.

FIGURE 5.6:   Different versions of the BLAST program.

we can also compare a DNA sequence with a protein sequence. Figure 5.6 summarizes five different ways of applying the BLAST algorithm to search DNA and protein databases.

5.4.3   BLAST1 versus BLAST2

BLAST1 spends 90 percent of its time on the extension of the hits (Step 3) which is the most computationally intensive step of BLAST1. BLAST2 is about three times faster than BLAST1. The speed up is attributed to the "two-hit requirement", which reduces the number of extensions significantly. The "two-hit requirement" is formulated based upon the observation that a HSP of interest is much longer than a single word pair (matching w-tuples), and may therefore entail multiple hits on the same diagonal within a relatively short distance of each another.

5.4.4   BLAST versus FastA

Both BLAST and FastA are significantly faster than the Smith-Waterman algorithm for sequence alignment. BLAST is an order of magnitude faster than FastA.

For sensitivity, BLAST also achieves better sensitivity for protein. However, for DNA, BLAST is less sensitive than FastA since it uses a long word size 11. BLAST is less effective than FastA in identifying important sequence matches with lower similarities, particularly when the most significant alignments contain spaces.

In practice, for most applications, the reduced sensitivity of BLAST over FastA is not an issue while the increased speed is (due to the number of queries made daily), so most biological databases use BLAST (and its variants) for finding local alignment.

5.4.5 Statistics for Local Alignment

Both BLAST and FastA find local alignments whose scores cannot be improved by extension. For local alignments without gap, such alignments are called high-scoring segment pairs or HSPs in BLAST. To determine the significance of those local alignments, BLAST and FastA use E-value and bit score. Below, we give a brief discussion of them.

All the statistics assume the expected score for aligning a random pair of residues/bases is negative. Otherwise, the longer the alignment, the higher the alignment score independent of whether the segments aligned are related or not.

First, we discuss E-value. E-value is the expected number of alignments having an alignment score $> S$ at random. Let m and n be the lengths of the query sequence and the database sequence, respectively. When both m and n are sufficiently long, the expected number E of HSPs with a score at least S follows the extreme distribution (Gumbel distribution). We have $E = Kmne^{-\lambda S}$ for some parameters K and λ which depend on the scoring matrix and the expected frequencies of the residues/bases. We can justify the equation intuitively. First, doubling the length of either sequence will double the expected number of HSPs; thus, we anticipate E is proportional to mn. Second, doubling the score S will exponentially reduce the expected number of HSPs; thus, we anticipate E is proportional to $e^{-\lambda S}$.

In BLAST, the HSP will be reported when its E-value is small. By default, we only show HSPs with an E-value at most 10.

Another statistic shown by BLAST and FastA is the bit score. Note that any alignment score S of an alignment depends on the scoring system. Without knowing the scoring system, the alignment score is meaningless. The bit score is defined as a normalized score independent of the scoring system, which is defined as follows.

$$S' = \frac{\lambda S - \ln K}{\ln 2}.$$

Note that $E = mne^{-S'}$. Hence, when S' is big, the HSP is significant.

The p-value is usually not shown in BLAST and FastA. To define the p-value, observe that the number of random HSPs with score $\geq S$ follows a

Poisson distribution. Thus, $Pr(\text{exactly } x \text{ HSPs with score} \geq S) = e^{-E}E^x/x!$, where $E = Kmne^{-\lambda S}$ is the E-score. Hence, the p-value $= Pr(\text{at least 1 HSP with score} \geq S) = 1 - e^{-E}$.

Note that when the E-value increases, the p-value is approaching 1. When $E = 3$, the p-value is $1 - e^{-3} = 0.95$. When $E = 10$, the p-value is $1 - e^{-10} = 0.99995$. When $E < 0.01$, the p-value is $1 - e^{-E} \approx E$. Hence, in BLAST, the p-value is not shown since we expect the p-value and the E-value are approximately the same when $E < 0.01$ while the p-value is almost 1 when $E > 10$.

Unlike local alignments without gap, there is no solid theoretical foundation for local alignment with gaps. Moreover, experimental results suggested that the theory for ungapped local alignment can be applied to the gapped local alignment as well.

5.5   Variations of the BLAST Algorithm

As BLAST is still among the best heuristics (in terms of striking an acceptable balance between speed and sensitivity) available for the problem of finding local alignments in a large set of biological sequences, it is natural that researchers have extended the original BLAST algorithm to deal with various specialized situations (in particular, making tradeoff between sensitivity and running time). Some of the interesting variants are described below.

5.5.1   MegaBLAST

MegaBLAST [317] is a greedy algorithm used for the DNA gapped sequence alignment search. MegaBLAST is designed to work only with DNA sequences. To improve efficiency, MegaBLAST uses longer w-tuples (by default, $w = 28$). But this reduces the sensitivity of the algorithm. As such, MegaBLAST is used to search larger databases where standard BLAST would be too slow.

MegaBLAST takes as input a set of DNA query sequences. The algorithm concatenates all the query sequences together and performs a search on the obtained long single query sequence. After the search is done, the results are re-sorted by query sequences.

Unlike BLAST, MegaBLAST by default uses linear gap penalty. To use the affine gap penalty, advanced options should be set. It is not recommended to use the affine version of MegaBLAST with large databases or very long query sequences.

5.5.2 BLAT

BLAT [174] stands for "BLAST-like alignment tool". It only works for DNA. Similar to BLAST, BLAT locates all hits and extends them into high-scoring segment pairs (HSPs).

Moreover, BLAT differs from BLAST in some significant ways. First, recall that BLAST builds the lookup table for the query and scan the database to find hits. BLAT does the reverse. It creates the lookup table for the database and scan the query to find hits. This approach speeds up the running time by a lot since it avoids the time consuming step of scanning the database. However, since the database is big, this approach takes up a lot of memory. To reduce the memory usage, BLAT creates the lookup table for the non-overlapping k-mers of the database instead of all k-mers. Although some hits may be missed due to this approach, BLAT finds that the sensitivity will not be reduced by a lot.

Second, BLAT also allows triggering an extension when 2 or more hits occur in close proximity. By default, BLAT uses the "two-hit requirement" and $w = 11$. It has been noted that BLAT is less sensitive than BLAST, but it is still more sensitive than MegaBLAST. Also, it is fast. In fact, in some cases it's claimed to be over 100 times faster.

Third, while BLAST returns each pair of homologous sequences as separate local alignments, BLAT stitches them together into a larger alignment. Such an approach effectively handles introns in RNA/DNA alignments. Unlike BLAST, which only delivers a list of exons, BLAT effectively "unsplices" mRNA onto the genome — giving a single alignment that uses each base of the mRNA only once, and which correctly positions the splice sites.

5.5.3 PatternHunter

PatternHunter [201] is designed to work for DNA only. It is a Java program which finds homologies between two DNA sequences with high sensitivity. PatternHunter can achieve an accuracy similar to BLAST and a speed similar to MegaBLAST. Its accuracy is contributed by its patented technology "spaced seed" and its efficiency is due to the use of a variety of advanced data structures including priority queues, a variation of red-black tree, queues and hash tables.

Below, we describe the concept of spaced seed. A seed of weight w and length m is a binary string $BS[1..m]$ with w 1's and $(m - w)$ 0's. Two DNA sequences $A[1..m]$ and $B[1..m]$ are matched according to the seed BS if $A[i] = B[i]$ for all positions i where $BS[i] = 1$. Figure 5.7 shows three examples of aligning DNA sequences according to two seeds. Note that BLAST uses a length-11 unspaced seed (a seed without 0). PatternHunter observed that spaced seeds (seeds with 0's) are better than unspaced seeds in general. The advantage of spaced seeds is that they can significantly increase hits to true homologous regions and reduce false hits as compared to using unspaced

11111111111	111010010100110111	111010010100110111
AGCATTCAGTC	ACTCCGATATGCGGTAAC	ACTCCAATATGCGGTAAC
|||||||||||	|||-|--|-|--||-|||	|||-|--|-|--x|-|||
AGCATTCAGTC	ACTTCACTGTGAGGCAAC	ACTCCAATATGCAGTAAC
(a)	(b)	(c)

FIGURE 5.7: This figure shows three alignments according to two seeds. (a) An example of aligning two length-11 strings according to the unspaced seed 11111111111. (b) An example of aligning two length-18 strings according to the space seed 111010010100110111. Note that, even though there are some mismatches for positions with 0's, we consider the two strings are matches according to the space seed. (c) An example not matched according to the space seed pattern (since there is one mismatch at position 13).

11111111111	111010010100110111
11111111111	111010010100110111
(a)	(b)

FIGURE 5.8: (a) Two adjacent unspaced 11-tuples. They share 10 symbols. (b) Two adjacent spaced 11-tuples. They share 5 symbols only.

w-tuples. Among all possible spaced seeds, PatternHunter selected an optimized spaced seed which maximizes the number of true hits and minimizes the number of false hits. Their default spaced seed is a length-18 weight-11 seed 111010010100110111.

The reason for improving the sensitivity is that the hits generated by the spaced seed model are more independent when compared to the unspaced seed model. For example, as shown in Figures 5.8, two adjacent unspaced 11-tuples share 10 positions while two adjacent spaced 11-tuples share just five positions. In other words, unspaced seeds will generate more hits per homologous region. If both spaced seeds and unspaced seeds generate a similar number of hits, we will expect the hits of the spaced seed to cover more homologous regions and thus, the spaced seed gives a better sensitivity.

Spaced seeds also reduce the number of hits, which implies a smaller number of false hits. The expected number of hits in a region can be easily calculated using Lemma 5.1. Based on the lemma, the longer the seed, the fewer the number of hits generated. This means that spaced seeds can improve efficiency.

LEMMA 5.1

The expected number of hits of a weight-w length-m seed model within a length L region with similarity $p(0 \leq p \leq 1)$ is $(L - m + 1)p^w$.

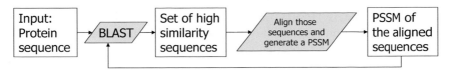

FIGURE 5.9: The flowchart for PSI-BLAST.

PROOF   For each possible position within the region, the probability of having w specific matches is p^w. Since there are $L - m + 1$ possible positions within the region, the expected number of hits is $(L - m + 1)p^w$.   ☐

For example, consider a region of length 64 in the database which has a similarity 0.7 with the query sequence. For the expected number of hits, by Lemma 5.1, BLAST has 1.07, while PatternHunter has 0.93. So, the expected number of hits decreases by 14%.

In addition to PatternHunter, the same group of authors developed a newer version of PatternHunter called PatternHunter II [187]. This version attempts to reach a sensitivity similar to that of the Smith-Waterman algorithm while matching the speed of BLAST. The main change is the use of multiple optimal spaced seeds which allows PatternHunter II to achieve a higher sensitivity.

5.5.4   PSI-BLAST (Position-Specific Iterated BLAST)

Position-specific iterated BLAST (PSI-BLAST) [10] is an application of BLAST to detect distant homology. It automatically combines statistically significant alignments produced by BLAST into a position-specific score matrix (PSSM). By replacing the query sequence with the constructed PSSM, PSI-BLAST can execute BLAST iteratively to search for more distant related sequences until no new significant alignment is found. Hence, PSI-BLAST is much more sensitive to weak but biologically relevant sequence similarities (but it is also slower than the standard BLAST).

Figure 5.9 shows the flowchart for PSI-BLAST. Given a query protein sequence, it has two main steps: (1) find homologous sequences with BLAST and (2) align the homologous sequences to form a PSSM profile. The two steps are iterated until no new significant sequence is found. We detail the two steps below.

Given an input protein sequence, Step 1 of PSI-BLAST executes a BLAST search with the query protein and identifies all database sequences with E-scores below a threshold (by default, 0.01). When more than one identified sequences have identity $> 98\%$, we only keep one copy of these sequences.

Then, Step 2 aligns the sequences identified in Step 1 using the query protein as the template. All gap characters inserted into the query sequence are ignored. From the alignment, we generate a position-specific score matrix (PSSM) profile. Precisely, if the query protein is of length n, the position-specific score matrix (PSSM) profile M is a $20 \times n$ matrix where each row represents one amino acid residue and each column represents a position. For each position i and each residue a in the profile, $M(a, i)$ equals the log-odds score $\log(O_{ia}/P_a)$ where O_{ia} is the observed frequency of the reside a at position i and P_a is the expected frequency of the residue a.

After we generate the PSSM profile, we reiterate Steps 1 and 2 using the PSSM profile. Moreover, we need to modify the BLAST search in Step 1. When we compare a position of the PSSM and a residue in the database, we cannot use the BLOSUM score since the PSSM is not a protein sequence. Instead, we replace the BLOSUM score by the corresponding log-odds score in that position.

5.6 Q-gram Alignment based on Suffix ARrays (QUASAR)

In 1999, Stefan Burkhardt, Andreas Crauser, Paolo Ferragina, Hans-Peter Lenhof, Eric Rivals, and Martin Vingron proposed a new database searching algorithm called "Q-gram Alignment based on Suffix ARrays (QUASAR)". QUASAR was designed to quickly detect sequences with strong similarity, especially if the searches are conducted on one database [44]. Note that QUASAR is an exact method for detecting ungapped local alignments.

5.6.1 Algorithm

QUASAR is an approximate matching algorithm. Formally, the problem is defined as follows.

- Input: a database D, a query S, the difference k, and the window size w

- Output: a set of (X, Y) where X and Y are length-w substrings in D and S, respectively, such that the edit distance of X and Y is at most k.

A pair of substrings X and Y with the above properties is called an *approximate match*.

The idea of QUASAR is to reduce the approximate matching problem to an exact matching problem of short substrings of length q (called q-grams). The approach is based on the following observation: if two sequences have an edit distance below a certain bound, one can guarantee that they share a certain number of q-grams. This observation allows us to design a filter that selects candidate positions from the database which possibly have a small edit distance with the query sequence. The precise filtering criteria is stated in the following lemma.

LEMMA 5.2

Given two length-w sequences X and Y, if their edit distance is at most k, then they share at least t common q-grams (length-q substrings) where $t = w + 1 - (k + 1)q$.

PROOF Let (X', Y') be an optimal alignment of X and Y, r be the number of differences between $X', Y' (r \leq k)$, L be the length of X' and Y' $(L \geq w)$. Consider the $L + 1 - q$ pairs of q-grams of X', Y' starting at the same position. Each difference in X', Y' can create at most q such pairs to be different, thus X' and Y' must have at least $L + 1 - q - rq \geq w + 1 - (k + 1)q$ common q-grams. Since any common q-gram for X', Y' is also common for X, Y, this completes the proof. ☐

The threshold stated in Lemma 5.2 is tight. For example, consider two strings ACAGCTTA and ACACCTTA. They have an edit distance of 1 and have $8 - 3(1 + 1) + 1 = 3$ common q-grams: ACA, CTT, and TTA.

Lemma 5.2 gives a necessary condition for any substring in D to be a candidate for an approximate match with $S[i..i + w - 1]$. If $S[i..i + w - 1]$ and $D[j..j + w - 1]$ share less than t q-grams where $t = w + 1 - (k + 1)q$, the edit distance between $S[i..i + w - 1]$ and $D[j..j + w - 1]$ is more than k and we can filter it away. Hence, we have the basic q-gram filtration method.

- First, the method finds all matching q-grams between the pattern S and the database D. In other words, it finds all pairs (i, j) such that the q-gram at position i in S is identical to the q-gram at position j in D. Such a pair is called a hit.

- Second, it identifies the database regions $D[j..j + w - 1]$ that have more than $w + 1 - (k + 1)q$ hits. These regions are then verified to see if the edit distance is at most k.

Figure 5.10 shows the algorithm formally.

To have a fast implementation of the filtration method, we need to find hits efficiently. QUASAR preprocesses the entire database D using the suffix array SA. It also stores an auxiliary array $idx[]$ where $idx[Q]$ is the start of the hit list for Q in SA. The suffix array allows us to locate all hits of any q-gram in D without scanning the database D. Figure 5.11 demonstrates how we find the list of hits for a q-gram.

The basic QUASAR algorithm

Require: The database D and the query S

Ensure: Report all position pairs (i, j) such that the edit distance of $D[j..j + w - 1]$ and $S[i..i + w - 1]$ is at most k.

1: **for** every $X = S[i..i + w - 1]$ **do**
2:     For every length-w substring Y in D, associate a counter with it and initialize it to zero.
3:     **for** each q-gram Q in X **do**
4:         Find the hit list, i.e., the list of positions in D so that Q occurs.
5:         Increment the counter for every length-w substring Y in D, which contains Q.
6:     **end for**
7:     For every length-w substring Y in D with $counter > w+1-(k+1)q$, X and Y are the potential approximate match. This will be checked with a sequence alignment algorithm.
8: **end for**

FIGURE 5.10: The basic QUASAR algorithm for finding potential approximate matches.

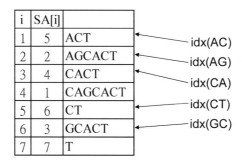

FIGURE 5.11: Consider a database $D = CAGCACT$. $SA[1..7]$ is the corresponding suffix array. idx is an auxiliary index which indicates the start positions of each q-gram. From the suffix array, we can read the hit list. For example, for the q-gram CA, the hit list is $SA[idx(CA)..idx(CT) - 1] = SA[3..4] = (4, 1)$.

5.6.2 Speeding Up and Reducing the Space for QUASAR

5.6.2.1 Window Shifting

The algorithm described in Figure 5.10 needs to build the counter list for every $S[i..i + w - 1]$, where $i = 1, 2, \ldots, (n - w + 1)$. For each i, we need to check $w - q + 1$ q-grams, which is a time consuming process. Window shifting reduces the running time.

The window shifting trick is as follows. Given the number of q-grams shared between $D[j..j + w - 1]$ and $S[i..i + w - 1]$ is c, we observe that the number of q-grams shared between $D[j + 1..j + w]$ and $S[i + 1..i + w]$ equals $c - \delta(D[j..j + q - 1], S[i..i + q - 1]) + \delta(D[j + w - q + 1..j + w], S[i + w - q + 1..i + w])$ where $\delta(x, y) = 1$ if they are the same and 0, otherwise.

Suppose we know the counters of all length-w windows in D for $S[i..i + w - 1]$. We can apply the window shifting trick to update the counters for $S[i + 1..i + w]$. First, we use the suffix array to find all occurrences of the q-gram $S[i..i + q - 1]$ and decrement the counter values of all windows that contain the q-gram $S[i..i + q - 1]$. Then, we use the suffix array to find all occurrences of the "new" q-gram $S[i + w - q + 1..i + w]$ and increment the corresponding window counters.

5.6.2.2 Block Addressing

Maintaining a counter for every length-w window of the database D requires a lot of space. Block addressing reduces the amount of space by storing fewer counters. The database D is conceptually divided into non-overlapping blocks of fixed size b ($b \geq 2w$). A counter is assigned to each block instead of each length-w window. These counters are stored in an array of size $|D|/b$. The counter of each block stores the number of q-grams in $S[1..w]$ which occur in the block. If a block contains more than t q-grams, this block potentially contains some approximate matches and we find those matches using a sequence alignment algorithm. Otherwise, we can filter out such a block.

Note that the blocks are non-overlapping. If an approximate match exists which crosses the block boundary of two adjacent blocks B_1 and B_2, we will miss them. To avoid this problem, a second block decomposition of the database, i.e., a second block array, is used. The second block decomposition is shifted by half of the length of a block ($b/2$). Then, if the situation as described above occurs, there exists a block, which overlaps with both B_1 and B_2, contains the potential candidate.

5.6.3 Time Analysis

The database preprocessing phase constructs the suffix array of the database D, which takes $O(|D|log|D|)$ time [204].

In the query phase, QUASAR generates the hit list for $O(|S|)$ q-grams. Since the expected number of hits per q-gram is at most $\frac{|D|}{4^q}$, the hit list

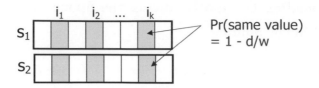

FIGURE 5.12:   LSH Function: 2 w-mers s_1 and s_2.

can be generated in $O(\frac{|S| \cdot |D|}{4^q})$ expected time. If c blocks have more than $w + 1 - (k + 1)q$ hits, the alignment takes $O(c \cdot b^2)$ time. In total, the query time is $O(\frac{|S||D|}{4^q} + cb^2)$.

QUASAR takes a lot of memory space. The space complexity of QUASAR is dominated by the space used for the suffix array, which is $O(|D|log|D|)$. More precisely, the algorithm requires $9|D|$ bytes space in the database pre-processing phase and $5|D|$ bytes space in the query phase.

5.7   Locality-Sensitive Hashing

LSH-ALL-PAIRS is a randomized search algorithm designed to find ungapped local alignments in a genomic sequence with up to a specified fraction of substitutions. The algorithm finds ungapped alignments efficiently using a randomized search technique called locality-sensitive hashing (LSH). It is efficient and sensitive enough to find local similarities with as little as 63 percent identity in mammalian genomic sequences with tens of megabases [42].

Consider a w-mer (a length-w string) s; a LSH function is a function which maps s to a sequence with k characters. Precisely, the LSH function chooses k indices $i_1, i_2, ..., i_k$ uniformly at random from the set $\{1, 2, ..., w\}$. The LSH function $\pi(s)$ is defined to be $(s[i_1], s[i_2], ..., s[i_k])$.

Now, consider two w-mers s_1 and s_2 as shown in Figure 5.12. If s_1 and s_2 are more similar, we have a higher probability that $\pi(s_1) = \pi(s_2)$. More precisely, if the Hamming distance of s_1 and s_2 equals d,

$$\Pr\left(\pi(s_1) = \pi(s_2)\right) = \prod_{j=1,...k} \Pr\left(s_1[i_j] = s_2[i_j]\right) = \left(1 - \frac{d}{w}\right)^k$$

In other words, if $\pi(s_1) = \pi(s_2)$, s_1 and s_2 are likely to be similar. Hence, by finding all w-mers s which have the same $\pi(s)$, it is likely that we identify most of the w-mers in a database D which are similar. However, such an approach may generate false positives and false negatives.

The LSH-ALL-PAIRS algorithm

Require: The DNA sequence database D

Ensure: Report all w-mer s and t in D such that the hamming distance of s and t is at most d.

1: Generate m random locality-sensitive hash functions $\pi_1(), \pi_2(), \dots, \pi_m()$;

2: For every w-mer s in D, compute $\pi_1(s), \dots, \pi_m(s)$;

3: **for** every pair of w-mers s and t such that $\pi_j(s) = \pi_j(t)$ for some j **do**

4:     **if** the hamming distance between s and t is at most d **then**

5:         Report the pair (s, t);

6:     **end if**

7: **end for**

FIGURE 5.13: The LSH-ALL-PAIRS algorithm.

- False positive: s_1 and s_2 are dissimilar but $\pi(s_1) = \pi(s_2)$.

- False negative: s_1 and s_2 are similar but $\pi(s_1) \neq \pi(s_2)$.

False positives can be avoided by validating the actual hamming distance between s_1 and s_2 where $\pi(s_1) = \pi(s_2)$. False negatives cannot be detected. Moreover, the number of false negatives can be reduced by repeating the test using different $\pi()$ functions. By the above observations, Buhler [42] suggested the LSH-ALL-PAIRS algorithm for identifying all w-mers in a database D which look similar. The algorithm is described in Figure 5.13.

In contrast to heuristic algorithms like FastA and BLAST presented earlier, LSH-ALL-PAIRS can guarantee the returned results fall within a bounded error probability. For any pair of w-mers s_1 and s_2 in the database D with hamming distance less than d, the chance that the algorithm does not report it is bounded above by

$$\Pi_{i=1}^{m} (1 - Pr(\pi_i(s_1) = \pi_j(s_2))) = \left(1 - \left(1 - \frac{d}{w} \right)^k \right)^m$$

For the time analysis, in the worst case, $O(N)$ w-mers might be hashed to the same LSH value, either because they are all pairwise similar or because the hash functions $\pi()$ yielded many false positives. The number of string comparisons performed can therefore in theory be as high as $O(N^2)$. On average, we expect $\frac{N}{4^w}$ w-mers are hashed to the same bucket. Hence, the expected number of string comparisons is $4^w \left(\frac{N}{4^w} \right)^2 = \frac{N^2}{4^w}$. This implies that the expected running time is $O(\frac{N^2(km+w)}{4^w})$, which is efficient in practice.

5.8 BWT-SW

As discussed previously, the optimal local alignment between a query sequence and a database can be computed using the Smith-Waterman dynamic programming. This approach, however, is slow. On the other hand, heuristic solutions like FastA, BLAST, BLAT, and PatternHunter are much faster; however, they cannot guarantee finding the optimal local alignment. This section discusses a method called BWT-SW [182] which uses a suffix tree or an FM-index to speed up the Smith-Waterman dynamic programming. BWT-SW is efficient and guarantees finding the optimal local alignment.

5.8.1 Aligning Query Sequence to Suffix Tree

Consider a text $S[1..n]$ and a query $Q[1..m]$. The problem of finding the optimal local alignment of Q and S can be rephrased as follows:

1: **for** $k = 1$ to n **do**
2:     Compute the optimal global alignments between any substring of Q and any prefix of $S[k..n]$;
3: **end for**
4: Report the highest scoring global alignment among all the best global alignments found.

The above procedure is correct since any substring of S is some prefix of $S[k..n]$ for some k. Hence, the above procedure finds the best global alignment between any substring of Q and any substring of S, i.e., the best local alignment between Q and S.

For a particular suffix $S' = S[k..n]$ of S, the optimal global alignment between any substring of Q and any prefix of S' can be found by dynamic programming. Let $V(i,j)$ be the optimal alignment score between any suffix of $Q[1..i]$ and the string $S'[1..j]$. Then, the best global alignment score between any substring of Q and any prefix of S' is $max_{0 \le i \le m, 0 \le j \le |S'|} V(i,j)$.

We devise the recursive formulae for $V(i,j)$ depending on three cases: (1) $j = 0$, (2) $i = 0$, and (3) both $i > 0$ and $j > 0$.

For case (1), when $j = 0$, the best suffix of $Q[1..i]$ which aligns with an empty string is an empty string. Hence, $V(i,0) = 0$.

For case (2), when $i = 0$, we need to introduce j spaces when we align an empty string with $S'[1..j]$. Thus, $V(0,j) = V(0, j-1) + \delta(\_, S'[j])$.

For case (3), when both $i > 0$ and $j > 0$, the last pair of aligned characters in the optimal alignment should be either match/mismatch, delete, or insert. Hence, we have

$$V(i,j) = \max \begin{cases} V(i-1, j-1) + \delta(Q[i], S'[j]) \\ V(i, j-1) + \delta(\_, S'[j]) \\ V(i-1, j) + \delta(Q[i], \_) \end{cases}$$

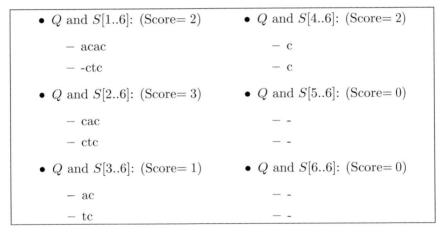

FIGURE 5.14: Dynamic programming to find the optimal alignment between any substring of $Q = ctc$ and any prefix of $S = acacag$.

	Q and S[1..6]: (Score= 2)	Q and S[4..6]: (Score= 2)
	– acac	– c
	– -ctc	– c
	Q and S[2..6]: (Score= 3)	Q and S[5..6]: (Score= 0)
	– cac	– -
	– ctc	– -
	Q and S[3..6]: (Score= 1)	Q and S[6..6]: (Score= 0)
	– ac	– -
	– tc	– -

FIGURE 5.15: For $i = 1, 2, \ldots, 6$, the figure shows the optimal alignment between any substring of $Q = ctc$ and any prefix of $S[i..6]$ where $S = acacag$.

For example, assume $match = 2$ and $mismatch/insert/delete = -1$. Consider $S = acacag$ and $Q = ctc$. Figure 5.14 shows an example for finding the optimal alignment between any substring of $Q = ctc$ and any prefix of $S = acacag$. The optimal alignment score is 2 with alignment $(acac, -ctc)$.

To generate the optimal local alignment, the same computation is repeated for all suffixes of S, that is, $acacag$, $cacag$, $acag$, cag, ag, and g. We get six alignments, as shown in Figure 5.15. Among all the alignment scores, the best alignment score is 3, which is the optimal local alignment score. This provides an alternative way to find the optimal local alignment.

However, the alternative way to find the optimal local alignment proposed above is slow. To speed it up, we can perform the dynamic programming between Q and all suffixes of S together using a suffix tree. Let T be the suffix tree for S. Note that every suffix of S corresponds to a path in the

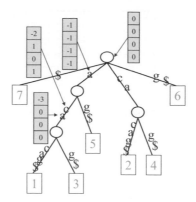

FIGURE 5.16:   Let $Q = ctc$ be the query. The figure shows the suffix tree for the sequence $S = acacag$. We also show the dynamic programming table columns when we go down the tree along the path aca. Note that it is the same as the first 4 columns in the dynamic programming table in Figure 5.14. Furthermore, observe that aca is the common prefix of the two suffixes $S[1..6]$ and $S[3..6]$. We can avoid filling in these four columns twice when we fill in the dynamic programming along the paths in the suffix tree.

suffix tree T. We can fill in the dynamic programming table for any suffix of S along the corresponding path in the suffix tree. When two suffixes share a common prefix, the suffix tree can avoid redundant table filling, which is illustrated in Figure 5.16.

How deep should we go when we align a pattern to a suffix trie? The depth of the suffix trie is n. However, we do not need to go down to depth n. If the pattern Q is of length m, we only need to go down the tree by at most cm characters, for some constant c depending on the scoring matrix. For instance, for our scoring matrix (score 2 for match and -1 for mismatch/insert/delete), we claim that we need to go down the suffix trie to depth at most $3m$. To prove the claim, observe that within every alignment, let x be the number of match/mismatch positions and y be the number of indel positions. Using our scoring matrix, the alignment score should be at most $2x - y \le 2m - y$. Since every alignment has a non-zero score, we have $y \le 2m$. Hence, we need to go down the suffix trie at most $x + y$ characters, which is at most $3m$. The claim is proved.

Figure 5.17 shows the algorithm for local alignment using a suffix trie. Below, we analyze the time complexity. Let L be the number of paths in T with depth cm. The number of nodes in those paths is at most cmL. For each node, we need to compute a dynamic programming column of size $m+1$. Hence the running time is $O(cm^2L)$. Note that $L = O(min\{n, \sigma^{cm}\})$. Hence, when m is not big, the worst case running time is faster than $O(nm)$.

Optimal local alignment using a suffix trie
Require: The suffix trie T of the string S and the query Q of length m
Ensure: The optimal local alignment score between Q and S
1: $CurScore = -\infty$;
2: **for** each node in T of depth at most cm visited in DFS order **do**
3:    When we go down the trie T by one character, we fill in one additional column of the DP table.
4:    When we go up the trie T by one character, we undo one column of the DP table.
5:    If any score s in the column is bigger than $CurScore$, set $CurScore = s$
6: **end for**
7: Report $CurScore$;

FIGURE 5.17:   Algorithm for computing the optimal local alignment using a suffix trie.

5.8.2   Meaningful Alignment

Can we do better? This section proposes another concept called meaningful alignment which enables effective pruning. Hence we can improve the running time by a lot in practice. Consider an alignment A of a substring X of S and a substring Y of the query Q. If the alignment score of some prefix X' of X and some prefix Y' of Y is less than or equal to zero, we say A is a meaningless alignment; otherwise, A is a meaningful alignment. Note that if A is meaningless with alignment score C, then there exists a meaningful alignment between a suffix of X and a suffix of Y with a score of at least C. For example, for the alignment between $Q[1..3]$ and $S[1..4]$ in Figure 5.16 with score 2, it is a meaningless alignment since the alignment between $Q[1..2]$ and $S[1..3]$ has score zero. Their suffixes $Q[3..3]$ and $S[4..4]$ in fact form a meaningful alignment with score 2.

Based on the above discussion, the optimal local alignment score between Q and S equals the optimal meaningful alignment score between any substring of S and any substring of Q.

Similar to Section 5.8.1, the best meaningful alignment can be found using dynamic programming. For any suffix S' of S, we apply dynamic programming to find the best meaningful alignment between any substring of Q and any prefix of S'. Let $V(i, j)$ be the best meaningful alignment score between any suffix of $Q[1..i]$ and $S'[1..j]$. Then, the best meaningful alignment score between any substring of Q and any prefix of S' equals $max_{1 \leq i \leq m, 1 \leq j \leq n}\{V(i, j)\}$.

The recursive formula for $V(i, j)$ depends on three cases: (1) $i = 0$, (2) $j = 0$, and (3) both $i > 0$ and $j > 0$.

For case (1), $V(0, j) = -\infty$.

	_	c	a	c	a	g
_	0	-∞	-∞	-∞	-∞	-∞
c	0	2←1←0			-∞	-∞
t	0	-∞	1←0		-∞	-∞
c	0	-∞	-∞	3←2←1		

Corresponding alignment:

    cac
    ctc

FIGURE 5.18: Example for finding the best meaningful alignment.

For case (2), $V(i,0) = 0$.

For case (3),

$$V(i,j) = \max \begin{cases} -\infty \\ V(i-1,j-1) + \delta(Q[i], S'[j]) & \text{if } V(i-1,j-1) > 0 \\ V(i,j-1) + \delta(\_, S'[j]) & \text{if } V(i,j-1) > 0 \\ V(i-1,j) + \delta(Q[i], \_) & \text{if } V(i-1,j) > 0 \end{cases}$$

The dynamic programming for finding the optimal meaningful alignment between all substrings of Q and S' is illustrated in Figure 5.18. To generate the best local alignment, we repeat the same computation for all suffixes S' of S. Figure 5.19 shows the optimal meaningful alignments between $Q = ctc$ and all suffixes of $S = acacag$. Among all the meaningful alignment scores, the best alignment score is 3, which is the best local alignment score.

There are two advantages when we compute the optimal local alignment using the new formulation. First, we can perform the meaningful alignments for all suffixes of S on the suffix trie T of S, thus avoiding redundant computations. Second, in practice, many entries in the dynamic programming are either zero or $-\infty$ (see Figure 5.18 for example). This allows us to have two pruning strategies.

Pruning strategy 1: For a particular node in T, if all entries in the column are non-positive, we can prune the whole subtree (see Figure 5.20(a)).

Pruning strategy 2: When we go down the tree, we do not need $O(m)$ time to compute a column when the column has many non-positive entries (see Figure 5.20(b)).

The algorithm can be further extended to handle the affine gap penalty (see Exercise 11).

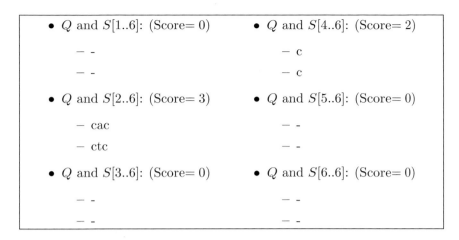

- Q and $S[1..6]$: (Score= 0)

  $-$ $-$

  $-$ $-$

- Q and $S[2..6]$: (Score= 3)

  $-$ cac

  $-$ ctc

- Q and $S[3..6]$: (Score= 0)

  $-$ $-$

  $-$ $-$

- Q and $S[4..6]$: (Score= 2)

  $-$ c

  $-$ c

- Q and $S[5..6]$: (Score= 0)

  $-$ $-$

  $-$ $-$

- Q and $S[6..6]$: (Score= 0)

  $-$ $-$

  $-$ $-$

FIGURE 5.19: For any $i = 1, 2, \ldots, 6$, the figure shows the optimal meaningful alignment between any substring of $Q = ctc$ and any prefix of $S[i..6]$, where $S = acacag$.

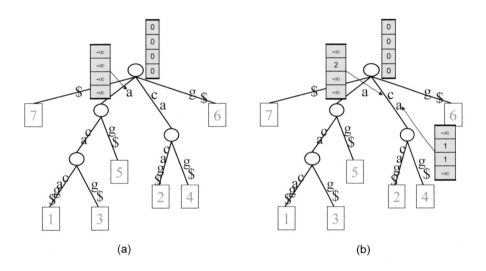

(a)   (b)

FIGURE 5.20: Figure (a) illustrates pruning strategy 1. Starting from the root and when we move down to a, all entries in the dynamic programming column are $-\infty$; in this case, we can prune the whole subtree. Figure (b) illustrates pruning strategy 2. Starting from the root and when we move down to c, only the second entry in the dynamic programming column is a positive integer. Hence, when we move further down to a, we only need to compute the values for the second and third entries. In this case, we can prune the computation for the first and the fourth entries.

5.9 Are Existing Database Searching Methods Sensitive Enough?

Given so many database searching methods, one would like to know if they are as accurate as Smith-Waterman local alignment method. In particular, many researchers use BLAST as the de facto standard for homology search. Will BLAST miss a lot of significant alignments?

To answer this question, Lam et al. [182] conducted an experiment using a set of 8,000 queries, which include 2,000 mRNAs from each of the 4 species: chimpanzee, mouse, chicken, and zebrafish. The length of the queries are in the range from 170 to 19,000. The average length is 2,700. Using the default setting of BLAST, the queries are aligned on the human genome. The significant of each alignment is evaluated using E-value (see Section 5.4.5).

Figure 5.9 shows the missing percentage of BLAST for different E-value threshold. When we focus on relatively significant alignments, says, with E-value less than or equal to 1×10^{-10}, BLAST only misses 0.06% when all 8,000 queries are considered together (precisely, 49 out of 81,054 alignments are missed by BLAST). We also observed that the missing percentage depends on the evolutionary distance between the species and human. For E-value less than or equal to 1×10^{-10}, the missing percentages for queries derived from chimpanzee, mouse, chicken, and zebrafish are 0.02%, 0.07%, 0.13%, and 0.39% respectively. This experiment concluded that BLAST is accurate enough, yet the few alignments missed could be critical for biological research.

5.10 Exercises

1. Go to the NCBI Genbank (http://www.ncbi.nlm.nih.gov/Genbank/) to extract the sequence NM_011443. This gene is called Sox2 in mice. Select a DNA segment and a corresponding amino acid segment of the gene and BLAST them using BLASTn and BLASTp, respectively. What are the results? Can you find the Sox2 gene in the other organisms?

2. Consider a query sequence AAACTGGGACATTT. Suppose the database has two sequences:

   Seq1 : CGTTTACGGGTACTTTGCAGTAGCTGATGGC
   Seq2 : ACGTAAATCGGGAGCATTTCGTACAGACT

   Assume the match score is +2, the indel score is −1, and the mismatch score is −1. Assume each hotspot is a 3-tuple.

E-value ($\leq$)	Chimpanzee Missing %	Mouse Missing %	Chicken Missing %	Zebrafish Missing %	All 4 species Missing %
1.0×10^{-16}	0.00	0.03	0.05	0.06	0.01
1.0×10^{-15}	0.00	0.03	0.05	0.06	0.02
1.0×10^{-14}	0.00	0.04	0.06	0.06	0.02
1.0×10^{-13}	0.00	0.03	0.07	0.14	0.02
1.0×10^{-12}	0.01	0.04	0.10	0.17	0.03
1.0×10^{-11}	0.02	0.05	0.11	0.28	0.05
1.0×10^{-10}	0.02	0.07	0.13	0.39	0.06
1.0×10^{-9}	0.03	0.09	0.16	0.60	0.08
1.0×10^{-8}	0.05	0.11	0.25	0.77	0.12
1.0×10^{-7}	0.10	0.19	0.31	0.81	0.18
1.0×10^{-6}	0.17	0.31	0.45	1.08	0.28
1.0×10^{-5}	0.32	0.47	0.70	1.45	0.45
1.0×10^{-4}	0.57	0.88	0.99	1.81	0.75
1.0×10^{-3}	0.99	1.36	1.25	2.25	1.17
1.0×10^{-2}	1.69	2.11	1.68	2.61	1.84
1.0×10^{-1}	2.70	2.97	2.33	2.86	2.76

FIGURE 5.21: This table shows the percentage of missing alignments by BLAST when performing local alignment of $2,000$ mRNAs from 4 species (chimpanzee, mouse, chicken, and zebrafish) on the human genome. The table is obtained from [182].

    (a) Can you identify the hotspots and the diagonal runs for both database sequences?

    (b) Can you compute the *init1* score for both database sequences?

    (c) Can you compute the *initn* score for both database sequences?

3. Consider the following query sequence Q = VPNMIHCTSAG. Can you perform the query preprocessing step of BLAST and generate the lookup table for all 2-tuples of Q? Assume the similarity threshold $T = 8$.

4. Consider a query sequence S=TCTCACCGTGCACGACATC and a database sequence T=CGGAACTGTGAACAATCCT. Assume the match score is $+5$, the mismatch score is -4, and the extension truncation threshold is $X = 13$. Suppose the word length is 3. One of the hit between S and T is GTG. Starting from the hit GTG, can you demonstrate the hit extension step in BLAST1? What is the MSP of this database sequence?

5. This question generalizes the two-hit requirement in BLAST to a three-hit requirement. Given a query $Q[1..m]$ and a database sequence $S[1..n]$. A hit (x, y) exists if $Q[x..x+w-1]$ looks similar to $S[y..y+w-1]$, where w is the word size. A hit (x_1, y_1) is said to satisfy the three-hit requirement if there exist $(x_2, y_2), (x_3, y_3)$ such that (1) $x_2 - x_1 = y_2 - y_1 > w$ and $x_3 - x_2 = y_3 - y_2 > w$, and (2) $x_3 - x_1 < A$ for some constant A. Suppose

you are given the set of hits D of size s. Can you give an efficient algorithm to identify all the hits satisfying the three-hit requirement? What is the running time?

6. Referring to the above question, generalize the definition of the three-hit requirement to the k-hit requirement. Can you give an efficient algorithm to identify all the hits satisfying the k-hit requirement? What is the running time?

7. Can you give the algorithm for score-limited dynamic programming for BLAST2?

8. Consider the following four seeds: 101010101, 11111, 111011, and 110111. Can you rank the goodness of the three seeds in term of getting at least one hit in the homologous region? Explain your answer. (You can assume the database is a randomly generated sequence.)

9. Design the shortest space seed of weight 4 so that every overlapping seed shares at most 1 position.

10. Consider a text $S[1..n]$ and a query $Q[1..m]$ where $n \gg m$. If the scores for match, mismatch, open gap, and space are 5, -4, -5, and -1, respectively, what is the maximum length of the optimal local alignment between S and Q? (State the answer in term of m.)

11. Describe how to extend the algorithm discussed in Section 5.8 to handle the affine gap penalty.

Chapter 6

Multiple Sequence Alignment

6.1 Introduction

Chapter 2 defined the alignment of two sequences and proposed methods for computing such alignment. This chapter extends the alignment concept to multiple sequences (that is, three or more sequences). Given a set of three or more DNA/RNA/protein sequences, multiple sequence alignment (MSA) aims to align those sequences by introducing gaps in each sequence. Figure 6.1 shows an example of a multiple sequence alignment of a set of HIV sequences.

Observe that the similarity among a set of sequences can be determined by pairwise alignment. Why do we need MSA? One reason is that pairwise alignment cannot identify regions which are conserved among all sequences. Multiple sequence alignment, on the other hand, can tell us such information. For example, biologists can align proteins of similar function to identify the functionally conserved regions of a set of proteins. Those conserved amino acid stretches in proteins are strong indicators of protein domains or preserved three-dimensional structures. Hence, multiple sequence alignment has been widely used in aiding structure prediction and characterization of protein families.

Below, we first formally define the multiple sequence alignment (MSA) problem. Then, we describe different methods for computing multiple sequence alignment, including the dynamic programming approach, the progressive approach, and the iterative approach.

6.2 Formal Definition of the Multiple Sequence Alignment Problem

Consider a set of k DNA/protein sequences $\mathcal{S} = \{S_1, S_2, \ldots, S_k\}$. A multiple alignment $\mathcal{M}$ of $\mathcal{S}$ is a set of k equal-length sequences $\{S_1', S_2', \ldots, S_k'\}$ where S_i' is a sequence obtained by inserting spaces into S_i. Figure 6.2 shows an example of a multiple alignment of 4 DNA sequences.

The multiple sequence alignment problem is to find an alignment $\mathcal{M}$ which

```
***  * * *      ** * * ******* * * *  **  *  *
CAAGAAAAAG___TATAAATATAGGACCAGGGAGAGCATTTTATACAACA
CAAGAAAAAG___TATACATATAGGACCAGGGAGAGCTTTTTATACAACA
CAAGRAAAAG___TATACMTATAGGACCAGGGAGAGCATTTTATACAACA
CAAGAAAAAG___TATAMMTATAAGACCAGGGAGAGCATTTTATACAACA
CAAGAAAAAG___GATACATATAGGACCAGGGAGAGCAGTTTATACAACA
CAAGAAGAAG___TATACMTTTMGGACCAGGGAAAGCATTTTATRC___A
CAAGAAGAAG___TATACATTTWGGACCAGGGAAAGCATTTTWTGC___A
CAAAAATAAGACATATACATATAGGACCAGGGAGACCATTTTATACAGCA
CAAGAAGARG___TRTACATATAGGACCAGGGARAGCATATTATAC___A
CAAGAAAAAG___TATACCTATAGGACCAGGGAGAGCMTTTTATRCAACA
CAAGAAAAAG___TGTACATATRGGACCAGGGAGAGCATATTAYAC___A
CAAGAAAAAG___TGTACATATAGGACCAGGGAAAGCATATTATAC___A
CAAGAAAAAG___TATACATATAGGACCAGGAARAGCATTTTATGCAACA
```

FIGURE 6.1: This is an alignment of a portion of 13 env genes of 13 different HIV strains from patients in Sweden. The '*' on the first line indicates the conserved bases. The alignment was generated by ClustalW [145, 287].

```
S1 = ACG__GAGA
S2 = _CGTTGACA
S3 = AC_T_GA_A
S4 = CCGTTCAC_
```

FIGURE 6.2: Multiple alignment of 4 DNA sequences.

maximizes a certain similarity function or minimizes a certain distance function. One popular similarity function is the Sum-of-Pair (SP) score, which is the function $\sum_{1 \leq i < j \leq k} align\_score_{\mathcal{M}}(S'_i, S'_j)$ where $align\_score_{\mathcal{M}}(S'_i, S'_j)$ is the alignment score between S'_i and S'_j. Precisely, $align\_score(S'_i, S'_j)$ equals the sum of the similarity score of every aligned pair of residues, minus gap penalties.

For example, assume the mismatch and match scores are 2 and -2, respectively. Also, assume the gap opening and gap extension scores are -2 and -1, respectively. Then, for the multiple alignment in Figure 6.2, the alignment scores for (S_1, S_2), (S_1, S_3), (S_1, S_4), (S_2, S_3), (S_2, S_4), and (S_3, S_4) are 3, 4, -5, 2, 6, and -6, respectively. Hence, the SP score of the multiple alignment is $3 + 4 - 5 + 2 + 6 - 6 = 4$.

By symmetry, Sum-of-Pair (SP) distance is a distance function. The Sum-of-Pair distance is defined as $\sum_{1 \leq i < j \leq k} d_{\mathcal{M}}(S'_i, S'_j)$ where $d_{\mathcal{M}}(S'_i, S'_j)$ equals the sum of the distance score of every aligned pair of residues.

The problem of finding a multiple alignment which maximizes the SP score (or minimizes the SP distance) is known to be NP-hard [164, 301]. In other words, it is unlikely to have a polynomial time solution for this problem.

6.3   Methods for Solving the MSA Problem

Roughly speaking, multiple sequence alignment methods can be classified as four types: global optimization methods, approximation algorithms, heuristic methods, and probabilistic methods.

For global optimization methods, the basic approach is dynamic programming (which is described in Section 6.4). However, the dynamic programming takes exponential time. To improve the efficiency, search-based approaches like A* and branch-and-bound are applied to prune the search space. MSA [126, 192] and DCA [274] are two examples of methods based on this strategy. We can also achieve optimal solution using strategies like simulated annealing (for example, MSASA [175]) and genetic algorithm (for example, SAGA [223]). However, the global optimization methods are still too slow for big datasets.

To reduce the running time, we can rely on approximation algorithms. Approximation algorithms report a multiple alignment with 'performance guarantee' (that is, an alignment whose score is within certain bound of the optimal score). For example, Gusfield [127] described the center star alignment method which guarantees to report a multiple alignment whose SP distance is at most twice that of the optimal distance. Later, Pevzner [236] and Bafna et al. [15] further improved the approximation ratio.

Apart from approximation algorithms, we can also apply heuristics to compute a multiple alignment. Although heuristics have no performance guarantee, they can report good multiple alignments most of the time. Many heuristics have been proposed. They can be broadly classified into two approaches: progressive approach (for example, ClustalW [145, 287]) and iterative approach (for example, MUSCLE [88, 89]).

The above three approaches tried to compute a multiple alignment which optimizes certain score. They do not try to explicitly model the mutation/indel probability and the evolution process. Probabilistic methods, on the other hand, assume certain model and try to learn the model parameters; then, a multiple alignment, which fits the model, is identified. Methods in this class include Handel [150, 151] and profile hidden Markov model (profile HMM) [269]. People also try to combine heuristic methods and probabilistic methods. Some example include ProAlign [198], COACH [91], SATCHMO [92], PRANK [197], PROBCONS [80], and MUMMALS [232].

The rest of this chapter is organized as follows. Section 6.4 first gives a dynamic programming solution to compute the optimal MSA in exponential time. Then, we describe an approximation algorithm with approximation ratio 2. Last, we discuss two heuristic solutions, ClustalW (progressive approach) and MUSCLE (iterative approach).

6.4 Dynamic Programming Method

Before we describe the dynamic programming algorithm for aligning k sequences, we first revisit the dynamic programming algorithm in Section 2.2.1 for aligning 2 sequences $S_1[1..n_1]$ and $S_2[1..n_2]$. In such a solution, we define $V(i_1, i_2)$ to be the alignment score (which is the same as the SP score) of the optimal alignment A_{opt} of $S_1[1..i_1]$ and $S_2[1..i_2]$. We observe that the optimal alignment has three cases depending on whether the last column of the alignment is (1) match/mismatch, (2) insert, or (3) delete. Hence, the recursive equation for $V(i_1, i_2)$ is derived as follows.

$$V(i_1, i_2) = \max \begin{cases} V(i_1 - 1, j_2 - 1) + \delta(S_1[i_1], S_2[i_2]) \\ V(i_1 - 1, j_2) + \delta(S_1[i_1], \_) \\ V(i_1, j_2 - 1) + \delta(\_, S_2[i_2]) \end{cases}$$

In fact, we can describe the three cases using another way. For each $j = 1, 2$, let b_j be an indicator variable which equals 0 if the last column of the alignment A_{opt} is a space in S_j; and 1, otherwise. Then, $(b_1, b_2) = (1, 1)$ if the last column is match/mismatch; $(b_1, b_2) = (0, 1)$ if the last column is insert; and $(b_1, b_2) = (1, 0)$ if the last column is delete. For technical reasons, we define $S_1[0] = S_2[0] = \_$. Then, the recursive equation for V can be rewritten as:

$$V(i_1, i_2) = \max_{(b_1, b_2) \in \{0,1\}^2 - \{(0,0)\}} \{V(i_1 - b_1, i_2 - b_2) + \delta(S_1[i_1 b_1], S_2[i_2 b_2])\}$$

We can generalize the above recursive equation to compute a multiple alignment for k sequences which maximizes the SP score. Consider a set of sequences $\mathcal{S} = \{S_1[1..n_1], S_2[1..n_2], \ldots, S_k[1..n_k]\}$. Assume $S_j[0] = \_$ for all $j \leq k$. Define $V(i_1, i_2, \ldots, i_k)$ to be the SP score of the optimal alignment of $\mathcal{S}$. Then, the maximum SP score among all possible multiple alignments of $\mathcal{S}$ is $V(n_1, n_2, \ldots, n_k)$.

We initialize the base case $V(0, \ldots, 0) = 0$. The recursive case is

$$V(i_1, i_2, \ldots, i_k) =$$
$$\max_{(b_1, \ldots, b_k) \in \{0,1\}^k - \{0^k\}} \{V(i_1 - b_1, \ldots, i_k - b_k) + \text{SP-score}(i_1 b_1, i_2 b_2, \ldots, i_k b_k)\}$$

where $\text{SP-score}(i_1, i_2, \ldots, i_k) = \sum_{1 \leq p < q \leq k} \delta(S[i_p], S[i_q])$.

Using the above recursive equation, the optimal SP score $V(n_1, n_2, \ldots, n_k)$ can be computed by filling in the dynamic programming table V. Then, by back-tracing starting from the entry $V(n_1, n_2, \ldots, n_k)$ (similar to the back-tracing described in Section 2.2.1), we can recover the optimal multiple alignment.

For time complexity, since we need to fill in $n_1 n_2 \ldots n_k$ entries and each entry can be computed in $O(k^2 2^k)$ time, the time required is $O(k^2 2^k n_1 n_2 \ldots n_k)$. For space complexity, since we need to store the table V, the space required is $O(n_1 n_2 \ldots n_k)$.

6.5 Center Star Method

Given a set of k strings $\mathcal{S} = \{S_1, \ldots, S_k\}$, computing the optimal multiple alignment of $\mathcal{S}$ takes exponential time. Can we compute a good enough multiple alignment in polynomial time which guarantee certain upper bound of the score? This section describes an approximation algorithm called center star method which constructs a multiple alignment for $\mathcal{S}$ whose sum-of-pair (SP) distance is at most twice that of the optimal multiple alignment.

First we define the center string of $\mathcal{S}$. Let $D(X, Y)$ be the pairwise optimal alignment distance between any strings X and Y. A string $S_c \in \mathcal{S}$ is called a center string if it minimizes $\sum_{i=1}^{k} D(S_c, S_i)$. The center star method tries to identify the center string S_c from $\mathcal{S}$. Then, it generates a multiple alignment $\mathcal{M}$ for $\mathcal{S}$ so that, for every $1 \leq i \leq k$, the alignment distance $d_{\mathcal{M}}(S_c, S_i)$ equals $D(S_c, S_i)$. The detail of the algorithm is given in Figure 6.3. Figures 6.4 and 6.5 demonstrate the execution of the center star method.

Center_Star_Method

Require: A set $\mathcal{S}$ of sequences

Ensure: A multiple alignment of M with sum of pair distances at most twice that of the optimal alignment of $\mathcal{S}$

1: Find $D(S_i, S_j)$ for all i, j.

2: Find the center sequence S_c which minimizes $\sum_{i=1}^{k} D(S_c, S_i)$.

3: For every $S_i \in \mathcal{S} - \{S_c\}$, choose an optimal alignment between S_c and S_i.

4: Introduce spaces into S_c so that the multiple alignment $\mathcal{M}$ satisfies the alignments found in Step 3.

FIGURE 6.3:   Center star method.

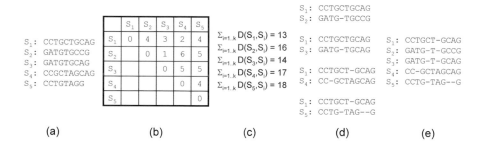

FIGURE 6.4:   (a) 5 DNA sequences. Assuming mismatch/indel scores 1, (b) shows the pairwise alignment distance between every pair of sequences. (c) computes $\sum_{i=1}^{k} D(S_c, S_i)$ for $1 \leq c \leq k$. S_1 minimizes $\sum_{i=1}^{k} D(S_c, S_i)$ and it is denoted as the center string. (d) shows the pairwise alignments between S_1 and S_i for $2 \leq i \leq k$. (e) shows the multiple alignment which is consistent with all the pairwise alignments in (d).

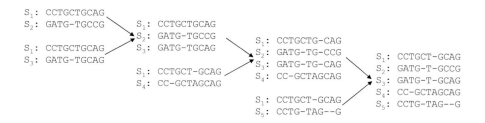

FIGURE 6.5:   This figure demonstrates how to convert the pairwise alignments in Figure 6.4(d) to the multiple alignment in Figure 6.4(e).

Below, we show that the multiple alignment generated by the center star method has the sum of pair distance at most twice the optimal sum of pair distance. Our proof relies on the fact that, for any multiple alignment $\mathcal{M}$, the distance score $d_{\mathcal{M}}()$ satisfies the triangle inequality, that is, $d_{\mathcal{M}}(X,Y) \leq d_{\mathcal{M}}(X,Z) + d_{\mathcal{M}}(Z,Y)$ for any three sequences X, Y, and Z in the multiple alignment $\mathcal{M}$.

Let $\mathcal{M}$ be the multiple alignment generated by the center star method. Let $\mathcal{M}^*$ be the optimal multiple alignment which minimizes the sum of pair distance. By the definition of the center string S_c, for any i, $\sum_{j=1}^{k} D(S_i, S_j) \geq \sum_{j=1}^{k} D(S_c, S_j)$. Hence, the sum of pair distance for M^*

$$
\begin{aligned}
&= \sum_{1 \leq i < j \leq k} d_{\mathcal{M}^*}(i,j) \\
&\geq \sum_{1 \leq i < j \leq k} D(S_i, S_j) \\
&= \tfrac{1}{2} \sum_{i=1}^{k} \sum_{j=1}^{k} D(S_i, S_j) \\
&\geq \tfrac{1}{2} \sum_{i=1}^{k} \sum_{j=1}^{k} D(S_c, S_j) \\
&= \tfrac{k}{2} \sum_{j=1}^{k} D(S_c, S_j)
\end{aligned}
$$

Thus, the sum of pair distance for $\mathcal{M}^*$ is at least $\tfrac{k}{2} \sum_j D(S_c, S_j)$.

Then, by the triangle inequality, the sum of pair distance for $\mathcal{M}$

$$
\begin{aligned}
&= \sum_{1 \leq i < j \leq k} d_{\mathcal{M}}(i,j) \\
&= \tfrac{1}{2} \sum_{i=1}^{k} \sum_{j=1}^{k} d_{\mathcal{M}}(i,j) \\
&\leq \tfrac{1}{2} \sum_{i=1}^{k} \sum_{j=1}^{k} [D(S_c, S_i) + D(S_c, S_j)] \\
&= \tfrac{k}{2} \sum_{i=1}^{k} D(S_c, S_i) + \tfrac{k}{2} \sum_{j=1}^{k} D(S_c, S_j) \\
&= k \sum_j D(S_c, S_j)
\end{aligned}
$$

Hence, the sum of pair distance of $\mathcal{M}$ is at most twice the optimal sum of pair distance for $\mathcal{S}$. We have the following lemma.

LEMMA 6.1
Let $\mathcal{M}$ be the multiple alignment of $\mathcal{S} = \{S_1, \ldots, S_k\}$ computed by the center star method. The sum of pair distance of $\mathcal{M}$ is at most twice that of the optimal alignment.

Last, we analyze the time complexity. Suppose all sequences in $\mathcal{S}$ are of length $O(n)$. Then, Step 1 takes $O(k^2 n^2)$ time to compute the optimal distance $D(S_i, S_j)$ for all $1 \leq i < j \leq k$. Step 2 takes $O(k^2)$ time to find the center string. Step 3 generates the pairwise alignment between S_c and S_i for every $S_i \in \mathcal{S} - \{S_c\}$. It takes $O(kn^2)$ time. Step 4 introduces spaces into the multiple alignment, which takes $O(k^2 n)$ time. Thus, the center star method runs in $O(k^2 n^2)$ time.

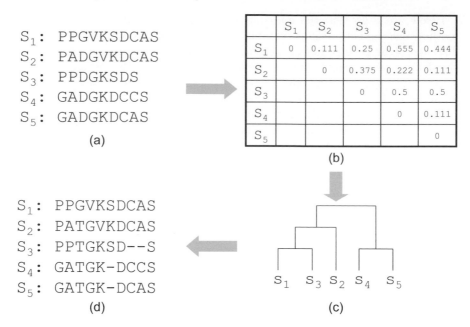

S$_1$: PPGVKSDCAS
S$_2$: PADGVKDCAS
S$_3$: PPDGKSDS
S$_4$: GADGKDCCS
S$_5$: GADGKDCAS

(a)

	S$_1$	S$_2$	S$_3$	S$_4$	S$_5$
S$_1$	0	0.111	0.25	0.555	0.444
S$_2$		0	0.375	0.222	0.111
S$_3$			0	0.5	0.5
S$_4$				0	0.111
S$_5$					0

(b)

S$_1$: PPGVKSDCAS
S$_2$: PATGVKDCAS
S$_3$: PPTGKSD--S
S$_4$: GATGK-DCCS
S$_5$: GATGK-DCAS

(d)

S$_1$ S$_3$ S$_2$ S$_4$ S$_5$

(c)

FIGURE 6.6:   The three steps of ClustalW (a progressive alignment meth-ods). Five input sequences are given in (a). Step 1 computes the pairwise distance scores for these five sequences (see (b)). Then, Step 2 generates the guide tree such that similar sequences are grouped together first (see (c)). Step 3 aligns the sequences one by one according to the branching order of the guide tree, yielding the multiple alignment of all input sequences (see (d)).

6.6   Progressive Alignment Method

The progressive alignment method was first described by Feng and Doolit-tle [104]. It is a heuristic to create multiple sequence alignment. Its basic idea is to utilize the pairwise alignments to guide us to successively construct the multiple alignment. This method first aligns the two most closely related sequences; then it progressively aligns the next most closely related sequences until all sequences are aligned.

In general, a progressive alignment method consists of three steps:

1: Compute pairwise distance scores for all pairs of sequences,
2: Generate the guide tree which ensures similar sequences are nearer in the tree, and
3: Align the sequences one by one according to the guide tree.

Figure 6.6 illustrates the three steps. Below, we give a detailed descrip-

```
S1: PPGVKSDCAS
S3: PPDGKSD--S
```

FIGURE 6.7:   The figure shows the alignment of S_1 and S_3 in Figure 6.6. There are eight non-gap positions and six identical positions. Hence, the distance is $1 - \frac{6}{8} = 0.25$.

tion of ClustalW [145, 287], which is one of the representative progressive alignment methods.

6.6.1   ClustalW

Given a set of sequences $\mathcal{S} = \{S_1, S_2, \ldots, S_k\}$, each of length $O(n)$. ClustalW [145, 287] performs the following three steps.

First, ClustalW computes the optimal global alignment for every pair of sequences S_i and S_j (see Section 2.2.1); then, the distance score of S_i and S_j is set to be $1 - \frac{y}{x}$ where x and y are the number of non-gap positions and the number of identical positions, respectively, in the alignment between S_i and S_j. Figure 6.7 shows an example to compute the distance between S_1 and S_2 in Figure 6.6(a). The pairwise distance matrix for the five sequences is shown in the table in Figure 6.6(b).

Second, ClustalW builds the guide tree from the distance matrix using the neighbor joining algorithm which is discussed in Section 7.3.4. The guide tree is shown in Figure 6.6(c).

Third, according to the guide tree, we perform a series of alignments to align larger and larger groups of sequences. For the sequences in Figure 6.6(a), according to the guide tree in Figure 6.6(b), we perform 4 stages of alignment: (a) align S_1 with S_3, (b) align S_4 with S_5, (c) align the alignment of (S_1, S_3) with S_2, and (d) align the alignment of (S_1, S_2, S_3) with the alignment of (S_4, S_5). At each stage, two sets of alignments are aligned by introducing new gaps. Moreover, gaps that are present in the original alignments will not be deleted. Such alignment is known as profile-profile alignment and its detail is described in Section 6.6.2.

After the three steps, ClustalW reports the final multiple alignment. For time complexity, Step (1) performs k^2 global alignments, which takes $O(k^2 n^2)$ time. Step (2) builds the guide tree using neighbor joining, which takes $O(k^3)$ time. For Step (3), each profile-profile alignment takes $O(kn+n^2)$ time. Since the guide has at most k internal nodes, Step (3) takes $O(k^2 n + kn^2)$ time. In total, the running time of ClustalW is $O(k^2 n^2 + k^3)$.

6.6.2   Profile-Profile Alignment

Step 3 of ClustalW computes the alignment of two alignments. The input consists of two sets of sequences $\mathcal{S}_1$ and $\mathcal{S}_2$ and the corresponding multiple alignments A_1 and A_2. The profile-profile alignment introduces gaps into A_1 and A_2 to generate the multiple alignment for $\mathcal{S}_1 \cup \mathcal{S}_2$. For example, in Figure 6.8, the two alignments A_1 and A_2 are aligned through introducing a gap of two spaces between columns 5 and 6 of A_2.

First, we need a scoring function to determine the similarity between a column i of A_1 and a column j of A_2. To maximize the SP score, a natural score is

$$PSP(A_1[i], A_2[j]) = \sum_{x,y \in \mathcal{A}} g_x^i g_y^j \delta(x, y)$$

where $\mathcal{A}$ is the alphabet of 20 amino acids or 4 nucleotides, g_x^i is the observed frequency of character x in position i, and $\delta(x, y)$ is the similarity between characters x and y. This score was suggested in ClustalW [145, 287].

Suppose A_1 and A_2 are the alignments of k_1 and k_2 sequences, respectively. Furthermore, suppose A_1 and A_2 have n_1 and n_2 columns, respectively. An alignment of A_1 and A_2 is formed by introducing new gaps into the two alignments so that they are of the same length. Each new gap introduced is given a full gap opening and extension penalties. Each aligned position pair of A_1 and A_2 is scored by the PSP score. The profile-profile alignment seeks to maximize the total score minus the gap penalty.

Similar to global alignment, the profile-profile alignment can be computed using dynamic programming. When there is no gap penalty, the alignment can be found as follows. Let $V(i, j)$ be the score of the best alignment between $A_1[1..i]$ and $A_2[1..j]$. We have $V(0, 0) = 0$, $V(i, 0) = V(i-1, 0) + PSP(A_1[i], \_)$ for $0 < i \le n_1$ $V(0, j) = V(0, j-1) + PSP(\_, A_2[j])$ for $0 < j \le n_2$. When $i, j > 0$, we have

$$V(i, j) = \max \begin{cases} V(i-1, j-1) + PSP(A_1[i], A_2[j]) \\ V(i, j-1) + PSP(\_, A_2[j]) \\ V(i-1, j) + PSP(A_1[i], \_) \end{cases}$$

Once we compute $V(n_1, n_2)$, by back-tracing, we can recover the optimal profile-profile alignment.

When the affine gap penalty is applied, the profile-profile alignment can be computed using dynamic programming similar to the global alignment with affine gap penalty.

The lemma below shows that the running time of the profile-profile alignment is $O(k_1 n_1 + k_2 n_2 + n_1 n_2)$.

LEMMA 6.2
Assuming that the size of $\mathcal{A}$ is a constant, the profile-profile alignment of A_1 and A_2 can be computed in $O(k_1 n_1 + k_2 n_2 + n_1 n_2)$ time.

PROOF   We first need to compute g_x^i for any $x \in \mathcal{A}$ and any column i in A_1. This takes $O(k_1 n_1)$ time. Similarly, we can compute g_y^j for any $y \in \mathcal{A}$ and

S1:	PPGVKSEDCAS			S1:PPGVKSEDCAS
S2:	PATGVKEDCAS			S2:PATGVKEDCAS
S3:	PPDGKSED--S			S3:PPDGKSED--S
		S4:	GATGKDCCS	S4:GATGK--DCCS
		S5:	GATGKDCAS	S5:GATGK--DCAS
	(a)		(b)	(c)

FIGURE 6.8: (a) Alignment A_1 of S_1, S_2, S_3. (b) Alignment A_2 of S_4, S_5. (c) Optimal alignment of A_1 and A_2, which is formed by introducing a gap of two spaces between columns 5 and 6 of A_2. Every column of the alignment corresponds to a column in A_1 and a column in A_2. For instance, for column 10 of the alignment, it corresponds to column 10 of A_1, which has amino acids $A, A, -$, and column 8 of A_2, which has amino acids C, A. The score of the alignment of A_1 and A_2 equals the PSP scores for columns $1, 2, 3, 4, 5, 8, 9, 10, 11$ minus the penalty of introducing a gap in columns 6 and 7. For column 10, the PSP score is $PSP(A_1[10], A_2[8]) = (2\delta(A, C) + 2\delta(A, A))$.

any column j in A_2. This takes $O(k_2 n_2)$ time. Finally, we perform dynamic programming to fill in the $n_1 n_2$ entries of the table V. Since each entry takes constant time to compute, the lemma follows. ⬚

6.6.3 Limitation of Progressive Alignment Construction

Since the progressive method is a heuristic which does not realign the sequences, the multiple sequence alignment is bad if we have a poor initial alignment. This also means that the method is sensitive to the distribution of the sequences in the sequence set. Another limitation is that progressive alignment does not guarantee convergence to the global optimal.

6.7 Iterative Method

To avoid the limitation inherent in progressive methods, iterative methods are introduced. The basic steps of iterative methods are as follows. First, an initial multiple alignment is generated; then, the multiple alignment is improved iteratively.

PRRP [122], MAFFT [170, 171], and MUSCLE [88, 89] are some examples of iterative methods. Below, we discuss the details of MUSCLE.

6.7.1   MUSCLE

Consider a set of sequences $\mathcal{S} = \{S_1, S_2, \ldots, S_k\}$. MUSCLE (multiple sequence comparison by log-expectation) is a heuristic that finds a multiple alignment of $\mathcal{S}$ maximizing the sum-of-pairs (SP) score (allowing an affine gap penalty). It consists of three stages: (1) draft progressive, (2) improved progressive, and (3) refinement.

Stage (1) generates a multiple alignment of the set of sequences $\mathcal{S}$ based on some progressive alignment method like ClustalW. Precisely, it computes the pairwise similarity among sequences in $\mathcal{S}$, then builds the guide tree, and finally aligns the sequences according to the branching orders of the guide tree. Moreover, MUSCLE has the following few changes. To improve accuracy, when two profiles are aligned, MUSCLE uses the log-expectation score (see Section 6.7.2) instead of the PSP score. To improve efficiency, instead of using the alignment score to compute the similarity of two sequences, MUSCLE uses k-mer distance, which is the number of k-mers shared by the two sequences. In addition, instead of using neighbor joining (which runs in $O(k^3)$ time) to construct the guide tree, MUSCLE uses UPGMA (which runs in $O(k^2)$ time. See Section 7.3.2).

Stage (2) improves the multiple alignment of $\mathcal{S}$ by performing the progressive alignment again. Using the multiple alignment generated in Stage (1), for every pair of aligned sequences, we can compute a more accurate pairwise distance as follows. First we compute the fraction D of identical bases between the pair of aligned sequences; then, their Kimura distance is computed, which is $-\ln(1 - D - D^2/5)$. Based on this distance matrix, MUSCLE rebuilds the guide tree and realigns the sequences according to the branching order of the guide tree.

To improve efficiency, Stage (2) only performs realignments for subtrees whose branching orders are changed relative to the original guide tree.

Stage (3) refines the multiple alignment to maximize the SP score using the tree-dependent restricted partitioning technique. By deleting any edge from the guide tree, the tree and the corresponding sequences are partitioned into two disjoint sets. The multiple alignment of each subset can be extracted from the multiple alignment of the previous iteration. Then, the two multiple alignments of the two subsets are realigned by means of profile-profile alignment. If the SP score of the new multiple alignment is improved, we keep the new multiple alignment; otherwise, we discard the new multiple alignment. The process is repeated until all edges are visited without a change in SP score or until a user-defined maximum number of iterations. In MUSCLE, the edges are visited in decreasing distance from the root, which has the effect of first realigning individual sequences, then realigning closely related groups.

6.7.2 Log-Expectation (LE) Score

The PSP score (see Section 6.6.2) ignores the spaces in the columns. Hence, it may favor more spaces in the columns. To resolve the problem, MUSCLE proposed a log-expectation (LE) score. The LE score for aligning column i of A_1 and column j of A_2 is defined as

$$LE(A_1[i], A_2[j]) = (1 - f_G^i)(1 - f_G^j) \log \sum_{x,y \in \mathcal{A}} f_x^i f_y^j \delta(x,y)$$

where f_G^i is the observed frequency of gaps in column i and f_x^i is the normalized observed frequency of character x in column i, that is, $g_x^i/(\sum_{x \in \mathcal{A}} g_x^j)$. In MUSCLE, the score $\delta(x,y)$, for any pair of amino acids x and y, is defined using the 240 PAM VTML matrix [215].

6.8 Further Reading

The multiple sequence alignment methods covered in this chapter just touch the tip of the iceberg. In-depth survey on multiple sequence alignment can be found in [79, 86, 90, 221, 222, 288, 298].

Apart from the four main methods in Section 6.3, other methods exist like ensemble methods and database-aided methods. Ensemble methods assemble a multiple alignment from a set of alignments made by different prediction techniques. The resulting alignment is expected to be more accurate. M-Coffee[299] and the meta_align program in MUMMALS[232] belong to this class of methods.

Database-aided methods add external information to help to improve and resolve ambiguities in the multiple alignment. For example, when the number of input sequences is small, we can perform database search to include homologous sequences of the input sequences. Addition of the homologous sequences has been shown to improve the alignment [264, 265]. We can also improve the alignment by adding known or predicted secondary structure (like PSI-PRALINE[264] and SPEM[318]) or structural information (like 3D-Coffee[226]).

Some multiple sequence alignment software is available on the web. For example, ClustalW can be found at http://www.ebi.ac.at/clustalw/, MUSCLE can be found at http://www.dive5.com/muscle, and ProbCons can be found at http://probcons.stanford.edu/. A comparison of different multiple sequence alignment software can be found in [30].

Benchmark datasets for evaluating multiple sequence alignment software are also available. They can be found in MDSA [52], BAliBASE [289], OXBench [241], SMART [240], PREFAB [89], and SABmark.

As a final remark, this section described global multiple sequence alignment methods. They require the sequences to be related over their whole length (or at least most of it). With inappropriate input sequences, the output is nonsense even though MSA methods will still produce alignment.

To perform local multiple alignment, a number of methods have been proposed, including DiAlign [277], Align-m [300], ProDa [238], SATCHMO [92], IterAlign [35], POA [185], and EulerAlign [316].

6.9   Exercises

1. Extract some sequences from BAliBASE. (The website of BAliBASE is `http://www-bio3d-igbmc.u-strasbg.fr/balibase/`.) Can you align those sequences using ClustalW and MUSCLE? Do the results look similar to the answers in BAliBASE?

2. Consider the following set of sequences.

$$S_1 = \texttt{ACTCTCGATC}$$
$$S_2 = \texttt{ACTTCGATC}$$
$$S_3 = \texttt{ACTCTCTATC}$$
$$S_4 = \texttt{ACTCTCTAATC}$$

   Can you compute their multiple sequence alignment using the center star method? Please show the steps.

3. Referring to previous question, can you compute the multiple sequence alignment of the 4 sequences using ClustalW? Please show the steps. (You can assume that match scores 1 and mismatch/indel scores -1.)

4. Given a set of k sequences $S_1, S_2, \ldots, S_k$, we would like to find k substrings $T_1, T_2, \ldots, T_k$ of $S_1, S_2, \ldots, S_k$, respectively, such that the optimal SP score of the multiple sequence alignment of $T_1, T_2, \ldots, T_k$ is maximized. Can you propose a dynamic programming algorithm to solve this problem? What is the running time? (Note that when $k = 2$, this problem is the same as the pairwise local alignment problem.)

5. Consider three sequences S_1, S_2, and S_3 of length n_1, n_2, and n_3, respectively. Assume $n_1 \geq n_2 \geq n_3$. We would like to find the multiple sequence alignment of S_1, S_2, and S_3 which optimizes the SP score. Can you propose an $O(n_1 n_2 n_3)$-time algorithm which uses $O(n_2 n_3)$ space?

6. Consider k sequences $S_1, S_2, \ldots, S_k$. Give an efficient dynamic programming algorithm to find an alignment $S_1', S_2', \ldots, S_k'$ which maximizes

the sum-of-pair score allowing an affine gap penalty. (The sum-of-pair score is defined as $\sum_{1 \leq i < j \leq k} score(S_i', S_j')$. To compute $score(S_i', S_j')$, we first remove all columns with a pair of spaces. Then, $score(S_i', S_j') = \sum_p \delta(S_i'[p], S_j'[p]) + h \cdot no\_of\_gaps$ where $\delta(x, y)$ is the similarity score between two residues, $\delta(x, \_) = \delta(\_, x) = s$, and no_of_gaps is the number of gaps in the alignment. Note that h is the open gap penalty, and s is the penalty of each space.)

7. Consider three sequences $S_1[1..n_1]$, $S_2[1..n_2]$, and $S_3[1..n_3]$. Assume their alignment contains at most d spaces. Furthermore, we assume the match is score 1 and the mismatch and indel scores are -1. Can you give an efficient algorithm to compute the maximum sum-of-pair score for the global alignment of the three sequences? What is the time complexity?

8. Assume that we compare the sequences by counting the shared number of 3-mers. For the following three sequences, which pair is more similar?

   - $S_1 = $ ACGTCGATC
   - $S_2 = $ CCGGCGTCA
   - $S_3 = $ ACGCTCGAT

Chapter 7

Phylogeny Reconstruction

7.1 Introduction

DNA is commonly known to be responsible for encoding the information of life. Through sexual reproduction, DNA is passed on as hereditary material to offspring. During the process, 'mistakes' might sometimes occur. These mistakes cause the DNA to be a bit different from its parent DNA (one or more of the bases might have been changed). Such 'mistakes' are known as mutations. Due to the mutations of many generations of reproduction, different taxa (or species) emerge (or evolve). Phylogenetics is the tool for studying the evolutionary relationship among different taxa.

7.1.1 Mitochondrial DNA and Inheritance

Understanding the evolution of modern man is one of the interesting topics in phylogenetics. In particular, people are interested in studying the evolution of mitochondrial DNA (mtDNA). mtDNA is located in the mitochondria, which are organelles located in the cytoplasm of the cell. These organelles are responsible for energy transfer and are basically the 'powerhouses'. There are about 1,700 mitochondria in every human cell; each includes an identical circular DNA of length about 16,000 base pairs and contains 37 genes. Whenever an egg cell is fertilized, nuclear DNA of a sperm cell enters the egg cell and combines with the nuclear DNA of an egg cell, producing a mixture of both parents' genetic code. The mtDNA from the sperm cell, however, is left behind, outside of the egg cell. So the fertilized egg contains a mixture of the father and mother's nuclear DNA and an exact copy of the mother's mtDNA, but none of the father's mtDNA. In other words, mtDNA is passed on only along the maternal line.

7.1.2 The Constant Molecular Clock

Even though all humans inherit the mtDNA from one person who lived a long time ago, our mtDNA is not exactly alike due to mutation. For example, let's say that 10,000 years after the most recent common ancestor, one of the mtDNA branches experienced a mutation. From that point on, that line of mtDNA would include that alteration. Another branch might experience a mutation in a different location. This alteration would also be passed on. By looking at the similarities and differences of the mtDNA of present day individuals, researchers try to reconstruct where the branching took place. In an original 1987 *Nature* article [48], Rebecca Cann, Mark Stoneking, and Allan Wilson studied the mtDNA of 147 people from continents around the world (though for Africans, they relied on African Americans). With the help of a computer program, they put together a sort of family tree by grouping those with the most similar DNA together, then grouping the groups, and then grouping the groups of groups. The resulting tree showed that one of the two primary branches consisted only of African mtDNA and the other branch consisted of mtDNA from all over the world, including Africa.

From the tree, they inferred that the most recent common mtDNA ancestor was an African woman, who appears roughly 100,000–200,000 years ago. This study was done under the assumption of a constant molecular clock. Such a theory assumes mutations accumulate at an approximately constant rate as long as the DNA sequence retains its original functions. The difference between the DNA sequences (or proteins) in two taxa would then be proportional to the time since the taxa diverged from a common ancestor (coalescence time). The time is measured in arbitrary units and then it can be calibrated in millions of years for any given gene if the fossil record of that taxa happens to be rich. Moreover, such a theory does not hold in general. Neutral mutations do not yield literally constant rates of molecular change but are expected to yield constant average rates of change over a long period of time (known as 'stochastically constant' rates). Hence, Cann et al.'s clock is not accurate and their analysis only supported the common ancestor of modern man who appeared from 100,000 to 1 million years ago.

7.1.3 Phylogeny

In Cann et al.'s project, the tree they built is known as phylogeny. Phylogeny, also known as a cladogram or a dendrogram, is usually represented by a leaf-labeled tree where the internal nodes refer to the hypothetical ancestors and the leaves refer to the existing taxa. Two taxa that look alike will be represented as neighboring external branches and will be joined to a common parent branch. The objective of phylogenetic analysis is to analyze all the branching relationships in a tree and their respective branch lengths.

As an example, the phylogeny of lizards is illustrated in Figure 7.1. In the figure, time is the vertical dimension with the current time at the bottom and earlier times above it. There are five extant taxa (taxa currently living) with their respective names stated. The lines above the extant taxa represent the same taxa, just in the past. When two lines converge to a point, that should be interpreted as the point when the two taxa diverge from a common ancestral taxon. Eventually, until some time in the past, all taxa were derived from just one taxon, the one displayed as the top point.

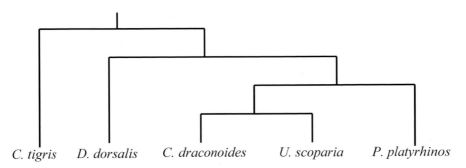

FIGURE 7.1: Phylogeny of lizards.

When performing analysis of phylogenetic trees, one has to understand that the genomes of many organisms have a complex origin. Through reproduction of taxa, some parts of the genome are passed on by a vertical descent. Other parts may have arisen by horizontal transfer of genetic material between taxa through a virus, DNA transformation, symbiosis (living together), or some other horizontal transfer mechanism. In such a case, we may need to represent the evolution of the taxa using multiple phylogenetic trees or using a phylogenetic network (see Section 8.5).

7.1.4   Applications of Phylogeny

Aside from the more obvious applications of understanding the history of life and analyzing rapidly mutating viruses such as HIV, phylogeny has several other applications. One example is multiple sequence alignment (see Chapter 6). Most multiple sequence alignment programs used in practice rely on a phylogenetic tree to guide the computation. In addition, phylogeny also helps to predict the structure of proteins and RNAs, helps to explain and predict gene expression, helps to discover a transcription factor binding motif, and helps to design enhanced organisms such as rice and wheat.

Below, we describe a successful application of phylogeny. Influenza is a fast evolving virus. To prepare a vaccine to protect against the influenza season every year, we need to predict accurately the correct strains of flu virus every year. Bush et al.[46] studied the evolution of the human influenza A (subtype H3) virus. They reconstructed the phylogenetic tree for the human influenza A isolates from 1983 to 1994. Bush et al. observed that the evolution of influenza was under a strong selection stress. The tree is skewed and there exists a main evolutionarily linkage. From this linkage, we are able to predict what will be the future influenza. Nowadays, the influenza vaccine is prepared based on such a prediction.

7.1.5   Phylogenetic Tree Reconstruction

To perform phylogenetic analysis, the first step is to reconstruct the phylogenetic tree. Given a collection of species/taxa, the problem is to reconstruct or to infer the ancestral relationships among the species/taxa, i.e., the phylogenetic tree among the species/taxa.

As a matter of fact, most methods for inferring phylogeny yield unrooted trees, that is, the common ancestor is unknown. Thus from a tree itself, it is impossible to tell which taxa branched off before the others. Examples of a rooted and an unrooted tree are illustrated in Figure 7.2.

To root a tree one should add an outgroup to the data set. An outgroup is a taxon for which external information (for example, paleontological information) is available that indicates that the outgroup branched off before all other taxa. Using more than one outgroup (they must not be closely related), we can improve the estimate of the final tree topology.

However, in the absence of a good outgroup, the root may be found by assuming approximately equal evolutionary rates over all the branches. In this way the root is put at the midpoint of the longest path between two taxa. This way of rooting is called mid-point rooting.

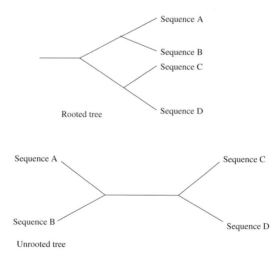

FIGURE 7.2:   Rooted and unrooted phylogenetic trees.

Depending on the types of input data, there are two classes of methods: character-based methods and distance-based methods. We will discuss the two classes of methods in Sections 7.2 and 7.3.

7.2 Character-Based Phylogeny Reconstruction Algorithm

We first study the character-based methods. Consider a set of taxa S. Every taxon can be described by a set of characters such as the number of fingers, the presence or the absence of a certain protein, the nucleotide at a particular location in the genome, etc. Each character has a finite number of states. By contrasting the difference and similarity of the characters of the taxa in S, we can reconstruct the phylogenetic tree for S. Such a method is called a character-based phylogeny reconstruction algorithm.

Precisely, the input to a character-based phylogeny reconstruction algorithm is the character-state matrix M, which is an $n \times m$ matrix where M_{ij} corresponds to the state of the j-th character of the i-th taxon. For example, Figure 7.3 shows the character-state matrix M for a set of four taxa W, X, Y, and Z, each described by four characters. Each character has four different states A, C, G, and T.

The aim of the character-based methods is to reconstruct a phylogeny for S which best explains M. Depending on the optimization criteria, this section describes three different character-based methods for reconstructing phylogeny given M. They are: (1) parsimony, (2) compatibility, and (3) maximum likelihood.

7.2.1 Maximum Parsimony

The principle of parsimony states that "if there exist two possible answers to a problem or a question, then a simpler answer is more likely to be correct". When applying parsimony to the phylogeny reconstruction problem, we aim to build the phylogeny with the fewest point mutations.

The input of the parsimony problem is an $n \times m$ character-state matrix M for a set S of n taxa. A tree T is said to be a valid phylogeny for M if every leaf of T is labeled by a unique taxa in S. The parsimony length of T, denoted as $L(T)$, is defined to be the minimum number of mutations required to explain the tree T. For example, for the phylogeny T in Figure 7.3, we can label the internal nodes of T so that the total number of mutations among all edges is 3. As we cannot reduce the number of mutations to 2, the parsimony length of T is $L(T) = 3$.

The aim of the parsimony problem is to compute a phylogeny T which minimizes $L(T)$. Such a phylogeny T is called the most parsimonious tree for M. In fact, the tree T in Figure 7.3 is the most parsimonious tree for M. Note that the most parsimonious tree for M is not unique. There may be more than one most parsimonious tree for the matrix M.

There are two computational problems for the parsimony problem.

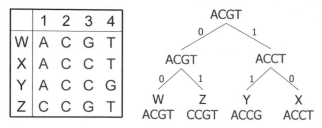

FIGURE 7.3: Example of the most parsimonious tree for four taxa W, X, Y, Z, each described by four characters. The parsimony length is 3.

- The Small Parsimony problem: Given the tree topology T, the problem asks for the parsimony length $L(T)$ and the corresponding assignment of the states to the internal nodes.

- The Large Parsimony problem: Given the character-state matrix M, the problem asks for the most parsimonious tree for M.

Below, we will discuss the details of both problems.

7.2.1.1 Small Parsimony Problem

Given a character-state matrix M for a set of taxa S and a rooted or unrooted tree T leaf-labeled by S, the small parsimony problem asks for the assignment of the states to the internal nodes of T to minimize the total number of mutations. A polynomial time solution to this problem was proposed by Fitch [108] and independently by Hartigan [138]. Sankoff [256] generalized Fitch's solution to handle mutations with weighted penalties. In this section, we just discuss Fitch's algorithm. For simplicity, the discussion assumes the tree T is a rooted binary tree.

We first consider the simple case when M has only one column. In other words, every taxon in S is described by exactly one character c. Let Σ be the set of possible states for the character c. Every leaf v in T is associated with a state $c(v) \in \Sigma$. Our aim is to compute the parsimony length $L(T)$ and to infer a state for each internal node of T so that the total number of state changes in the edges is $L(T)$.

Fitch showed that $L(T)$ can be computed by dynamic programming. For any node u in T and any $\sigma \in \Sigma$, we denote T^u as the subtree of T rooted at u and $L(T^u, \sigma)$ as the minimum total number of state changes in the edges of T^u when u is labeled by σ. Note that $L(T^u) = \min_{\sigma \in \Sigma} L(T^u, \sigma)$. Let R_u be the set of states for u such that $L(T^u, \sigma) = L(T^u)$. In other words, $R_u = \{\sigma \in \Sigma \mid L(T^u, \sigma) = L(T^u)\}$. The following lemma states the recursive formula for computing R_u and $L(T^u)$ for all nodes u in T.

LEMMA 7.1

For every leaf u in T, we have

$$R_u = \{c(u)\}, L(T^u) = 0.$$

For every internal node u in T where v and w are its children, let $\delta = 1$ if $R_v \cap R_w = \emptyset$; and 0 otherwise. We have

$$L(T^u) = L(T^v) + L(T^w) + \delta, R_u = \begin{cases} R_v \cap R_w & \text{if } \delta = 0 \\ R_v \cup R_w & \text{otherwise} \end{cases}$$

PROOF   The state of every leaf u is fixed to be $c(u)$ by the input. Hence, $R_u = \{c(u)\}$. Since there is no mutation, $L(T^u) = 0$.

For every internal node u in T, there are two cases depending on whether $\delta = 0$ or 1.

- When $\delta = 0$, $R_v \cap R_w \neq \emptyset$. We claim that $L(T^u) = L(T^v) + L(T^w)$. Observe that the parsimony length of T^u is at least $L(T^v) + L(T^w)$. Thus, $L(T^u) \geq L(T^v) + L(T^w)$. By labeling u, v, w by any $\sigma \in R_v \cap R_w$, both edges (u, v) and (u, v) have no state change. Thus, we have

$$L(T^u) \leq L(T^u, \sigma) \leq L(T^v, \sigma) + L(T^w, \sigma) = L(T^v) + L(T^w)$$

  Hence, the claim follows. Note that the equality satisfies when u is labeled by a state in $R_v \cap R_w$. This means that $R_u = R_v \cap R_w$.

- When $\delta = 1$, $R_v \cap R_w = \emptyset$. We claim that $L(T^u) = L(T^v) + L(T^w) + 1$. Since there is no common state in R_v and R_w, the parsimony length of T^u is strictly larger than $L(T^v) + L(T^w)$. Thus, $L(T^u) \geq L(T^v) + L(T^w) + 1$. By label v and w by $\sigma_1 \in R_v$ and $\sigma_2 \in R_w$, respectively, and u by σ where σ is either $sigma_1$ or $sigma_2$, we have

$$\begin{aligned} L(T^u) &\leq L(T^u, \sigma) \\ &\leq H(\sigma, \sigma_1) + H(\sigma, \sigma_2) + L(T^v, \sigma_1) + L(T^w, \sigma_2) \\ &= 1 + L(T^v) + L(T^w) \end{aligned}$$

  Hence, the claim follows. Note that the equality satisfies when u is labeled by a state in $R_v \cup R_w$. This means that $R_u = R_v \cup R_w$.

☐

By the above lemma, we can determine the parsimony length of a tree T and report the corresponding character-state assignment using a two phase algorithm. The first phase visits the nodes u in T in the bottom-up order and computes $L(T^u)$ and R_u by Lemma 7.1. The second phase performs back-tracing and generates the character-state assignment for all internal nodes of T. The algorithm is formally stated in Figure 7.4.

Fitch's algorithm

1: Initialization: $R_u = \{c(u)\}$ for all leaves u
2: **for** every internal node u in postorder traversal **do**
3:     Let v and w be the children of u.
4:     Set $\delta = 1$ if $R_v \cap R_w = \emptyset$; 0, otherwise.
5:     Set $L_u = L_v + L_w + \delta$
6:     Set $R_u = \begin{cases} R_v \cap R_w & \text{if } \delta = 0 \\ R_v \cup R_w & \text{otherwise} \end{cases}$.
7: **end for**
8: **for** every internal node u in preorder traversal **do**
9:     Set $c(u) = \begin{cases} c(v) & \text{if } c(v) \in R_u \\ \text{arbitrary state in } R_v & \text{otherwise} \end{cases}$ where v is u's parent.
10: **end for**

FIGURE 7.4: Algorithm for computing the parsimony length and the corresponding character-state assignment of a tree T.

Figure 7.5 shows an example for the implementation of the above algorithm. In the left phylogeny, we label every internal node u by R_u in the bottom-up order using the recursive formula in Lemma 7.1. In the right phylogeny, we traverse the phylogeny in top-down order to generate the label for each internal node. The time complexity of the algorithm is $O(n|\Sigma|)$ where n is the number of taxa. Then, the parsimony length equals the number of state changes in T.

Now, we consider the case when M has m columns, that is, every taxon is represented by m characters. We note that the i-th character and the j-th character are independent for any i and j. Thus, the problem can be solved using m instances of the simple case problem. Then, the parsimony length of T equals the total number of changes over all m characters. For each character, the running time is $O(n|\Sigma|)$. For m characters, the time complexity would be $O(mn|\Sigma|)$.

In the above discussion, we assume the input tree T is rooted. However, the parsimony length is independent of where the root is located. When T is unrooted, its parsimony length can be computed by rooting the tree at any edge and running Fitch's algorithm. The above discussion also assumes the tree is binary. The solution can be generalized to a non-binary tree case (see Exercise 6).

7.2.1.2 Large Parsimony Problem

Given the character-state matrix M for a set S of n taxa, the large parsimony problem asks for a tree T which is leaf-labeled by S and minimizes $L(T)$. Such a tree is called the most parsimonious tree for the set of taxa S.

The large parsimony problem is NP-hard [74, 110], that is, the problem

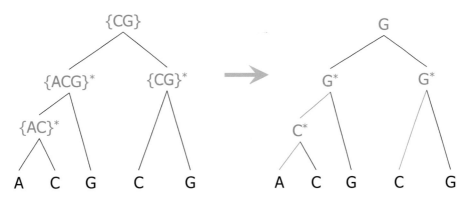

FIGURE 7.5:   Illustration of Fitch's algorithm for solving the small parsimony problem. On the left, we demonstrate the computation of R_u for all internal nodes u in bottom-up order. Each asterisk (*) indicates a mutation is required for one of its child edges. On the right, we demonstrate how to label the internal nodes in top-down order.

is unlikely to have a polynomial time solution. The simple solution is to exhaustively enumerate and compute the parsimony length of all possible trees using Fitch's algorithm. Since there are $(2n - 5)!!$ possible unrooted degree-3 trees for n taxa where $p!! = 1 \cdot 3 \cdot 5 \cdot \ldots \cdot p$, such an enumeration method can only work when n is small. People have suggested a number of solutions to resolve the problem, including branch and bound algorithms, integer linear programming, approximation algorithms, and heuristics algorithms. In this section, we will discuss a branch and bound algorithm by Hendy and Penny [140] and a 2-approximation algorithm for the large parsimony problem.

Branch and bound algorithm: Hendy and Penny proposed a branch and bound algorithm to compute the most parsimonious tree for a set of taxa $S = \{a_1, a_2, \ldots, a_n\}$. Before we detail the solution, we first describe a way to enumerate all possible unrooted trees. We start with the tree with three leaves $\{a_1, a_2, a_3\}$. This is a star tree with three edges and we can compute its parsimony length. Then, incrementally, given any tree T for leaves $\{a_1, a_2, \ldots, a_i\}$, we extend T to form a tree with leaves $\{a_1, a_2, \ldots, a_{i+1}\}$ for $i = 3, 4, \ldots, n - 1$. Precisely, since T is of degree 3 and has $(2i - 3)$ edges, by adding an extra edge with a leaf labeled a_{i+1} to any of these $(2i-3)$ edges, we generate $(2i - 3)$ trees with leaves $\{a_1, a_2, \ldots, a_{i+1}\}$. By this method, we can generate all $(2n - 5)!!$ trees and identify the tree with the minimum parsimony length.

Hendy and Penny observed that when we add a new leaf a_{i+1} to a tree T, the parsimony length is monotonic increasing. Hence, if the parsimony length of the current incomplete tree T is larger than the parsimony length of some complete tree seen so far, then we do not generate or search the trees extended from T. This is the pruning strategy used by Hendy and Penny to reduce the enumeration time.

The efficiency of this method relies on whether we know some complete tree with small enough parsimony length. Hence, it is suggested that we can use some efficient phylogenetic tree construction algorithm like neighbor-joining (see Section 7.3.4) to generate a good enough tree first. The parsimony length of this tree helps to improve the effectiveness of the pruning strategy.

Approximation algorithm: Instead of finding an exact solution, there exists a polynomial time approximation algorithm to solve the large parsimony problem. The best known solution has an approximation ratio of 1.55 [7]. In this section, we present a simple polynomial time algorithm with approximation ratio 2, that is, we can always find a tree whose parsimony length is at most two times worse than that of the most parsimonious tree. The solution is based on the transformation to the minimum spanning tree problem.

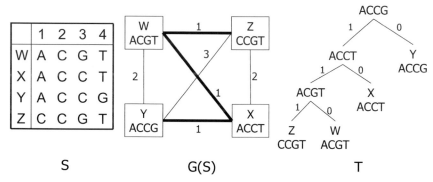

FIGURE 7.6:   An example demonstrating the 2-approximation algorithm for the large parsimony problem. Given the set of 4 taxa S, we construct the graph $G(S)$ and compute the minimum weight spanning tree T' (bold edges in $G(S)$). From T', we construct a phylogenetic tree leaf-labeled by the 4 taxa.

The 2-approximation algorithm is as follows. From the character-state matrix M of a set S of n taxa, we create a weighted complete graph $G(S)$ where (1) the vertex set is S and (2) the weight of every edge (a_i, a_j) equals the Hamming distance (that is, the number of changes) between the characters of a_i and a_j where $a_i, a_j \in S$. Let T' be the minimum weight spanning tree of $G(S)$.[1]   We then convert T' to a phylogenetic tree T by replacing every

---

[1]A spanning tree of $G(S)$ is a tree which connects all nodes in $G(S)$. The minimum spanning tree is a spanning tree of the minimum weight. Such a tree can be found in polynomial time. The current best solution runs in $O(n\alpha(m, n))$ time, where n and m are the number of nodes and edges, respectively, in $G(S)$. Note that $\alpha(m, n)$ is the Ackermann function, which is a very slow growing function.

internal node labeled by taxa a with an unlabeled node and attaching a leaf labeled by a as its child. Figure 7.6 shows an example. The running time of this algorithm is dominated by the construction of the graph $G(S)$, which takes $O(n^2m)$ time. The lemma below shows that the approximation ratio of this algorithm is 2.

LEMMA 7.2
Let T' be a minimum spanning tree of $G(S)$. Then, the parsimony length of T' is at most twice that of the most parsimonious tree.

PROOF   Let T^* be the most parsimonious tree. Let C be a Euler cycle of T^*. (Note that in the Euler cycle C of T^*, every edge in T^* appears twice.) For any graph H, define $w(H)$ be the sum of the weights of all its edges. Then, $w(C) = 2w(T^*)$. Let P be the path containing all nodes in $G(S)$ ordered by their first occurrences in C. We have $w(P) \leq w(C)$ since Hamming distance satisfies the triangle inequality. Since P is not a minimum weight spanning tree, $w(P) \geq w(T')$. In total, $w(T') \leq w(P) \leq w(C) = 2w(T^*)$.   ⧠

Although maximum parsimony is a popular method, it suffers from the problem that it is statistically inconsistent [102], that is, it is not guaranteed to produce the true tree with high probability, given sufficient data. Moreover, parsimony works well in practice. Some people argue that trees on which parsimony is clearly inconsistent are biologically unrealistic [243].

7.2.2   Compatibility

Compatibility is another method to reconstruct a phylogenetic tree. It assumes that mutation is rare and most of the characters mutate at most once in the history. More precisely, compatibility finds a phylogenetic tree that maximizes the number of characters which have at most one mutation in the tree. The following discussion focuses on binary characters, that is, each character has two states 0 and 1 only.

We first give some basic definitions. A binary character c is said to be compatible to a leaf-labeled tree T if and only if there exists an assignment of states to the internal nodes of T such that at most one edge in T has state change. There is another way to express the compatibility property. If a character c is compatible to a tree T, then there exists an edge (u, v) in T such that the deletion of (u, v) partitions T into two subtrees where all leaves in one subtree have state 0 while all leaves in another subtree have state 1. Figure 7.7 shows two examples of leaf-labeled trees where one is compatible while another one is incompatible.

Based on the concept of compatibility, perfect phylogeny can be defined as follows. Consider a set S of n taxa; each is characterized by m binary characters. The input can be expressed as an $n \times m$ character-state matrix

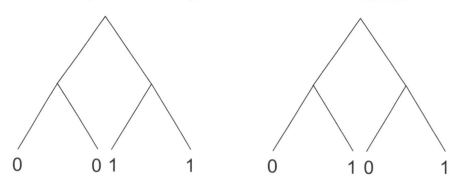

FIGURE 7.7: The binary character in the left tree is compatible since we can label the internal nodes so that there is only one state change. For the right tree, the binary character is not compatible since there are at least two state changes.

M where M_{ij} is the state of character j of taxon i. M is said to admit a perfect phylogeny if and only if a rooted tree T for the n taxa exists where all m characters are compatible.

Figure 7.8 shows an example input matrix M. As shown in the figure, the matrix M admits a perfect phylogeny since X_1, X_2, and X_3 are all compatible to the tree T.

There are two computational problems related to compatibility.

- Perfect Phylogeny Problem: This problem checks if M admits a perfect phylogeny. If yes, we need to report the tree. This problem is equivalent to asking whether there exists a tree T such that all m characters are compatible to the tree T.

- Large Compatibility Problem: Supposing M cannot admit a perfect phylogeny, this problem aims to find the maximum set of mutually compatible characters and recover the corresponding tree.

To ease our discussion, we assume, for each character c in the character-state matrix M, the number of state 1 is smaller than or equal to the number of state 0. If some characters do not satisfy this constraint, we can simply flip those states without affecting the compatibility result.

7.2.2.1 Perfect Phylogeny Problem

Consider a set S of n taxa, each characterized by m characters. This input can be expressed as an $n \times m$ character-state matrix M where M_{ij} is the state of character j of taxon i. For every character i, let O_i be the set of taxa with state 1, that is, $O_i = \{j \mid M_{ij} = 1\}$. For example, in Figure 7.8, $O_1 = \{1, 5\}$, $O_2 = \{1\}$, and $O_3 = \{2, 4\}$. For any pair of characters i and j, they are pairwise compatible if O_i and O_j are disjoint or one of them contains

M	X_1	X_2	X_3
Species 1	1	1	0
Species 2	0	0	1
Species 3	0	0	0
Species 4	0	0	1
Species 5	1	0	0

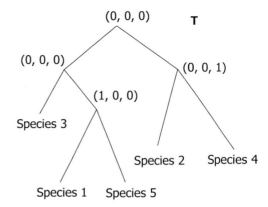

FIGURE 7.8: An example matrix M which admits a perfect phylogeny. Note that all three characters are compatible.

the other. For example, in Figure 7.8, X_1 and X_2 are pairwise compatible since O_1 contains O_2. X_1 and X_3 are pairwise compatible since O_1 and O_3 are disjoint. Below is the key lemma for determining compatibility.

LEMMA 7.3
M admits a perfect phylogeny if and only if every pair of characters i and j are pairwise compatible.

PROOF ($\Rightarrow$) Given that M admits a perfect phylogeny. Note that $|O_i| \leq n/2$ for every character i. In contrast, assume that some characters i and j are not pairwise compatible, i.e., there exist three taxa X, Y, Z such that $X \in O_j - O_i$, $Y \in O_i - O_j$, and $Z \in O_i \cap O_j$. Since $O_i \cap O_j$ is non-empty, $|O_i \cup O_j| < n$. Thus, there exists a taxon $W \notin O_i, O_j$. Since character i is compatible in T, there exists an edge e in T which separates Y and Z from X and W. Since character j is compatible in T, there exists an edge e' in T which separates X and Z from Y and W. T cannot contain both edges e and e'. Thus, a contradiction occurs. Therefore, i and j are pairwise compatible.

($\Leftarrow$) Suppose that every characters i and j are pairwise compatible. Without loss of generality, we reorder the characters so that $|O_i| \geq |O_j|$ for $i < j$. Below, we show that we can construct a perfect phylogeny iteratively in m phases. For each phase i, let $M^{(i)}$ be a submatrix of M for the characters $1, 2, \ldots, i$. Note that multiple taxa may have the same states for characters $1, 2, \ldots, i$. We merge them into one row, says row r, in $M^{(i)}$ and denote the corresponding set of taxa as $N^{(i)}(r)$. The aim of phase i is to build a perfect phylogeny for the matrix $M^{(i)}$.

In phase 1, $M^{(1)}$ has only the character 1, which has two states 0 and 1. The perfect phylogeny for $M^{(1)}$ consists of one root attaching two leaves, where one leaf has state 0 while another leaf has state 1. By induction, suppose T is a perfect phylogeny for $M^{(i-1)}$ in phase $i-1$. We claim that O_i should be a subset of $N^{(i-1)}(r)$ for some row r in $M^{(i-1)}$. Then, the perfect phylogeny for $M^{(i)}$ is formed by splitting the leaf r in T into two nodes u and v such that $N^{(i)}(u) = O_i$ and $N^{(i)}(v) = N^{(i-1)}(r) - O_i$.

What remains is to prove the claim. Assume the taxa in O_i occur in both $N^{(i-1)}(r)$ and $N^{(i-1)}(s)$ for some rows r and s in $M^{(i-1)}$. Since rows r and s are different in $M^{(i-1)}$, there exists some character $j < i$ such that $M_{rj}^{(i-1)} = 0$ and $M_{sj}^{(i-1)} = 1$. Hence, some taxa in O_i have state 0 while some taxa in O_i have state 1 for the character j. Thus, $O_i \not\subseteq O_j$. On the other hand, since $|O_j| \geq |O_i|$, $O_j \not\subseteq O_i$. This means that O_i and O_j are neither disjoint nor one contains the other. This contradicts the fact that i and j are pairwise compatible. $\square$

Below, we first describe a method to check if the input character-state matrix M admits a perfect phylogeny. Then, supposing M admits a perfect phylogeny, we give an algorithm to recover the corresponding tree.

Checking whether M admits a perfect phylogeny: By Lemma 7.3, M admits a perfect phylogeny if all $\binom{m}{2}$ pairs of characters are pairwise compatible. Since containment or disjointedness can be checked in $O(n)$ time, $O(n\binom{m}{2}) = O(m^2 n)$ time is sufficient to determine if M admits a perfect phylogeny.

The bottleneck in the $O(m^2 n)$-time algorithm is the way that checking is done for containment and disjointedness. This section describes how to speed up the checking and gives an $O(mn)$-time algorithm. The idea is to construct a matrix L from M which allows us to check for containment and disjointedness efficiently. L is defined as follows. First we relabel the characters in M so that $|O_i| \geq |O_j|$ if $i < j$. Then, the matrix L is defined so that $L_{ij} = 0$ if $M_{ij} = 0$; otherwise, $L_{ij} = \max\{-1, \max\{k < j \mid M_{ik} = 1\}\}$. Figure 7.9 shows the matrix L derived from M. For example M_{11} has value 1, but there does not exist a smaller value of k such that $M_{1k} = 1$, thus $L_{11} = -1$. As another example, L_{13} has value 1 because $M_{11} = 1$ and $M_{13} = 1$. (Note that the value 1 for L_{13} comes from the column index of M_{11}, which is 1.) According to the algorithm, $L_{ij} = 0$ when $M_{ij} = 0$.

The following lemma shows how to use the matrix L to check if M admits a perfect phylogeny.

LEMMA 7.4

If there exists some column j in L with two different nonzero entries, then M does not admit a perfect phylogeny. Otherwise, M admits a perfect phylogeny.

PROOF   Suppose $L_{ij} = x$ and $L_{kj} = x'$ where both x and x' are nonzero. Without loss of generality, $x > x'$. Since L_{ij} and L_{kj} are non-empty,

M	X1	X2	X3
Species 1	1	0	1
Species 2	0	1	0
Species 3	0	0	0
Species 4	0	1	0
Species 5	1	0	0

L	X1	X2	X3
Species 1	-1	0	1
Species 2	0	-1	0
Species 3	0	0	0
Species 4	0	-1	0
Species 5	-1	0	0

FIGURE 7.9: An example demonstrates how to construct the matrix L from M. Note that we arrange the characters in M so that $|O_i| \geq |O_{i+1}|$.

by definition, $M_{ij} = M_{kj} = 1$. Since $L_{ij} = x$ and $L_{kj} > x$, $M_{ix} = 1$ and $M_{kx} = 0$. (See figure below.)

Thus, O_j contains taxa i and k and O_x contains taxon i, but not taxon k. It means that (1) $O_j \cap O_x \neq \phi$, (2) O_j is not a subset of O_x. Note that $j > x$. Thus, $|O_x| \geq |O_j|$. As $k \notin O_x$, O_x should contain some taxon which does not appear in O_j. So, (3) O_x is not a subset of O_j. By (1) to (3) and applying Lemma 7.3, M does not admit a perfect phylogeny.

Otherwise, for every character j, all nonzero entries L_{ij} have the same value. We claim that every pair of characters is pairwise compatible and hence, by Lemma 7.3, M admits a perfect phylogeny. Below, we prove the claim. For a particular character j, let k be the value of a nonzero entry L_{ij}. Since all nonzero entries L_{ij} have the same value k, by definition, $O_j \subseteq O_k$ and $O_p \cap O_j = \emptyset$ for $k < p < j$. Hence, character j is pairwise compatible with any character p where $k \leq p < j$. If $k > 0$, we apply the same principle and set k' as the value of a nonzero entry L_{ik}. By the same argument, character j is pairwise compatible with any character p where $k' \leq p < k$. By applying this argument repeatedly until $k = -1$, we show that character j is pairwise compatible with any character p where $p < j$. □

Our algorithm is as follows. Given a matrix M, we first construct the matrix L from it. Then, for every character j, we search for two taxa i and k such that $L_{ij} \neq L_{kj}$ and $L_{ij}, L_{kj} \neq 0$. If such a pair of taxa exists, we report M

Algorithm   Perfect   Binary   Phylogeny   Construction

Let T be a tree containing a single root node r;
Let $N(r)=\{1, \ldots, n\}$;
for $j=1$ to m **do**
    By examining the leaves in T, we find a leaf v such that $N(v)$ can
    be partitioned into two non-empty sets N_0 and N_1 where $N_s = \{x \in$
    $N(v) \mid M_{xj} = s\}$ for $s = 0, 1$;
    Create two children v_0 and v_1 for v;
    Set $N(v_0) = N_0$, $N(v_1) = N_1$;
end for
for every leaf $v \in T$ **do**
    If $|N(v)| > 1$, assign the taxa in $N(v)$ as the children of v;
    If $|N(v)| = 1$, node v represents the taxa in $N(v)$;
end for

FIGURE 7.10:   Algorithm for constructing a perfect phylogeny.

does not admit a perfect phylogeny. Otherwise, we report M admits a perfect phylogeny.

The correctness of the algorithm follows from Lemma 7.4. For the time complexity, L can be constructed in $O(mn)$ time. Then, for each of the m characters, it takes $O(n)$ time to check if there exist two different nonzero entries. Thus, $O(mn)$ time is sufficient to decide whether M admits a perfect phylogeny or not.

Reconstruct Perfect Phylogeny from M: Suppose M admits a perfect phylogeny. This section describes an $O(nm)$-time algorithm to construct the perfect phylogeny. Without loss of generality, we assume the columns of M are arranged such that $|O_i| \geq |O_j|$ for $1 \leq i < j \leq m$. The construction algorithm incrementally builds a perfect phylogeny T_i for characters $\{1, 2, \ldots, i\}$, for $i = 1, 2, \ldots, m$. For each leaf v in T, we maintain $N(v)$ to be the set of taxa which share the same states for characters 1 to i. Initially, the construction algorithm starts with a tree T_0 with a single node r where $N(r) = \{1, 2, \ldots, n\}$. Then, we process the characters one by one to refine the tree. Since M admits a perfect phylogeny, for every character i, T_i is constructed by splitting at most one leaf in T_{i-1} into two leaves. Figure 7.10 details the algorithm.

As an example, consider the character-state matrix M for the 5 taxa in Figure 7.9. According to the previous section, M admits a perfect phylogeny. Figure 7.11 illustrates the steps to construct the corresponding perfect phylogeny.

For the time complexity, observe that, for every character j, it takes $O(n)$ time to identify a leaf and to split it into two leaves. Thus, the total time taken is $O(nm)$.

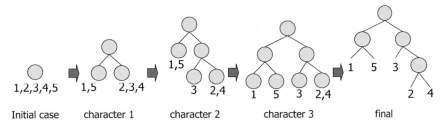

Initial case   character 1   character 2   character 3   final

FIGURE 7.11: Given the character-state matrix M in Figure 7.9, this figure illustrates the reconstruction of the perfect phylogeny from M. Initially, all five taxa are represented in a node. By character X_1, the taxa are split into two groups $\{1, 5\}$ and $\{2, 3, 4\}$. Then, the taxa are further split based on characters X_2 and X_3.

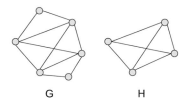

G   H

FIGURE 7.12: The clique problem.

7.2.2.2 Large Compatibility Problem

This section considers the large compatibility problem, which identifies a maximum subset of taxa in M so that the submatrix admits a perfect phylogeny. We show that this problem is equivalent to the clique problem, which is a classical NP-hard problem in computer science. Hence, the large compatibility problem is NP-hard [132]. Furthermore, all results applied to the clique problem are also applicable to the large compatibility problem.

We first formally define the clique problem. Given a graph G, a clique is an induced subgraph of G which is a complete graph. The clique problem asks for the clique of maximum size. For example, in Figure 7.12, the maximum clique of G is in fact H.

Below two lemmas show that the clique problem can be transformed to the large compatibility problem in polynomial time, and vice versa.

LEMMA 7.5

The clique problem can be transformed to the large compatibility problem in polynomial time.

PROOF   Given a graph $G = (V, E)$ with $|V| = m$, we construct a character-state matrix M as follows. M consists of m characters, each corresponds to a vertex in V. M has $3\binom{m}{2}$ taxa; every pair of vertices corresponds to 3 taxa. For every $(u,v) \notin E$, let x, y, z be the three corresponding taxa. We set $M_{xu} = 0$, $M_{yu} = 1$, $M_{zu} = 1$, $M_{xv} = 1$, $M_{yv} = 1$, and $M_{zv} = 0$. By this setting, we ensure characters u and v are not pairwise compatible. All remaining entries in M are set to 0.

We show that $V' \subseteq V$ is a clique if and only if all characters u and v are pairwise compatible for $u, v \in V'$. If V' is a clique, by definition, for every $u, v \in V'$, $(M_{xu}, M_{xv}) \in \{(0,0),(0,1),(1,0)\}$ for any taxa x. We have $O_u \cap O_v = \emptyset$. Thus, u and v are pairwise compatible. On the other hand, if u and v are pairwise compatible for all $u, v \in V'$, then $(u,v) \in E$ for all $u, v \in V'$. Thus, V' forms a clique. □

LEMMA 7.6
The large compatibility problem can be transformed to the clique problem in polynomial time.

PROOF Given a character-state matrix M, we can transform the large compatibility problem to the clique problem as follows: We define a graph $G = (V, E)$ where every character i in M corresponds to a vertex i in G and an edge (i, j) is in G if characters i and j are pairwise compatible in M. The transformation takes $O(n^2 m)$ time. It can easily be shown that G contains a clique of size c if and only if M contains a subset of compatible characters whose size is c. □

7.2.2.3 Compatibility for Characters with k Possible States

Now we generalize the problem for non-binary characters. A character c with k possible states is said to be compatible to a leaf-labeled tree T if and only if there exists an assignment of states to the internal nodes of T such that the total number of state changes is at most $k - 1$.

Let's revisit the two computational problems for compatibility for characters with multiple states.

- Perfect Phylogeny Problem: When the number of states is constant, a polynomial time algorithm is still feasible [3]. When the number of states is variable, the problem is NP-hard.

- Large Compatibility Problem: The problem is NP-hard.

7.2.3 Maximum Likelihood Problem

J. Felsenstein, A. Edwards, and E. Thompson proposed to apply maximum likelihood to infer phylogeny. It assumes the observed character-state matrix M of a set of n taxa is generated from some model of evolution. Based on such an evolution model, the maximum likelihood method tries to estimate the phylogeny which best explains M.

This section will first study the evolution model. Then, we discuss two computational problems related to phylogeny inference using maximum likelihood.

7.2.3.1 Cavender-Felsenstein Model

A number of evolution models have been proposed, including the Cavender-Felsenstein model (or called Cavender-Farris model)[53] and the Jukes-Cantor model [163]. This section focuses on the Cavender-Felsenstein (CF) model, which is the simplest possible Markov model of evolution for binary characters. The model is described by a weighted tree $\mathcal{T} = (T, \{p_e \mid e \in T\})$ where T is the phylogeny for the n taxa and, for every edge $e = (u, v)$ in T, p_e is the probability that the states of the character at u and v are different. The CF model also assumes the root has equal probability of having state 0 or 1.

We can access if a particular phylogenetic tree satisfies a given CF model. For example, consider a CF model $\mathcal{T} = (T, \{p_e | e \in T\})$ where the tree T is shown in Figure 7.13 and the mutation probabilities of the edges $p_{(r,a)}, p_{(r,u)}, p_{(u,b)}, p_{(u,c)}$ are $0.2, 0.1, 0.3, 0.4$, respectively. Then, the chance that $\{r = 1, u = 0, a = 1, b = 0, c = 1\}$ is

$$Pr(r = 1, u = 0, a = 1, b = 0, c = 1 \mid \mathcal{T})$$
$$= Pr(r = 1)Pr(a = 1|r = 1)Pr(u = 0|r = 1)Pr(b = 0|u = 0)Pr(c = 1|u = 0)$$
$$= \qquad 0.5(1 - 0.2)(0.1)(1 - 0.3)(0.4) = 0.0112$$

Formally, consider a CF model $\mathcal{T} = (T, \{p_e | e \in T\})$. Suppose, for every node u in T, its state is i_u where i_u is either 0 or 1. Then, $Pr(u = i_u \forall u \in T | \mathcal{T})$ can be computed as follows.

$$Pr(u = i_u \forall u \in T | \mathcal{T})$$
$$= Pr(r = i_r) \prod_{(u,v) \in T} \left[\delta(i_u, i_v)(1 - p_{(u,v)}) + (1 - \delta(i_u, i_v))p_{(u,v)} \right] \quad (7.1)$$

where r is the root of T and $\delta(x, y) = 1$ if $x = y$ and 0 otherwise.

In reality, we do not observe the states of the characters for the taxa in the internal nodes of T. Let $L(T)$ and $I(T)$ be the set of leaves and the set of internal nodes, respectively, in T. Suppose the observed state of every leaf $u \in L(T)$ is i_u. We can estimate $Pr(u = i_u \forall u \in L(T) \mid \mathcal{T})$ using the following formula.

$$Pr(u = i_u \forall u \in L(T) \mid \mathcal{T}) = \sum_{i_v \in \{0,1\} \forall v \in I(T)} Pr(u = i_u \forall u \in T \mid \mathcal{T})$$

For example, for the tree in Figure 7.13, if the observed states of a particular

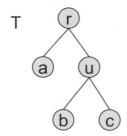

T

FIGURE 7.13:   Cavender-Felsenstein model.

character for the three taxa are $a = 1, b = 0, c = 1$, we have

$$Pr(a = 1, b = 0, c = 1 \mid T) = \sum_{i,j \in \{0,1\}} Pr(r = i, u = j, a = 1, b = 0, c = 1 \mid T)$$

The CF model also assumes all characters are independent. Hence, for a character-state matrix M, the likelihood of T is

$$Pr(M|T) = \prod_{i=1}^{m} Pr(M_i|T). \tag{7.2}$$

where $M_i[j]$ is the state for character i of the taxa j.

There are two computational problems for the Cavender-Felsenstein model.

- Computing the likelihood of a model: Given any character-state matrix M and a model $T = (T, \{p_e \mid e \in T\})$, we try to compute $Pr(M \mid T)$;

- Find a model with maximum likelihood: Given any character-state matrix M, we try to find a model $T = (T, \{p_e \mid e \in T\})$ which maximizes $Pr(M \mid T)$.

7.2.3.2   Likelihood of a Model

Give a model $T = (T, \{p_e \mid e \in T\})$ and any character-state matrix M, this section describes a polynomial time algorithm to compute $Pr(M \mid T)$.

It is obvious that the likelihood $Pr(M \mid T)$ can be computed using Equation (7.2), but it will take exponential time. This section describes a recursive formula which allows us to compute the likelihood by recursion. Let T^v be the subtree of T rooted at v. Let T^v be $(T^v, \{p_e \mid e \in T^v\})$. For any character i and any node v in T, we denote $L_i(v)$ and $L_i(v, s)$ as $Pr(M_i \mid T^v)$ and $Pr(M_i, v = s \mid T^v)$, respectively. Let r be the root of T. Our aim is to compute $Pr(M \mid T)$, which equals

$$\prod_{i=1}^{m} L_i(r) = \prod_{i=1}^{m} \left(\frac{1}{2} L_i(r, 0) + \frac{1}{2} L_i(r, 1) \right) \tag{7.3}$$

Algorithm ComputeLikelihood(i, u)
  if u is a leaf **then**
    Compute and return $\{L_i(u, 0), L_i(u, 1)\}$ using Equation 7.4;
  else
    Let v and w be the two children of u;
    Call ComputeLikelihood(i, v) to compute $\{L_i(v, 0), L_i(v, 1)\}$;
    Call ComputeLikelihood(i, w) to compute $\{L_i(w, 0), L_i(w, 1)\}$;
    Compute and return $\{L_i(u, 0), L_i(u, 1)\}$ using Equation 7.5;
  end if

FIGURE 7.14: Algorithm ComputeLikelihood(i, u) for computing $\{L_i(u, 0), L_i(u, 1)\}$.

Below, we derive the recursive formula for $L_i(v, s)$ depending on whether v is a leaf or is an internal node.

Consider any leaf u in T and any state $s \in \{0, 1\}$. We have

$$L_i(u, s) = \begin{cases} 1 \text{ if } M_{ui} = s \\ 0 \text{ otherwise.} \end{cases} \tag{7.4}$$

Consider any internal node u in T and any state $s \in \{0, 1\}$, let v and w be the children of u. We have

$$L_i(u, s) = \tag{7.5}$$
$$\left[\sum_{y \in \{0,1\}} L_i(v, y) Pr(v_i = y | u_i = s) \right] \left[\sum_{x \in \{0,1\}} L_i(w, x) Pr(w_i = x | u_i = s) \right]$$

Based on Equations 7.4 and 7.5, we apply the algorithm in Figure 7.14 to compute $\{L_i(r, 0), L_i(r, 1)\}$ for every character i. Then, by Equation 7.3, we compute $Pr(M \mid \mathcal{T})$.

Now we analyze the time complexity. To compute $\{L_i(r, 0), L_i(r, 1)\}$, the algorithm in Figure 7.14 needs to evaluate $\{L_i(u, 0), L_i(u, 1)\}$ for every node u according to Equation 7.5; each takes $O(1)$ time. Since there are n nodes and m characters, $\{L_i(u, 0), L_i(u, 1)\}$ for all nodes $u \in T$ can be computed in $O(mn)$ time. Then, $Pr(M \mid \mathcal{T})$ can be computed in $O(m)$ time using Equation 7.3. The total running time is $O(mn)$.

7.2.3.3 Find a Model with Maximum Likelihood

Given the character-state matrix M, this section discusses how to find the best model $\mathcal{T} = (T, \{p_e \mid e \in T\})$ which maximizes $Pr(M \mid \mathcal{T})$. Roch [247] showed that this problem is NP-hard.

This section presents one popular heuristic method DNAml proposed by Felsenstein [103]. Note that we will construct an unrooted tree as the placement of the root of the tree will not affect the maximum likelihood (pulling

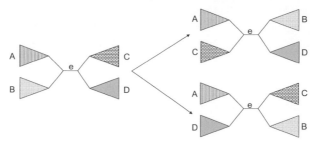

FIGURE 7.15: The two possible nearest neighbor interchange (NNI) operations across the edge e.

Algorithm DNAml

Let $S = \{u_1, u_2, \ldots, u_n\}$ be the set of taxa.
Build the tree T for species $\{u_1, u_2\}$
for $k = 3$ to n **do**
    Among all $(2k - 5)$ ways to insert u_k into T, we choose the way with the best likelihood.
    if $k \geq 4$ **then**
        while there exists a nearest neighbor interchange (NNI) operation which can improve the likelihood of T **do**
            We apply such NNI on T;
        end while
    end if
end for

FIGURE 7.16: Algorithm for DNAml.

principle, [103]). The method starts with a tree T with two taxa u_1 and u_2. Then, we attach taxa $u_3, u_4, \ldots, u_n$ into T one by one. When we attach the k-th taxon u_k, the tree T has $2k - 5$ edges. We try to attach u_k to each of these $2k - 5$ edges and evaluate the likelihood (the likelihood is computed based on Section 7.2.3.2 and the mutation probability of an edge is estimated using Lemma 7.8). The attachment yielding the highest likelihood is accepted. Then, local rearrangements are carried out in T to see if any of these improves the likelihood of the tree. The local rearrangement performed is the nearest neighbor interchange (NNI). It exchanges two subtrees incident across an internal edge in T. Figure 7.15 gives an example of two possible NNI operations across an edge. If any NNI operation improves the likelihood, it is accepted and the local rearrangement process continues until a tree is found in which no local rearrangement can improve the likelihood. The detail of the algorithm is presented in Figure 7.16.

In the algorithm in Figure 7.16, we need to estimate the mutation probability of an edge linking two subtrees (a subtree may be just a leaf). Consider

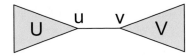

FIGURE 7.17: A tree formed by connecting the subtrees U and V through the edge (u, v).

two rooted phylogenetic trees U and V for two disjoint sets of taxa rooted at u and v, respectively. Let T be the tree formed by connecting U and V through the edge (u, v) (see Figure 7.17). Suppose $L_i(U, s)$ and $L_i(V, s)$ are given for any character i and any $s \in \{0, 1\}$. The lemma below expresses the maximum likelihood of T in term of $p_{(u,v)}$. Then, the next lemma states how to estimate the mutation probability $p_{(u,v)}$.

LEMMA 7.7
Given the character-state matrix M and a model $\mathcal{T} = (T, \{p_e | e \in T\})$ where T is as shown in Figure 7.17, the likelihood of $\mathcal{T}$, $Pr(M|\mathcal{T})$, equals

$$\prod_{i=1}^{m} \left[\sum_{s \in \{0,1\}} L_i(U, s) L_i(V, s)(1 - p_{(u,v)}) + \sum_{s \in \{0,1\}} L_i(U, s) L_i(V, 1 - s) p_{(u,v)} \right]$$

PROOF   Suppose the character i of u and v have states s and s', respectively. Let M_i be the states of the taxa for character i. Then, $Pr(M_i|T)$ equals $\sum_{s,s' \in \{0,1\}} L_i(U, s) Pr(v_i = s | u_i = s') L_i(V, s')$, where $Pr(v_i = s | u_i = s') = p_{(u,v)}$ if $s' \neq s$ and $(1 - p_{(u,v)})$ otherwise. Since $Pr(M|T) = \prod_{i=1}^{m} Pr(M_i|T)$, the lemma follows.   ☐

LEMMA 7.8
Given the character-state matrix M and a model $\mathcal{T} = (T, \{p_e | e \in T\})$ where T is as shown in Figure 7.17, the mutation probability $p = p_{(u,v)}$ of the linking edge (u, v) in T satisfies the following equation.

$$f(p) = \sum_{i=1}^{m} \frac{A_i - B_i}{B_i(1 - p) + A_i p} = 0$$

where $A_i = \sum_{s \in \{0,1\}} [L_i(U, s) L_i(V, (1 - s))]$, $B_i = \sum_{s \in \{0,1\}} [L_i(U, s) L_i(V, s)]$. Since $f(p)$ is a decreasing function, $f(p) = 0$ can be solved by performing a binary search.

PROOF   By Lemma 7.7, $Pr(M|T) = \prod_{i=1}^{m} (B_i(1 - p) + A_i p)$. The log-likelihood is $\ln Pr(M|T) = \sum_{i=1}^{m} \ln (B_i(1 - p) + A_i p)$. To find p which maximizes the log-likelihood L, we need to identify p such that $\frac{dL}{dp} = 0$, that is, $\frac{dL}{dp} = \sum_{i=1}^{m} \frac{A_i - B_i}{B_i(1-p) + A_i p} = 0$. Since $\frac{d^2L}{dp^2} = -\sum_{i=1}^{m} \frac{(A_i - B_i)^2}{(B_i(1-p) + A_i p)^2} < 0$, $\frac{d^2L}{dp^2}$ is a decreasing function. As $0 \leq p \leq 1$, by binary search, we can identify p such that $\frac{dL}{dp} = 0$. ▯

7.2.3.4 Final Remark for Maximum Likelihood

Maximum likelihood is statistically consistent, i.e., given a long enough sequence, the maximum likelihood method is able to recover the correct tree with arbitrarily high probability.

After we obtain $\mathcal{T} = (T, \{p_e \mid e \in T\})$, we can estimate the evolution time λ_e for every edge $e \in T$. The relationship between p_e and λ_e is as follows. p_e is the chance that the state of the character is changed an odd number of times within the period λ_e. The lemma below mathematically describes the precise relationship between p_e and λ_e.

LEMMA 7.9
For any edge $e \in T$, $\lambda_e = -\ln(1 - 2p_e)$.

PROOF Let u_t be the state of the character after time t. Let α be the mutation probability at any point of time. Suppose $P_t = Pr(u_t = 1 | u_0 = 1)$. Then, $P_0 = 1$. We have $P_1 = (1 - \alpha)$ and $P_2 = (1 - \alpha)P_1 + \alpha(1 - P_1)$. In general, $P_{t+1} = (1 - \alpha)P_t + \alpha(1 - P_t)$. Thus, $P_{t+1} - P_t = \alpha(1 - 2P_t)$ and we have

$$\frac{dP_t}{dt} = \alpha(1 - 2P_t)$$

By solving the differential equation, we have $Pr(u_t = 1 | u_0 = 1) = P_t = \frac{1}{2}(1 - e^{-2\alpha t})$. Similarly, we can derive $Pr(u_t = 0 | u_0 = 0) = P_t = \frac{1}{2}(1 - e^{-2\alpha t})$. Hence, $Pr(\text{change} | t) = Pr(u_t = 1 - s | u_0 = s) = 1 - Pr(u_t = s | u_0 = s) = 1 - \frac{1}{2}(1 - e^{-2\alpha t})$. By solving the equation, we have $t = -\frac{1}{2\alpha} \ln(1 - 2Pr(\text{change} | t))$. As the constant $\frac{1}{2\alpha}$ is just a scaling factor, we eliminate it. The lemma follows.
▯

7.3 Distance-Based Phylogeny Reconstruction Algorithm

In addition to construct phylogeny using character-based data, phylogeny can be constructed based on the similarity among taxa/species. Intuitively, taxa which look similar should be evolutionarily more related. This motivates us to define the distance between two taxa to be the number of mutations required to change one species to another.

This section focuses on distance-based phylogenetic tree reconstruction methods. Given n taxa, the input is an $n \times n$ distance matrix M where M_{ij} is the mutation distance between taxa i and j. Our aim is to output a tree of degree 3 which is consistent with the distance matrix. Figure 7.18 illustrates an example of such a distance matrix and the corresponding phylogeny.

In the rest of this section, we first study what is a valid distance matrix. Then, we discuss three distance-based methods: the UPGMA method, an additive tree reconstruction method, and the neighbor-joining method.

7.3.1 Additive Metric and Ultrametric

The distance matrix M is not just an $n \times n$ matrix of real numbers. It needs to satisfy some constraints. The basic requirement is that M should be a metric. Precisely, a distance matrix M is said to be a metric if and only if it satisfies:

Symmetric: $M_{ij} = M_{ji}$ and $M_{ii} = 0$; and

Triangle Inequality: $M_{ij} + M_{jk} \geq M_{ik}$.

Look back at the matrix in Figure 7.18. It is symmetric and satisfies the triangle inequality. Thus, it is a metric. In the following discussion, we will further discuss two additional constraints for a distance matrix: additive and ultrametric.

7.3.1.1 Additive Metric

Recall that the distance matrix M describes the number of mutations between every pair of taxa. If x is the lowest common ancestor that evolves to two taxa i and j, we should expect $M_{ij} = M_{ix} + M_{xj}$. This motivates the additive metric. A distance matrix M for a set of species S is said to be additive if and only if there exists a phylogeny T for S such that

- Every edge (u, v) in T is associated with a positive weight d_{uv}; and

- For every $i, j \in S$, M_{ij} equals the sum of the edge weights along the path from i to j in T.

In this case, M is called an additive matrix and T is called an additive tree. Figure 7.18 shows an example of an additive matrix and the corresponding additive tree. Note that it is an unrooted phylogeny, since we do not know the root.

Below, we discuss how to decide if a distance matrix is additive. If M is a distance matrix for at most three taxa, we claim that M must be additive. When M has one taxon, the additive tree is just one node. When M has two taxa i and j, the additive tree is an edge with weight M_{ij}. When M has 3 taxa i, j and k, the additive tree's topology must be three leaves attaching

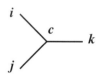

M	a	b	c	d	e
a	0	11	10	9	15
b	11	0	3	12	18
c	10	3	0	11	17
d	9	12	11	0	8
e	15	18	17	8	0

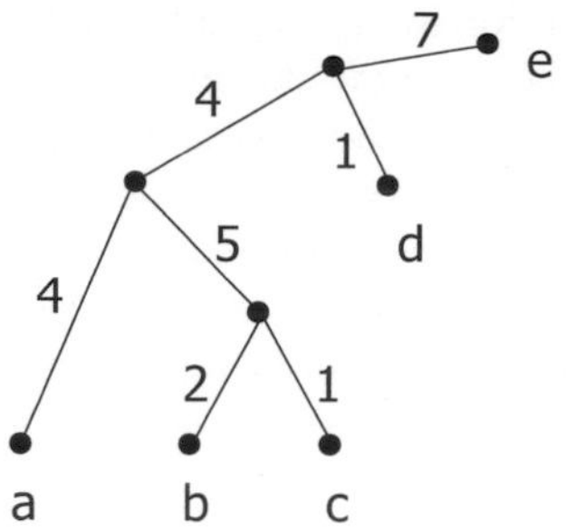

FIGURE 7.18: An example additive metric and the corresponding additive tree.

FIGURE 7.19: Additive tree for three taxa i, j, and k.

to a common center c (see Figure 7.19). The following lemma states the edge weight of $(c, i), (c, j), (c, k)$ can be uniquely determined.

LEMMA 7.10
Consider an additive matrix M. For any three taxa i, j and k, the corresponding additive tree is unique. Its topology is as shown in Figure 7.19 and the edge weights are:

$$d_{ic} = \frac{M_{ij} + M_{ik} - M_{jk}}{2} \tag{7.6}$$

$$d_{jc} = \frac{M_{ij} + M_{jk} - M_{ik}}{2} \tag{7.7}$$

$$d_{kc} = \frac{M_{ik} + M_{jk} - M_{ij}}{2} \tag{7.8}$$

PROOF   Since M is additive, we have $M_{ik} = d_{ic} + d_{ck}$, $M_{jk} = d_{jc} + d_{ck}$, and $M_{ij} = d_{ic} + d_{cj}$. By solving the three equations, we obtain Equations (7.6)–(7.8). Since M satisfies triangle inequality, d_{ic}, d_{jc}, and d_{kc} are positive.
▯

When M has at least 4 taxa, M may not be additive. Buneman proposed a 4-point condition [43] to determine if M is additive, which is stated in the following theorem.

THEOREM 7.1 (Buneman's 4-point condition)

M is additive if and only if the 4-point condition is satisfied, that is, for any four taxa in S, we can label them as i, j, k, l such that $M_{ik} + M_{jl} = M_{il} + M_{jk} \geq M_{ij} + M_{kl}$

PROOF ($\Rightarrow$) If M is additive, then there exists an additive tree T for S. For any four taxa, we can relabel the four taxa as i, j, k, l so that we get the subtree shown in Figure 7.20. Note that x and y may be identical (when $M_{xy} = 0$). We can easily verify that $M_{ik} + M_{jl} = M_{il} + M_{jk} \geq M_{ij} + M_{kl}$.

($\Leftarrow$) Consider $P(k)$ to be the assertion that "for any size-k subset S' of S, if the 4-point condition is satisfied for any taxa i, j, k, l in S', then there exists a unique additive tree for S'."

The assertions $P(1), P(2), P(3)$ are true since trees with at most three taxa must be additive (see Lemma 7.10).

Suppose $P(k-1)$ is true for some $k > 1$. We prove by induction that $P(k)$ is also true. Let S' be a size-k subset of S. Consider any $a, b \in S'$. Since $S' - \{a\}$ is of size $k - 1$, there exists a unique additive tree T' for $S' - \{a\}$. Since $S' - \{b\}$ is of size $k - 1$, there exists a unique additive tree T'' for $S' - \{b\}$. Denote by T a tree formed by overlaying T' and T''. As T' and T'' are isomorphic when restricting to the taxa $S' - \{a, b\}$, the overlaying should be feasible.

What remains is to show that T is an additive tree. For any taxa $(x, y) \in S \times S - \{(a, b)\}$, the path length between x and y in T equals the path length between x and y in either T' or T'', which equals M_{xy}. For (a, b), we identify a path between c and d such that both a and b attach to the path in T. Then, it can easily be checked that the path length between a and b in T equals $M_{ad} + M_{bc} - M_{cd}$. By the 4-point condition, $M_{ad} + M_{bc} - M_{cd} = M_{ab}$. Thus, T is an additive tree. $\qquad\Box$

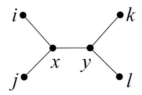

FIGURE 7.20: Buneman's 4-point condition.

As an example, let us choose a, b, c, d in Figure 7.18 as the four taxa. We have $M_{ac} + M_{bd} = 10 + 12 = 22; M_{ab} + M_{cd} = 11 + 11 = 22; M_{ad} + M_{bc} = 9 + 3 = 12$. So, $M_{ac} + M_{bd} = M_{ab} + M_{cd} \geq M_{ad} + M_{bc}$.

7.3.1.2   Ultrametric and constant molecular clock

The simplest evolution rate model is the constant molecular clock (see Section 7.1.2). This model assumes all taxa evolved at a constant rate from a common ancestor. The ultrametric matrix is an additive matrix which models the constant molecular clock. Precisely, a matrix M is ultrametric if there exists a tree T such that

1. M_{ij} equals the sum of the edge weights along the path from i to j in T; and

2. A root of the tree can be identified such that the distance to all leaves from the root is the same, that is, the length is a fixed value.

The rooted tree T corresponding to the ultrametric matrix M is called an ultrametric tree. An example of an ultrametric matrix and its corresponding ultrametric tree are given in Figure 7.21. Note that an ultrametric tree is rooted and any subtree of an ultrametric tree is also an ultrametric tree. Also, by definition, if M is ultrametric, M is also additive.

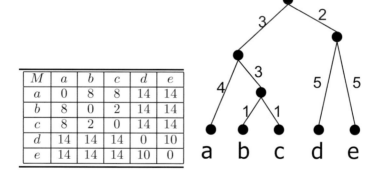

M	a	b	c	d	e
a	0	8	8	14	14
b	8	0	2	14	14
c	8	2	0	14	14
d	14	14	14	0	10
e	14	14	14	10	0

FIGURE 7.21:   An example of an ultrametric distance matrix and the corresponding phylogenetic tree.

For any node v in the ultrametric tree T, we define $height(v)$ to be the sum of the edge weights along the path from v to any descendant leaf. We have the following lemma.

LEMMA 7.11
Consider an ultrametric tree T corresponding to the ultrametric matrix M. For any pair of leaves i, j, let v be the least common ancestor of i and j in T. We have $height(v) = \frac{M_{ij}}{2}$.

PROOF Let M_{vi} and M_{vj} be the length of the paths from v to i and j, respectively. By definition, $height(v) = M_{vi} = M_{vj}$. Since T is additive, $M_{ij} = M_{vi} + M_{vj} = 2height(v)$. Hence, the lemma follows. □

Since an ultrametric matrix is an additive matrix, it satisfies the 4-point condition. The theorem below states that it also satisfies the 3-point condition.

THEOREM 7.2 (3-point condition)

M is ultrametric if and only if the 3-point condition is satisfied, that is, for any three taxa in S, we can label them i, j, k such that $M_{ik} = M_{jk} \geq M_{ij}$.

PROOF ($\Rightarrow$) If M is ultrametric, then there exists an ultrametric tree T for S. For any three taxa, we can relabel them as i, j, k so that we get the subtree shown in Figure 7.22 where y is the common ancestor of i and j while x is the common ancestor of i, j, k. Note that x and y may be identical (when the mutation distance between x and y is 0). We can easily verify that $M_{ik} = M_{jk} = 2 * (M_{iy} + M_{yx}) \geq 2M_{iy} = M_{iy} + M_{jy} = M_{ij}$.

($\Leftarrow$) Since M satisfies the 3-point condition, M also satisfies the 4-point condition. By Theorem 7.1, M is additive. Let T be the corresponding additive tree. Let a and b be taxa that maximize M_{ab}. Denote x as a point which partitions the path between a and b in T into two halves. We claim that the length of the path from x to s is $M_{ab}/2$ for all $s \in S$. Note that the length of the path from x to a or b equals $M_{ab}/2$. For any $s \in S$, without loss of generality, assume $M_{sb} \geq M_{sa}$. By the 3-point condition, we have $M_{sb} = M_{ab} \geq M_{sa}$. Then $M_{xs} = M_{sb} - M_{xb} = M_{ab} - M_{ab}/2 = M_{ab}/2$. Thus, T rooted at x is an ultrametric tree. The lemma is proved. □

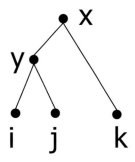

FIGURE 7.22: Ultrametric tree.

Based on the above discussion, there are three computational problems on the distance matrix. Let M be a distance matrix for a set of taxa S.

- If M is ultrametric, we want to reconstruct the corresponding ultrametric tree T in polynomial time.

- If M is additive, we want to reconstruct the corresponding additive tree T in polynomial time.

- If M is not exactly additive, we want to find the nearest additive tree T.

The rest of this chapter will discuss solutions for these three problems.

7.3.2   Unweighted Pair Group Method with Arithmetic Mean (UPGMA)

Given a distance matrix M of a set S of n taxa, this section presents the Unweighted Pair Group Method with Arithmetic Mean (UPGMA) for reconstructing the phylogenetic tree T for S.

The basic principle of UPGMA is that similar taxa should be closer in the phylogenetic tree. Hence, it builds the tree by clustering similar taxa iteratively. The method works by building the phylogenetic tree bottom up from its leaves. Initially, we have n leaves (or n singleton trees), each representing a taxon in S. Those n leaves are referred as n clusters. Then, we perform $n - 1$ iterations. In each iteration, we identify two clusters C_1 and C_2 with the smallest average distance $dist(C_1, C_2)$, that is,

$$dist(C_1, C_2) = \frac{\sum_{i \in C_1, j \in C_2} M_{ij}}{|C_1| \times |C_2|} \tag{7.9}$$

The two identified clusters C_1 and C_2 are linked with a new root r to form a bigger cluster C. Precisely, C consists of C_1, C_2, and two additional edges (r, r_1) and (r, r_2), where r_1 and r_2 are the roots of C_1 and C_2, respectively. We define $height(C) = dist(C_1, C_2)/2$. To ensure the average distance from r to the leaves in C equals $height(C)$, We set the edge weights $d(r, r_1) = height(C) - height(C_1)$ and $d(r, r_2) = height(C) - height(C_2)$.

In the algorithm, once two clusters C_1 and C_2 are merged to form a bigger cluster C, we need to compute $dist(C, C')$ for all other clusters C'. The simple method is to apply Equation 7.9 to compute $dist(C, C')$. However, it is time consuming. To speed up, UPGMA computes $dist(C, C')$ based on the following lemma.

LEMMA 7.12
Suppose C is a cluster formed by merging two clusters C_1 and C_2. For all

---

UPGMA algorithm

1: Set $\mathcal{C} = \{\{c_1\}, \{c_2\}, \ldots, \{c_n\}\}$ where $height(\{c_i\}) = 0$ for $i \in \{1, \ldots, n\}$;

2: For all $\{c_i\}, \{c_j\} \in \mathcal{C}$, set $dist(\{c_i\}, \{c_j\}) = M_{ij}$;

3: **for** $i = 2$ to n **do**

4:     Determine clusters $C_i, C_j \in \mathcal{C}$ such that $dist(C_i, C_j)$ is minimized;

5:     Let C_k be a cluster formed by connecting C_i and C_j to the same root;

6:     Let $height(C_k) = dist(C_i, C_j)/2$;

7:     Let $d(C_k, C_i) = height(C_k) - height(C_i)$;

8:     Let $d(C_k, C_j) = height(C_k) - height(C_j)$;

9:     $\mathcal{C} = \mathcal{C} - \{C_i, C_j\} \cup \{C_k\}$;

10:     For all $C_x \in \mathcal{C} - \{C_k\}$, define $dist(C_x, C_k) = dist(C_k, C_x) = \frac{|C_i|dist(C_i, C_x) + |C_j|dist(C_j, C_x)}{(|C_i| + |C_j|)}$;

11: **end for**

FIGURE 7.23: The UPGMA algorithm.

other clusters C',

$$dist(C, C') = \frac{|C_1|dist(C_1, C') + |C_2|dist(C_2, C')}{|C|}$$

PROOF $dist(C, C') = \frac{\sum_{i \in C, j \in C'} M_{ij}}{|C| \times |C'|} = \frac{\sum_{i \in C_1, j \in C'} M_{ij} + \sum_{i \in C_2, j \in C'} M_{ij}}{|C| \times |C'|}$. As $dist(C_1, C') = \frac{\sum_{i \in C_1, j \in C'} M_{ij}}{|C_1||C'|}$ and $dist(C_2, C') = \frac{\sum_{i \in C_2, j \in C'} M_{ij}}{|C_2||C'|}$, the lemma follows. $\Box$

Figure 7.23 shows the detailed algorithm. The execution of the algorithm is illustrated in Figure 7.24. The next lemma shows that if M is ultrametric, the tree reconstructed by the UPGMA algorithm is in fact an ultrametric tree.

LEMMA 7.13

Suppose M is ultrametric. For any cluster C created by the UPGMA algorithm, C is a valid ultrametric tree.

PROOF The UPGMA algorithm starts with a set of n leaves and each leaf forms a cluster. Then, at the i-th iteration, a cluster C_i is created by merging two clusters $C_{i,1}$ and $C_{i,2}$ created before the i-th iteration. We prove by induction that C_i is a valid ultrametric tree and $height(C_i) \geq height(C_{i-1})$ for any i.

M	a	b	c	d	e
a	0	8	8	14	14
b	8	0	2	14	14
c	8	2	0	14	14
d	14	14	14	0	10
e	14	14	14	10	0

FIGURE 7.24:   Illustration of the UPGMA algorithm.

At the 0-th iteration, since all n clusters are n leaves, every cluster is a valid ultrametric tree of height 0.

Suppose, at the i-th iteration, C_i is a valid ultrametric tree. We first show that $height(C_{i+1}) \geq height(C_i)$ by considering two cases.

Case 1: $C_i \notin \{C_{i+1,1}, C_{i+1,2}\}$. In this case, $C_{i+1,1}, C_{i+1,2}, C_{i,1}, C_{i,2}$ are clusters available at the i-th iteration. We select to merge $C_{i,1}$ and $C_{i,2}$ at the i-th iteration implies that $dist(C_{i+1,1}, C_{i+1,2}) \geq dist(C_{i,1}, C_{i,2})$. Hence, $height(C_{i+1}) \geq height(C_i)$.

Case 2: $C_i \in \{C_{i+1,1}, C_{i+1,2}\}$. Without loss of generality, assume $C_i = C_{i+1,1}$. Then, $C_{i+1,2}, C_{i,1}, C_{i,2}$ are clusters available at the i-th iteration. We select to merge $C_{i,1}$ and $C_{i,2}$ at the i-th iteration implies that $dist(C_{i,1}, C_{i+1,2}) \geq dist(C_{i,1}, C_{i,2})$ and $dist(C_{i,2}, C_{i+1,2}) \geq dist(C_{i,1}, C_{i,2})$. This implies that

$$
\begin{aligned}
dist(C_{i+1,1}, C_{i+1,2}) &= dist(C_i, C_{i+1,2}) \\
&= \frac{|C_{i,1}| dist(C_{i,1}, C_{i+1,2}) + |C_{i,2}| dist(C_{i,2}, C_{i+1,2})}{|C_{i,1}| + |C_{i,2}|} \\
&\geq \frac{|C_{i,1}| dist(C_{i,1}, C_{i,2}) + |C_{i,2}| dist(C_{i,2}, C_{i,2})}{|C_{i,1}| + |C_{i,2}|} \\
&= dist(C_{i,1}, C_{i,2}).
\end{aligned}
$$

Hence, $height(C_{i+1}) \geq height(C_i)$.

Next, we show that C_{i+1} is a valid ultrametric tree. C_{i+1} consists of a root r linking to the root r_1 of $C_{i+1,1}$ and the root r_2 of $C_{i+1,2}$. Since both $C_{i+1,1}$ and $C_{i+1,2}$ are valid ultrametric trees, proving that the weights of (r, r_1) and (r, r_2) are positive is sufficient to ensure C_{i+1} is a valid ultrametric tree. Since $height(C_{i+1}) \geq height(C_i)$, we have $height(C_{i+1}) \geq height(C_{i+1,1}), height(C_{i+1,2})$. Then, for $j = 1, 2$, the weight of the edge (r, r_j) is $height(C_{i+1}) - height(C_{i+1,j}) \geq 0$. The lemma follows. □

We now analyze the time complexity. UPGMA takes $n - 1$ iterations as mentioned above. Each iteration is dominated by the step to find the pair of clusters with the minimum distance (Step 4 in Figure 7.23). This step requires n^2 operations in the first iteration. Subsequently we maintain, for each i, an integer $\tau(i)$ such that $dist(C_i, C_{\tau(i)})$ is the minimum among the values in $\{dist(C_i, C_j) \mid j = 1, 2, \ldots, i - 1\}$. Then, Step 4 only needs n steps to find two clusters C_i and $C_j = C_{\tau(i)}$ which minimize $dist(C_i, C_j)$. In addition, it needs n steps to maintain the function τ, that is, to compute the pairwise distance between the new cluster C_k and the rest of the clusters. Therefore, each iteration takes $O(n)$ time. The total time complexity is $O(n^2)$.

The space usage is upper bounded by the distance matrix used, thus it is $O(n^2)$.

UPGMA is a robust algorithm. Even if the distance matrix M is slightly deviated from ultrametric, UPGMA can find an ultrametric tree which approximates M.

As a final note, UPGMA may not be able to return the correct tree since it is not true that the evolution rate is the same for every path from the root to the leaf. It is known that the evolutionary clock is running differently for different species and even for different genome regions of the same species.

7.3.3 Additive Tree Reconstruction

Suppose M is an additive matrix. This section presents a reconstruction algorithm which reconstructs the additive tree T in $O(n^2)$ time. As a side product, we also show that the additive tree is unique.

The algorithm works by inserting taxa into the tree one by one. Initially, by Lemma 7.10, we can uniquely determine the additive tree for three taxa i, j, k, which consists of a center c connecting to i, j, and k. To include another taxon h in the tree, we need to introduce one more internal node c' (see Figure 7.25). c' will split either (i, c), (j, c), or (k, c). To check whether h splits the edge (k, c), we apply Lemma 7.10 for taxa i, k, and h to create a new internal node c'. If $d_{kc'} < d_{kc}$, then we split (k, c) at c' and insert an edge from c' to h. Otherwise, we use the same approach to check whether c' splits the edge (i, c) or the edge (j, c). Note that since c' can only split exactly one edge, the additive tree for four species is unique.

In general, given an additive tree T' for $k - 1$ taxa, we can iteratively insert the k-th taxon into T' to reconstruct the additive tree as follows:

- For every edge of T', we check whether the k-th taxon splits the edge

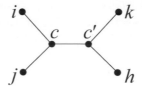

FIGURE 7.25:   Four species with two internal nodes.

using Lemma 7.10. Because the k-th taxon can only split exactly one edge of T', the additive tree for the k taxa is unique.

For example, consider an additive matrix M in Figure 7.26. First, we form an additive tree for two taxa a and b. Then, we choose a third taxon, says, c, and attach it to a point x on the edge between a and b. The edge weights d_{ax}, d_{bx}, d_{cx} can be computed by Lemma 7.10. Next, we would like to add the fourth taxon d to the additive tree. d can be attached to any of the three edges ax, bx, and cx. By applying Lemma 7.10 on the three edges, we identify that d should be attached to the edge ax. Finally, we add the last taxon e to the additive tree using the same method. The process is illustrated in Figure 7.26.

M	a	b	c	d	e
a	0	11	10	9	15
b	11	0	3	12	18
c	10	3	0	11	17
d	9	12	11	0	8
e	15	18	17	8	0

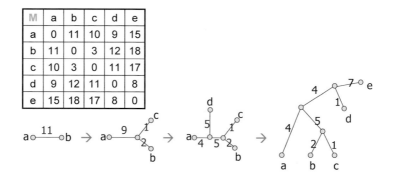

FIGURE 7.26:   Example of additive tree reconstruction.

We now analyze the time complexity. The time complexity for checking if a taxon should be attached to a particular edge requires $O(1)$ time based on Lemma 7.10. To attach a taxon into a tree with k leaves, we need to check $O(k)$ edges, which takes $O(k)$ time. Therefore, to construct an additive tree for n taxa, the time complexity is $O(1 + 2 + \ldots + n) = O(n^2)$.

Neighbor-Joining algorithm

1: Let $Z = \{\{1\}, \{2\}, \cdots, \{n\}\}$ be the set of initial clusters;
2: For all $\{i\}, \{j\} \in Z$, set $dist(\{i\}, \{j\}) = M_{ij}$;
3: **for** $i = 2$ to n **do**
4:    For every cluster $A \in Z$, set $u_A = \frac{1}{n-2} \sum_{D \in Z} dist(D, A)$;
5:    Find two clusters $A, B \in Z$ which minimizes $dist(A, B) - u_A - u_B$;
6:    Let C be a new cluster formed by connecting A and B to the same root r. Let r_A and r_B be the roots of A and B. The edge weights of (r, r_A) and (r, r_B) are $\frac{1}{2}dist(A, B) + \frac{1}{2}(u_A - u_B)$ and $\frac{1}{2}dist(A, B) + \frac{1}{2}(u_B - u_A)$, respectively;
7:    Set $Z = Z \cup \{C\} - \{A, B\}$;
8:    For any $D \in Z - \{C\}$, define $dist(D, C) = dist(C, D) = \frac{1}{2}(dist(A, D) + dist(B, D) - dist(A, B))$;
9: **end for**

FIGURE 7.27: Neighbor-joining algorithm.

7.3.4 Nearly Additive Tree Reconstruction

Consider a matrix M for a set S of n taxa. Suppose that M is not additive. This section discusses methods that are used to build a nearly additive tree. In other words, we want to find an additive matrix D that minimizes the sum of the squares of error $SSQ(D)$ where

$$SSQ(D) = \sum_{i=1}^{n} \sum_{j \neq i} (D_{ij} - M_{ij})^2.$$

The additive tree T corresponding to D is called the least square additive tree. Computing the least square additive tree is NP-hard [101]. People have applied heuristics algorithms to resolve this problem. The most popular heuristic is the neighbor-joining method.

The neighbor-joining (NJ) method [253, 275] was proposed in 1987. Its idea is similar to UPGMA. By hierarchical clustering, it iteratively constructs a larger cluster (say C) by merging two nearest clusters (say A and B). Moreover, NJ has an extra constraint that the distances from A and B to the remaining clusters should be as dissimilar as possible.

Figure 7.27 presents the NJ algorithm. Below, we briefly describe the neighbor-joining algorithm and analyze the time complexity. The n taxa are considered as n clusters in Z. We first initialize the distance for n^2 pairs of clusters, which takes $O(n^2)$ time. Then we perform $n - 1$ iterations, where each iteration merges two clusters. For each iteration, we aim to identify two clusters A and B such that $dist(A, B)$ is minimized and $u_A + u_B$ is maximized where $u_A = \frac{1}{n-2} \sum_{D \in Z} dist(D, A)$ and $u_B = \frac{1}{n-2} \sum_{D \in Z} dist(D, B)$. In other words, we hope to find A and B which minimize $dist(A, B) - u_A - u_B$. Note that u_A for all $A \in Z$ can be computed in $O(n^2)$ time. Also, A and B which minimize $dist(A, B) - u_A - u_B$ can be found in $O(n^2)$ time. Afterward, we merge A and B into a new cluster C and remove A and B from Z. Since C is the new cluster added to Z, we need to update the pairwise distances between C and all other clusters in Z. This step requires $O(n)$ time. So, the execution time of each iteration is $O(n^2)$. There are $n - 1$ iterations. Hence the time complexity of the neighbor-joining algorithm is $O(n^3)$.

Apart from the neighbor-joining method, the Fitch-Margoliash method [109] and a number of other methods also assume the nearly additive tree is the tree which minimizes the least square error. Moreover, there are other metrics for defining the nearly additive tree. One example is the L_∞-metric. The L_∞-metric for two distance matrices A and B is $L_\infty(A, B) = \max_{i,j} |A_{ij} - B_{ij}|$. Based on the L_∞ metric, for a given non-additive matrix M, we have to find an additive matrix E such that $L_\infty(M, E)$ is minimized. The problem of minimizing the L_∞ metric is also NP-hard. However, Agarwala et al. [3] have proposed a 3-approximation algorithm with respect to the L_∞-metric which runs in polynomial time.

7.3.5   Can We Apply Distance-Based Methods Given a Character-State Matrix?

Given a character-state matrix N, we can convert it into a distance matrix M and apply distance-based methods to construct a phylogeny.

However, how can we convert a character-state matrix to a distance matrix? A simple-minded method is that, for any two taxa i and j, we define M_{ij} to be the Hamming distance h_{ij} between the character-state vectors of taxa i and j. (For example, let's say taxon i is (1,0,1,1,0) and taxon j is (1,0,0,1,1). The Hamming distance is $h_{ij} = 2$.)

Unfortunately, the answer is "NO"! The Hamming distance fails to capture the "multiple" mutations on the same site. It does not satisfy the additive property of a distance matrix. For example, at the first position of taxa i and j, we cannot tell whether there is no mutation in the site or there are two mutations $(1 \to 0 \to 1)$.

The solution is to correct the distance based on some evolution model. Using the CF tree model described in Section 7.2.3.1, the corrected distance M_{ij} equals:

$$-\frac{1}{2} \log \left(1 - \frac{2h_{ij}}{m} \right)$$

As the number of characters increases, M converges to an additive matrix.

7.4 Bootstrapping

Given a character-state matrix N, the tree generated by character-based methods or distance-based methods may be unstable. In other words, removing or adding some characters may change the topology of the reconstructed phylogenetic tree. Bootstrapping is proposed to reduce the instability.

Bootstrapping means sampling with replacement from the input. Consider n taxa, each described by m characters. Through bootstrapping, we build a set of character matrices where each character matrix N_i is formed by m randomly selected characters (with replacement). For each N_i, a tree T_i can be reconstructed using either a character-based method or a distance-based method. For a character-based method, we directly reconstruct T_i from N_i using the character-based method. For a distance-based method, we first build a distance matrix M_i from N_i as in Section 7.3.5 and reconstruct a tree T_i using the distance-based method. By computing the consensus tree of all T_i (see Section 8.4), we get a stable phylogenetic tree. Figure 7.28 illustrates an example.

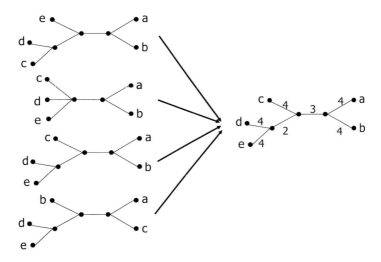

FIGURE 7.28: Example of bootstrapping. By randomly selecting characters with replacement, we generate four character matrices and reconstruct four trees. Then, we generate the consensus tree from the four trees. In the consensus tree, the number on each edge indicates the number of trees which have such edges (or splits).

7.5   Can Tree Reconstruction Methods Infer the Correct Tree?

This chapter introduced a number of phylogenetic tree reconstruction methods. Can those methods infer the correct tree?

A number of studies have tried to answer this important problem. In 1992, Hillis et al. [146] performs the following *in vitro* experiment to study the goodness of phylogenetic tree reconstruction methods. In the laboratory, bacteriophage T7 was propagated and split sequentially in the presence of a mutagen, where each lineage was tracked. A phylogeny of nine taxa was constructed. The authors also determined the occurrences of 34 restriction enzyme cleavage sites in each of these nine taxa. Five different phylogenetic methods (parsimony method, Fitch-Margoliash method, Cavalli-Sforza method, neighbor-joining method, and UPGMA method) were applied on the restriction site map data to reconstruct the phylogenetic tree. Out of 135,135 possible phylogenetic trees, all five methods could correctly determine the true tree in a blind analysis. This experiment showed that a phylogenetic tree reconstruction algorithm can correctly reconstruct the true phylogenetic tree.

The previous study was performed *in vitro* in the laboratory. Can we validate the performance of phylogenetic tree reconstruction algorithms using a real evolution? Leitner et al. [186] studied the evolution of HIV-1 samples in a period of 13 years (1981–1994). Since HIV evolves approximately one million times faster than the nucleic genomes of higher organisms, this study corresponds to approximately 13 million years of evolution in higher organisms. Such a time period is considerably longer than the age of the genus Homo, which includes humans.

The HIV-1 samples are collected from 11 HIV-1 infected individuals with known epidemiological relationships. Precisely, the relationship of the individuals is as follows. A Swedish male became HIV-1 infected in Haiti in 1980. Through sexual interaction, six females were infected. Two of the six females had sexual relationships with two other male partners and they were also infected. Two children of two females were also infected. Together with the information of the sexual interaction time, the authors got the actual phylogenetic tree of the collected HIV-1 samples.

On the other hand, Leitner et al. got DNA sequences from the collected HIV-1 samples. Then, seven tree reconstruction methods (Fitch-Margoliash (FM), neighbor-joining (NJ), minimum-evolution (ME), maximum-likelihood (ML), maximum-parsimony (MP), unweighted pair group method using arithmetic averages (UPGMA), and an FM method assuming a molecular clock (KITSCH)) were applied to those sequences. The result showed that FM, NJ, and ML perform the best; MP is in the middle; while UPGMA and KITSCH, which assume a constant molecular clock, perform the worst. All methods tended to overestimate the length of short branches and underestimate long branches.

The above results showed that phylogenetic reconstruction methods can correctly reconstruct the real evolution history. However, we still cannot estimate accurately the length of the branches.

7.6 Exercises

1. For the following matrix M, can you find the parsimony tree using the approximation algorithm? Is this the maximum parsimony tree? If not, what is the maximum parsimony tree?

M	C_1	C_2	C_3	C_4
S_1	1	0	1	1
S_2	1	1	0	1
S_3	0	1	1	0

2. Can you solve the large compatibility problem, that is, find the largest set of characters which admit the perfect phylogeny? Can you check if the following matrix has a perfect phylogeny? If yes, can you report the corresponding tree?

M	C_1	C_2	C_3	C_4
S_1	0	0	0	1
S_2	1	1	0	0
S_3	0	0	0	1
S_4	0	1	1	0
S_5	0	1	1	1

3. Consider the following tree topology for taxa $\{AC, CT, GT, AT\}$. Can you compute the parsimony length? Also, can you give the corresponding labeling for the internal nodes?

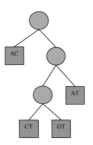

4. For the above question, do we have another labeling for the internal nodes such that we have the same parsimony length?

5. For the following character-state matrix M, can you describe how to compute construct a phylogenetic tree using the approximation algorithm in Section 7.2.1.2?

M	C_1	C_2	C_3
S_1	0	0	1
S_2	1	1	0
S_3	1	0	1
S_4	0	1	0
S_5	0	1	1

6. Consider a set L of n taxa. Let T be the non-binary tree T leaf-labeled by L and M be an $n \times m$ character-state matrix for L. Suppose each character has k possible states. Can you give an $O(nmk)$ time algorithm to compute the parsimony length of T?

7. Can you check if the following matrix has a perfect phylogeny? If yes, can you report the corresponding tree?

	C_1	C_2	C_3
S_1	0	0	1
S_2	1	1	0
S_3	0	0	1
S_4	0	1	0
S_5	0	1	1

8. Consider n species and m binary characters. Let $M[i, j]$ be the value of the j-th character for the i-th species. Assume that every column contains both 0 and 1. Can you propose an algorithm to check if there exists a tree T for M with the parsimony length at most m? If the parsimony length is m, can you propose a polynomial time algorithm to reconstruct T?

9. Let M be the following (incomplete) directed character state matrix for objects $\{A, B, C, D, E\}$:

	c_1	c_2	c_3	c_4	c_5
A	?	1	0	1	1
B	0	0	?	0	0
C	1	?	0	0	1
D	1	0	?	0	0
E	0	1	1	1	?

Prove that there is no way of filling in the missing entries of M with 0's and 1's so that the resulting matrix admits a perfect phylogeny.

10. Consider a binary $n \times m$ matrix M. For every column j, let O_j be $\{i \mid M_{ij} = 1\}$. Two columns j and k are pairwise compatible if O_j and O_k are disjoint or one of them contains the other.

    Show that if columns j and k are pairwise compatible for any j, k, then M admits a perfect phylogeny.

11. (a) Consider n species, each characterized by m binary characters. Such input can be represented using an $n \times m$ matrix M where M_{ij} is the state of character j for species i. What is the best lower bound of the time complexity to check whether M admits a perfect phylogeny?

    (b) Instead of storing the input using the matrix M, for every species i, we store the set of values $\{j \mid M_{ij} = 1\}$ using a linked list. Let x be the total length of the linked lists for all n species. Give an $O(n + m + x)$-time algorithm which can determine whether M admits a perfect phylogeny or not.

12. Consider the following tree topology T. The value on the edge is the mutation probability p. Can you compute (i) $Pr(a = 1, b = 0, c = 1, u = 1, r = 1 \mid T, p)$, (ii) $Pr(a = 1, b = 0, c = 1, u = 0, r = 0 \mid T, p)$, (iii) $Pr(a = 1, b = 0, c = 1, u = 1, r = 0 \mid T, p)$, (iv) $Pr(a = 1, b = 0, c = 1, u = 0, r = 1 \mid T, p)$, and (v) $Pr(a = 1, b = 0, c = 1 \mid T, p)$?

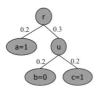

13. Consider the following tree topology T. The value on the edge is the mutation probability p. Can you compute $Pr(a = 10, b = 01, c = 11 | T, p)$?

14. Compute a phylogenetic tree for the following set of DNA sequences such that the total cost of your phylogenetic tree is at most twice that of the optimal solution.

$$S_1 = \text{ACCGT}$$
$$S_2 = \text{ACGTT}$$
$$S_3 = \text{CCGTA}$$
$$S_4 = \text{GTCCT}$$
$$S_5 = \text{AGCTT}$$

15. For the following matrix M, does it admit a perfect phylogeny? If not, can you find the minimum number of cells you need to flip so that M admits a perfect phylogeny?

M	1	2	3	4
A	1	1	1	0
B	0	0	1	1
C	1	1	0	0
D	1	0	0	0
E	0	1	0	1

16. For the following matrix M, is it an ultrametric matrix? Can you construct the corresponding ultrametric tree or the corresponding nearly ultrametric tree?

M	S_1	S_2	S_3	S_4	S_5
S_1	0	20	20	20	8
S_2		0	16	16	20
S_3			0	10	20
S_4				0	20
S_5					0

17. Is the following matrix additive? If not, give a reason. If yes, give the additive tree.

	S_1	S_2	S_3	S_4
S_1	0	3	8	7
S_2		0	7	6
S_3			0	5
S_4				0

18. (a) For the distance matrix below, is it an ultrametric? If yes, can you construct the corresponding ultrametric tree? If not, is it an additive matrix? If yes, can you construct the corresponding additive tree?

(b) Suppose, for every $j \neq 4$, $d(S_4, S_j)$ $(= d(S_j, S_4))$ is reduced by 1. Is it an ultrametric? If yes, can you construct the corresponding ultrametric tree? If not, is it an additive matrix? If yes, can you construct the corresponding additive tree?

	S_1	S_2	S_3	S_4	S_5	S_6
S_1	0	8	12	15	14	6
S_2		0	12	13	12	8
S_3			0	12	14	12
S_4				0	13	15
S_5					0	14
S_6						0

19. Refer to the matrix in the above question. Suppose $d(S_4, S_3)$ $(= d(S_3, S_4))$ is changed to 17. Is it ultrametric? If yes, can you construct the corresponding ultrametric tree? If not, is it an additive matrix? If yes, can you construct the corresponding additive tree?

20. Construct an additive tree for the following distance matrix.

	A	B	C	D	E
A	0	10	9	16	8
B		0	15	22	8
C			0	13	13
D				0	20
E					0

21. For the following matrix M, can you construct the nearly additive tree using neighbor joining?

M	S_1	S_2	S_3	S_4	S_5
S_1	0	7	11	13	15
S_2		0	12	14	18
S_3			0	8	10
S_4				0	5
S_5					0

22. Given an additive tree T for n species, describe an efficient algorithm to compute the additive distance matrix for T. What is the time complexity?

23. This exercise considers a simple alignment-free phylogeny reconstruction method. For each species, its sequence is mapped to a vector defined by the counts of k-mers. The distance between every pair of species is defined by the Euclidean distance of their vectors. Then, the phylogeny can be reconstructed by any distance-based reconstruction method.

Consider the following four sequences:

$$S_1 = \text{actgaatcgtact}$$
$$S_2 = \text{cctgaatcgtact}$$
$$S_3 = \text{actggtaatcact}$$
$$S_4 = \text{actggtaatccct}$$

(a) Can we generate the vector of 2-mers for each sequence?

(b) Can you construct the distance matrix for the set of sequences based on Euclidean distance of their vectors?

(c) Can you construct the phylogeny using neighbor joining method?

Chapter 8

Phylogeny Comparison

8.1 Introduction

A phylogenetic tree models the evolutionary history for a set of species/taxa. The previous chapter describes methods to construct phylogeny for the same set of taxa. However, different phylogenetic trees are reconstructed in different situations:

- Different kinds of data: The data used to represent the same taxa/species for tree reconstruction is not always the same, which will definitely lead to different results. For example, different segments of the genomes are used as the input to the tree reconstruction algorithm.

- Different kinds of models: Every phylogenetic tree reconstruction algorithm is based on some models (or assumptions), such as the maximum parsimony, the Cavender-Felsenstein model, the Jukes-Cantor model, etc. There is no consensus about which model is the best.

- Different kinds of algorithms: We use different kinds of reconstruction algorithms for a set of taxa. Various algorithms do not always give the same answer on the same input. For some reconstruction algorithms, different runs on the same input may give different answers.

Surely, the resulting trees may agree in some parts and differ in others. Tree comparison helps us to gain the similarity and dissimilarity information from multiple trees. There are three computational problems related to tree comparison.

- **Measure the similarity of trees**: The similarity measurement determines the common structure among the given trees. One popular similarity measure is the Maximum Agreement Subtree (MAST). Since information extracted agrees among the trees, the information is trustable to a certain extent.

- **Measure the difference of trees**: The dissimilarity measurement determines the difference/distance among the given trees. One popular dissimilarity measurement is the Robinson-Founds distance. Other well

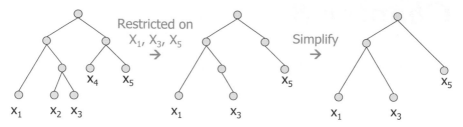

FIGURE 8.1:   Example of a restricted subtree

known examples are the nearest neighbor interchange (NNI) distance, the subtree-transfer (STT) distance, and the quartet distance.

- **Find the consensus of a set of trees**: The consensus of a set of trees is a tree which looks similar to all trees. Different consensus trees exist including strict consensus tree, majority rule consensus tree, median consensus tree, greedy consensus tree, and R* tree.

The rest of this chapter studies the three computational problems.

8.2   Similarity Measurement

Suppose two different methods report two different phylogenetic trees. If both methods say humans are more closely related to mice than to chickens, then we have higher confidence that such a relationship is true. This section discusses the maximum agreement subtree problem, which tries to extract such common relationships.

We need some definitions to define the maximum agreement subtree problem. Given a phylogenetic tree T leaf-labeled by a set of taxa S, a tree is called the restricted subtree of T with respect to $L \subseteq S$, denoted as $T|L$, if

1. $T|L$ is leaf-labeled by L and its internal nodes are the least common ancestors of such leaves, and

2. the edges of $T|L$ preserve the ancestor-descendant relationship of T.

To derive a restricted subtree $T|L$, we remove all leaves not labeled by L and contract all internal nodes with at most one child. Figure 8.1 illustrates this process.

Given two phylogenetic trees T_1 and T_2, T' is called an agreement subtree of T_1 and T_2 if T' is a common restricted subtree derived from both trees. A maximum agreement subtree (MAST) of T_1 and T_2 is an agreement subtree

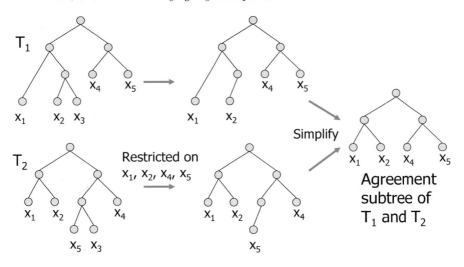

FIGURE 8.2: This is an agreement subtree of two phylogenetic trees T_1 and T_2. In fact, this is also a maximum agreement subtree of T_1 and T_2.

of the two trees with the largest possible number of leaves [166]. Figure 8.2 illustrates an example. Note that if the MAST of T_1 and T_2 has more leaves, T_1 and T_2 are considered to be more similar.

8.2.1 Computing MAST by Dynamic Programming

This section presents a dynamic programming algorithm to compute the MAST of two binary rooted trees. Let $MAST(T_1, T_2)$ be the number of leaves in the maximum agreement subtree of two binary rooted trees T_1 and T_2.

Let T^u be the subtree of T rooted at u for any $u \in T$. For any nodes $u \in T_1$ and $v \in T_2$, we define the recursive formula for $MAST(T_1^u, T_2^v)$ depending on two scenarios. For scenario (1), when either u or v is a leaf node, $MAST(T_1^u, T_2^v) = 1$ if both T_1^u and T_2^u contain a common taxon; and 0 otherwise.

For scenario (2), when both u and v are internal nodes, $MAST(T_1^u, T_2^v)$ can be computed by the following lemma.

LEMMA 8.1

Let u and v be internal nodes in T_1 and T_2, respectively. Let u_1 and u_2 be the *children of u in T_1 and v_1 and v_2 be the children of v in T_2. We have*

$$MAST(T_1^u, T_2^u) = max \begin{cases} MAST(T_1^{u_1}, T_2^{v_1}) + MAST(T_1^{u_2}, T_2^{v_2}) \\ MAST(T_1^{u_2}, T_2^{v_1}) + MAST(T_1^{u_1}, T_2^{v_2}) \\ MAST(T_1^{u_1}, T_2^v) \\ MAST(T_1^{u_2}, T_2^v) \\ MAST(T_1^u, T_2^{v_1}) \\ MAST(T_1^u, T_2^{v_2}) \end{cases}$$

PROOF   Let T' be a maximum agreement subtree of T_1^u and T_2^v. Depending on the distribution of the taxa of T' in T_1^u and T_2^v, we have the following 3 cases.

- If the taxa of T' appear in either $T_1^{u_1}$ or $T_1^{u_2}$, then $MAST(T_1^u, T_2^v)$ equals either $MAST(T_1^{u_1}, T_2^v)$ or $MAST(T_1^{u_2}, T_2^v)$. Therefore, we have

$$MAST(T_1^u, T_2^v) = \max\{MAST(T_1^{u_1}, T_2^v), MAST(T_1^{u_2}, T_2^v)\}$$

- If the taxa of T' appear in either $T_2^{v_1}$ or $T_2^{v_2}$, we have

$$MAST(T_1^u, T_2^v) = \max\{MAST(T_1^u, T_2^{v_1}), MAST(T_1^u, T_2^{v_2})\}$$

- Otherwise, the taxa of T' appear in both subtrees of T_1 and T_2. Then, the two subtrees of T' are the restricted subtrees of $T_1^{u_1}$ and $T_1^{u_2}$. Similarly, the two subtrees of T' are the restricted subtrees of $T_2^{v_1}$ and $T_2^{v_2}$. Hence, we have

$$MAST(T_1^u, T_2^v) = \max \begin{cases} MAST(T_1^{u_1}, T_2^{v_1}) + MAST(T_1^{u_2}, T_2^{v_2}) \\ MAST(T_1^{u_2}, T_2^{v_1}) + MAST(T_1^{u_1}, T_2^{v_2}) \end{cases}$$

☐

Now we analyze the time complexity of the algorithm. Suppose T_1 and T_2 are rooted phylogenies for n taxa. We have to compute $MAST(T_1^u, T_2^v)$ for every node u in T_1 and every node v in T_2. Thus, we need to fill in n^2 entries. Each entry can be computed in constant time. In total, the time complexity is $O(n^2)$.

In fact, we can compute the MAST of two degree-d rooted trees T_1 and T_2 with n leaves in $O(\sqrt{d} n log(\frac{n}{d}))$ time. For details, please refer to [167].

8.2.2   MAST for Unrooted Trees

In real life, we normally want to compute the MAST for unrooted trees. For unrooted degree-3 trees U_1 and U_2, $MAST(U_1, U_2)$ can be computed in $O(n \log n)$ time [165]. For unrooted trees U_1 and U_2 of unbounded degree,

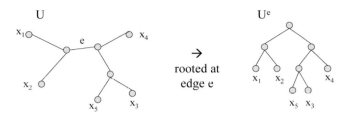

FIGURE 8.3: Converting an unrooted tree to a rooted tree U_e at edge e.

$MAST(U_1, U_2)$ can be computed in $O(n^{1.5} \log n)$ time [166]. This section shows how can we transform the problem of finding the MAST of two unrooted trees to the problem of finding the MAST of rooted trees.

For any unrooted tree U, for any edge e in U, we denote U^e to be the rooted tree rooted at the edge e. Figure 8.3 illustrates an example.

LEMMA 8.2

For any edge e of U_1, $MAST(U_1, U_2) = \max\{MAST(U_1^e, U_2^f) \mid f$ is an edge of $U_2\}$.

PROOF   See Exercise 4.   ☐

The above lemma reveals the relationship of the MAST between unrooted trees and rooted trees. Using this lemma directly, the MAST of U_1 and U_2 can be computed by executing the rooted MAST algorithm n times. Moreover, there is some redundant computation in this naïve method. By avoiding the redundant computation, we can compute the MAST of two unrooted trees in $O(n^{1.5} \log n)$ time [165].

8.3   Dissimilarity Measurements

The previous section discussed how to compute the similarity of two phylogenetic trees, including both unrooted and rooted trees. Now we discuss how to compute the dissimilarity of two phylogenetic trees.

We will discuss four dissimilarity measurements. They are the Robinson-Foulds distance, the nearest neighbor interchange distance, the subtree transfer distance, and the quartet distance.

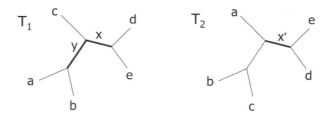

FIGURE 8.4:   Both the edge x in T_1 and the edge x' in T_2 form the same split $\{a, b, c\}|\{d, e\}$. Hence, both x and x' are good edges. On the other hand, y is a bad edge since there is no edge y' in T_2 such that both y and y' form the same split.

8.3.1   Robinson-Foulds Distance

One intuitive way to define the dissimilarity is based on counting the number of splits (see below for the definition) that are not shared by two trees, which is called the Robinson-Foulds distance [246].

Before we define the Robinson-Foulds distance, we first introduce the definition of a split. Given a set of taxa S, a split is a partition of the set S into A and $S - A$ for any subset $A \subseteq S$. As shorthand, a split is denoted as $A|S - A$.

Consider a tree T leaf-labeled by S. Every edge in T defines a split. For example, consider the tree T_1 in Figure 8.4. The edge x partitions the species into $\{a, b, c\}$ and $\{d, e\}$. Note that for any leaf $\ell \in S$, $\{\ell\}|S - \{\ell\}$ must be a valid split for T since the leaf edge connecting to ℓ always partitions S into $\{\ell\}$ and $S - \{\ell\}$.

Consider two unrooted trees T_1 and T_2. An edge x in T_1 is called a good edge if there exists an edge x' in T_2 such that both of them form the same split. Otherwise, the edge x is called a bad edge. Note that, by definition, every leaf edge is a good edge. Figure 8.4 illustrates an example.

Given two phylogenetic trees T_1 and T_2, the Robinson-Foulds (RF) distance between T_1 and T_2 is half the sum of the number of bad edges in T_1 with respect to T_2 and that in T_2 with respect to T_1. For example, in Figure 8.4, there is only one bad edge in each phylogenetic tree. By definition, the RF distance is $(1 + 1)/2 = 1$.

When the phylogenetic trees are of degree 3, the RF distance satisfies the following property.

LEMMA 8.3
When both T_1 and T_2 are of degree-3, the number of bad edges in T_1 with respect to T_2 is the same as the number of bad edges in T_2 with respect to T_1.

PROOF   As both trees are of degree 3, we know that they have the same number of edges. Also we know that the number of good edges in T_1 with respect to T_2 equals the number of good edges in T_2 with respect to T_1.

Since the number of bad edges equals the total number of edges minus the number of good edges, the lemma follows.   □

In the rest of this section, we first describe a brute-force algorithm to compute the RF distance. Then, we present Day's algorithm which computes the RF distance in linear time.

For every edge e in T_1, if the split formed by e is the same as the split formed by some edge e' in T_2, e is a good edge. Thus, a brute-force solution is to enumerate all edge pairs in T_1 and T_2 and report an edge pair (e, e') to be a good edge pair if the splits formed by both e and e' are the same. Then, based on the number of good edges, we can compute the number of bad edges and hence deduce the RF distance.

The correctness of the brute-force solution is straight forward. For the time analysis, note that for every edge e in T, the algorithm checks every edge e' in T' to determine whether e and e' give the same split. Such checking takes $O(n)$ time. Since there are n edges in T, the algorithm takes $O(n^2)$ time.

An $O(n^2)$-time algorithm for computing the RF distance may be a bit slow. In 1985, W. H. E. Day proposed an algorithm which can find the set of good edges in T with respect to T' using $O(n)$ time [73].

The algorithm first preprocesses a hash table which enables us to check if a particular split exists in T_1 using $O(1)$ time. Then, $O(n)$ time is sufficient to check the number of splits in T_2 which exist in T_1.

Below, we detail the steps. We root T_1 and T_2 at both leaves labeled by the same taxa. Also, we relabel the remaining $n-1$ leaves of T_1 in increasing order and change the labels of T_2 correspondingly. Figure 8.5 gives an example of the relabeling of T_1 and T_2. Such labeling has the following property.

LEMMA 8.4
The set of labels in any subtree of T_1 forms a consecutive interval.

After the preprocessing of T_1, Day's algorithm creates a hash table $H[1..n]$ to store the internal nodes of T_1. Precisely, for every internal node x in T_1, the corresponding interval $i_x..j_x$ is stored in either $H[i_x]$ or $H[j_x]$. The algorithm makes decisions as follows.

- If x is the leftmost child of its parent in T_1, $i_x..j_x$ is stored in $H[j_x]$.

- Otherwise, the interval $i_x..j_x$ is stored in $H[i_x]$.

Figure 8.6 shows the hash table for the internal nodes of T_1 in Figure 8.5.

The only issue in creating the hash table using the above rule is that two intervals may be stored in the same entry in the hash table H. Moreover, such a collision is impossible, as implied by the following lemma.

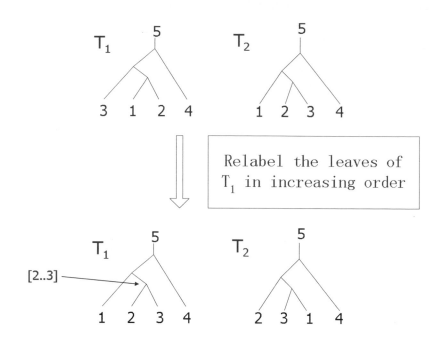

FIGURE 8.5: An example of relabeling all leaves of T_1 in increasing order.

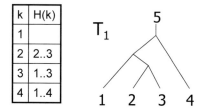

FIGURE 8.6: Hash table to store intervals for internal nodes in T_1.

LEMMA 8.5

If the intervals are stored in H according to the rules stated above, every entry in H stores at most one interval.

PROOF   By contrary, suppose $H[i]$ stores two intervals representing two internal nodes x and y. By definition, i should be the endpoints of the two intervals representing x and y. Thus, x and y should satisfy the ancestor-descendant relationship. Without loss of generality, assume x is the ancestor of y. Then y's interval should be the subinterval of x's interval. So, we have two cases.

Case 1: x's interval is $j..i$ and y's interval is $j'..i$ for $j < j'$; or

Case 2: x's interval is $i..j$ and y's interval is $i..j'$ for $j > j'$.

For case 1, y should be the rightmost child of its parent. Hence, $j'..i$ should be stored in $H[j']$ instead. So we get a contradiction! For case 2, y should be the leftmost child of x. Hence, $i..j'$ should be stored in $H[j']$ instead. We arrive at a contradiction similarly. In conclusion, we can store at most one interval in each entry in H.   ☐

The hash table H allows us to determine whether an interval $i..j$ exists in T_1 by checking $H[i]$ and $H[j]$. If either $H[i]$ or $H[j]$ equals $i..j$, then the interval $i..j$ does exist in T_1; otherwise, $i..j$ does not exist in T_1. Given the hash table H, we describe how to identify all good edges below. For each internal node u in T_2, we define

1. min_u and max_u to be the minimum and the maximum leaf labels in the subtree of T_2 rooted at u, respectively; and

2. $size_u$ to be the number of leaves in the subtree of T_2 rooted at u.

Figure 8.7 shows the results of computing min_u, max_u, and $size_u$ for the internal nodes of T_2 in Figure 8.5. If $max_u - min_u + 1 = size_u$, then we know that the leaf labels in the subtree of T_2 rooted at u form a consecutive interval $min_u..max_u$. Then we check whether $H[min_u]$ or $H[max_u]$ equals $min_u..max_u$. If yes, (u, v) is a good edge where v is the parent of u in T_2.

Figure 8.8 summarizes the pseudocode for Day's algorithm. For time complexity, Step 1 relabels the leaves, which takes $O(n)$ time. For Steps 2–6, the for-loop traverses $O(n)$ nodes in T_1 and stores the internal nodes of T_1 in the hash table $H[]$. The processing time per node is $O(1)$ time. Lastly, in Steps 7–12, the for-loop processes every internal node in T_2 and identifies good edges, which also requires $O(n)$ time. In total, the time complexity is $O(n)$.

size$_x$=max$_x$-min$_x$+1. Also, H[3]=1..3.
Thus, (x, z) is a good edge!

T_2

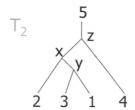

	min$_u$	max$_u$	size$_u$	max$_u$-min$_u$+1
x	1	3	3	3
y	1	3	2	3

FIGURE 8.7: Compute the minimum and maximum leaf labels and the number of leaves.

Day's algorithm

1: Root T_1 and T_2 at the leaf labeled by the same taxa. Then, the remaining $n - 1$ leaves are relabeled so that they are in increasing order in T_1. The leaves in T_2 are also relabeled correspondingly.
2: Initialize the hash table $H[1..n]$
3: **for** every internal node u in T_1 **do**
4:     Let $i_u..j_u$ be the interval corresponding to u
5:     If u is the leftmost child of its parent in T_1, set $H[j_u]$ to be $i_u..j_u$; otherwise, set $H[i_u]$ to be $i_u..j_u$.
6: **end for**
7: **for** every internal node u in T_2 **do**
8:     Compute min_u, max_u, and $size_u$
9:     **if** $max_u - min_u + 1 = size_u$ **then**
10:        if either $H[min_u]$ or $H[max_u]$ equals $min_u..max_u$, (u, v) is a good edge where v is the parent of u in T_2
11:     **end if**
12: **end for**

FIGURE 8.8: Day's algorithm.

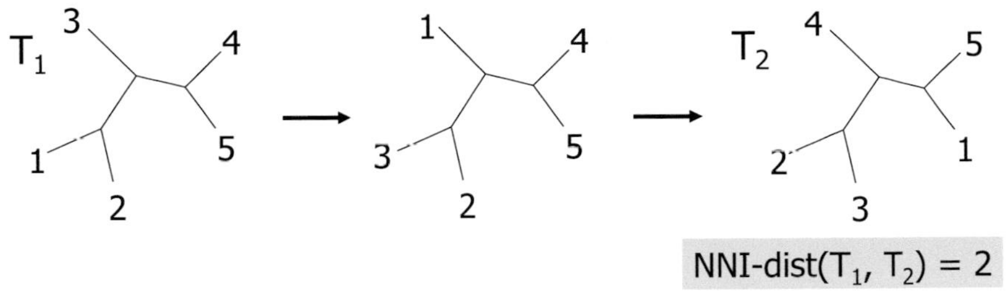

$$\text{NNI-dist(T}_1, \text{T}_2) = 2$$

FIGURE 8.9: The NNI distance of T_1 and T_2 is 2.

8.3.2 Nearest Neighbor Interchange Distance (NNI)

Given two unrooted degree-3 trees T_1 and T_2, the distance $NNI(T_1, T_2)$ between the two trees is the minimum number of nearest neighbor interchange (NNI) operations required to transform one tree into another [179, 213].

Below, we define the NNI operation and the NNI distance.

- NNI operation: A nearest neighbor interchange (NNI) is the interchange of two of the subtrees incident to an internal edge in a binary tree. Two such interchanges are possible for each internal edge. Figure 7.15 gives an example to illustrate the two possible NNI operations across an edge.

- NNI distance: Given two unrooted, degree-3 trees T_1 and T_2, the NNI distance is defined as the minimum number of NNI operations required to convert T_1 to T_2, which is denoted as NNI-dist(T_1, T_2). Figure 8.9 shows an example.

The following lemmas state some properties of NNI-dist.

LEMMA 8.6
$NNI\text{-}dist(T_1, T_2) = NNI\text{-}dist(T_2, T_1)$.

PROOF Since the nearest neighbor interchange (NNI) distance measures the minimum number of NNIs required to change T_1 into T_2, if one NNI is required to convert T_1 to T_2, then obviously one NNI is required to convert T_2 to T_1; hence, the lemma follows. ▯

LEMMA 8.7
$NNI\text{-}dist(T_1, T_2) \geq$ *number of bad edges in T_1 with respect to T_2*

PROOF To remove one bad edge, we require at least one NNI operation. Hence, the lemma follows. ▯

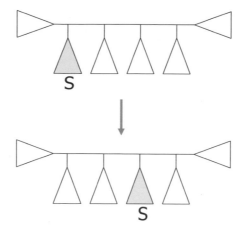

FIGURE 8.10:   The cost of the above STT operation is 2.

Computing the NNI-dist is NP-hard. However, there exists a polynomial time $O(\log n)$-approximation algorithm [72]. This approximation algorithm makes use of a certain merge sort technique, which is not covered in this book.

8.3.3   Subtree Transfer Distance (STT)

The subtree transfer (STT) distance is introduced in [139]. The STT operation and the STT distance are defined as follows.

- STT operation: Given an unrooted degree-3 tree T, a subtree transfer operation is the operation of detaching a subtree and reattaching it to the middle of another edge. There are a few different methods to charge the STT operation. For simplicity, we assume the STT operation is charged by the number of nodes the subtree is transferred. Figure 8.10 gives an example of the STT operation.

- STT distance: Given two unrooted degree-3 trees T_1 and T_2, the STT distance is defined as the minimum total cost of the STT operations required to transfer T_1 to T_2, which is denoted as STT-dist(T_1, T_2).

Below we show a property of the STT distance.

LEMMA 8.8
$STT\text{-}dist(T_1, T_2) = NNI\text{-}dist(T_1, T_2).$

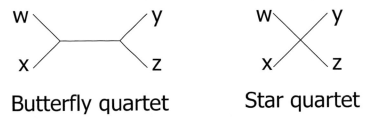

Butterfly quartet   **Star quartet**

FIGURE 8.11: Butterfly quartet and star quartet.

PROOF   First, STT-dist$(T_1, T_2) \leq$ NNI-dist(T_1, T_2) because each NNI operation is an STT operation.

Then, STT-dist$(T_1, T_2) \geq$ NNI-dist(T_1, T_2) because each STT operation of cost k can be simulated by k NNI operations. □

Based on the above lemma, the STT-distance equals the NNI distance. Therefore, computing the STT-dist(T_1, T_2) is also an NP-hard problem. However, there also exists a polynomial time $O(\log n)$ approximation algorithm to compute STT-dist(T_1, T_2).

8.3.4   Quartet Distance

A quartet is a phylogenetic tree with 4 taxa. Depending on the topology, the quartets can be classified into two types: butterfly quartet and star quartet. Figure 8.11 shows the two types of quartets. For the star quartet, all 4 taxa are attached to the same center. For the butterfly quartet, an internal edge exists which splits the taxa into two sets, say, $\{w, x\}$ and $\{y, z\}$. As a shorthand, we usually denote such a butterfly quartet as $wx|yz$.

Given two unrooted trees T_1 and T_2, Estabrook, McMorris, and Meacham [97] proposed to compute the distance between T_1 and T_2 based on the number of shared quartets. Precisely, the quartet distance is defined to be the number of sets of 4 taxa $\{w, x, y, z\}$ such that $T_1|\{w, x, y, z\} \neq T_2|\{w, x, y, z\}$, where $T_i|\{w, x, y, z\}$ represents the restricted subtree of T_i with respect to $\{w, x, y, z\}$.

For example, consider T_1 and T_2 in Figure 8.4 which are leaf-labeled by $\{a, b, c, d, e\}$. There are 5 different subsets of size 4.

- $\{a, b, c, d\}$: $T_1|\{a, b, c, d\} \neq T_2|\{a, b, c, d\}$

- $\{a, b, c, e\}$: $T_1|\{a, b, c, e\} \neq T_2|\{a, b, c, e\}$

- $\{a, b, d, e\}$: $T_1|\{a, b, d, e\} \neq T_2|\{a, b, d, e\}$

- $\{a, c, d, e\}$: $T_1|\{a, c, d, e\} \neq T_2|\{a, c, d, e\}$

- $\{b, c, d, e\}$: $T_1|\{b, c, d, e\} = T_2|\{b, c, d, e\}$

Hence, the quartet distance of T_1 and T_2 is 4.

Observe that $\binom{n}{4}$ equals the sum of the number of shared quartets and that of different quartets between T_1 and T_2. Hence, we have the following brute-force algorithm.

1: $count = 0$;
2: **for** every subset $\{w, x, y, z\}$ **do**
3:     **if** $T_1|\{w, x, y, z\} = T_2|\{w, x, y, z\}$ **then**
4:         $count = count + 1$;
5:     **end if**
6: **end for**
7: Report $\binom{n}{4} - count$;

The brute-force algorithm takes at least $O(n^4)$ time to compute the quartet distance. Efficient methods have been proposed. When both T_1 and T_2 are of degree 3, Steel and Penny [271] gave an algorithm which runs in $O(n^3)$ time. Bryant et al. [41] improved the time complexity to $O(n^2)$. Brodal et al. [36] gave the current best solution, which runs in $O(n \log n)$ time. When both T_1 and T_2 are arbitrary degree, Christiansen et al. [61] gave an $O(n^3)$-time algorithm. The running time can be further improved to $O(d^2 n^2)$ when the degree of both trees is bounded by d.

Below, we describe the $O(n^3)$-time algorithm proposed by Christiansen et al. [61]. We first give an observation. Consider three leaves x, y, and z. For each T_i, there exists a unique internal node c_i in T_i such that c_i appears in any paths from x to y, y to z, and x to z. We denote c_i as the center of x, y, and z in T_i. Let T_i^x, T_i^y, and T_i^z be the three subtrees attached to c_i which contain x, y, and z, respectively. Let T_i^{rest} be $T_i - T_i^x - T_i^y - T_i^z$. (See Figure 8.12 for an illustration.) We have the following observations.

- For every taxa $w \in T_i^x - \{x\}$, the quartet for $\{w, x, y, z\}$ in T_i is $wx|yz$.

- For every taxa $w \in T_i^y - \{y\}$, the quartet for $\{w, x, y, z\}$ in T_i is $wy|xz$.

- For every taxa $w \in T_i^z - \{z\}$, the quartet for $\{w, x, y, z\}$ in T_i is $wz|xy$.

- For every taxa $w \in T_i^{rest}$, the quartet for $\{w, x, y, z\}$ in T_i is a star quartet.

Based on the above observations, the number of shared butterfly quartets involving x, y, z is $|T_1^x \cap T_2^x| + |T_1^y \cap T_2^y| + |T_1^z \cap T_2^z| - 3$. The number of shared star quartets involving x, y, z is $|T_1^{rest} \cap T_2^{rest}|$. In total, the number of shared quartets involving x, y, z is $Count(x, y, z)$, which equals

$$|T_1^x \cap T_2^x| + |T_1^y \cap T_2^y| + |T_1^z \cap T_2^z| + |T_1^{rest} \cap T_2^{rest}| - 3 \qquad (8.1)$$

Hence, the total number of shared quartets is $\frac{1}{4} \sum_{\{x,y,z\} \subseteq S} Count(x, y, z)$.

To have an efficient algorithm, there are still two issues. The first issue is on the computation of $|T_1^{rest} \cap T_2^{rest}|$ in the above equation. We claim that it can be computed from $|T_1^a|$, $|T_2^a|$ and $|T_1^a \cap T_2^b|$ for all $a, b \in \{x, y, z\}$. Recall that

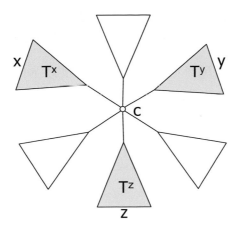

FIGURE 8.12: c is the center of x, y, and z. T^x, T^y, and T^z are the subtrees attached to c which contain x, y, and z, respectively. T^{rest} is $T - T^x - T^y - T^z$, which is represented by the uncolored part of T.

$T_i^{rest} = T_i - \cup_{a \in \{x,y,z\}} T_i^a$. Then, $|T_1^{rest} \cap T_2^{rest}| = n - |A_1 \cup A_2| + |A_1 \cap A_2|$ where $A_i = \cup_{a \in \{x,y,z\}} T_i^a$. Thus, $|T_1^{rest} \cap T_2^{rest}|$ can be computed in $O(1)$ time with the following equation.

$$n - \sum_{a \in \{x,y,z\}} (|T_1^a| + |T_2^a|) + \sum_{a \in \{x,y,z\}} \sum_{b \in \{x,y,z\}} |T_1^a \cap T_2^b| \qquad (8.2)$$

Second, to compute $Count(x, y, z)$, we require the values $|T_1^a \cap T_2^b|$ for all $a, b \in \{x, y, z\}$. Note that T_1^a and T_2^b are subtrees of T_1 and T_2, respectively. So, what we need is an efficient method to compute $|R_1 \cap R_2|$ for all subtrees R_1 of T_1 and all subtrees R_2 of T_2. Since each T_i has $O(n)$ subtrees, there are $O(n^2)$ pairs of subtrees (R_1, R_2). As every $|R_1 \cap R_2|$ can be computed in $O(n)$ time, we can compute $|R_1 \cap R_2|$ for all subtree pairs (R_1, R_2) in $O(n^3)$ time. In fact, the method can be improved to compute these values in $O(n^2)$ time (see Exercise 13).

Figure 8.13 summarizes the algorithm for computing the quartet distance between T_1 and T_2. First, Step 1 computes the size of $|R_1 \cap R_2|$ for all subtree pairs (R_1, R_2). This step takes $O(n^2)$ time. Then, Steps 3–12 compute $Count(x, y, z)$ for every three taxa x, y, z. Since there are n^3 sets of $\{x, y, z\}$ and $Count(x, y, z)$ can be computed in $O(1)$ time, the running time is $O(n^3)$. Last, we report the quartet distance in Step 13. The total running time of the algorithm is $O(n^3)$.

We can compute the distance of two rooted trees by counting the number of different triplets. Such a distance is called the triplet distance (see Exercise 10).

Algorithm Quartet_dist(T_1, T_2)

Require: two unrooted phylogenetic trees T_1 and T_2

Ensure: the quartet distance between T_1 and T_2

1: Compute $|R_1 \cap R_2|$ for any subtree R_1 of T_1 and any subtree R_2 of T_2;

2: $count = 0$;

3: **for** every $\{x, y, z\}$ **do**

4:    **for** $i = 1, 2$ **do**

5:       Let c_i be the center of x, y, and z in T_i;

6:       Let T_i^x, T_i^y, and T_i^z be the subtrees attached to c_i containing x, y, and z, respectively;

7:       Set $T_i^{rest} = T - T_i^x - T_i^y - T_i^z$;

8:    **end for**

9:    Compute $|T_1^{rest} \cap T_2^{rest}|$ based on Equation 8.2;

10:   Compute $Count(x, y, z)$ based on Equation 8.1;

11:   $count = count + Count(x, y, z)$;

12: **end for**

13: Report $\binom{n}{4} - count/4$;

FIGURE 8.13: Algorithm for computing the quartet distance of T_1 and T_2.

8.4   Consensus Tree Problem

Given a set S of n taxa. Let $\mathcal{T}$ be a set of trees $\{T_1, T_2, \ldots, T_m\}$ where the leaves of every tree T_i are labeled by S. The consensus tree problem asks for one tree (called the consensus tree) to summarize the information agreed by all trees in $\mathcal{T}$.

The consensus tree methods have been applied to find the bootstrapping tree (see Section 7.4). Another application is to infer the species tree from a set of gene trees.

The idea of a consensus tree was first described by Adams [2]. Then, a number of different definitions of a consensus tree were proposed. Margush and McMorris [205] defined the majority rule consensus tree. Around the same time, Sokal and Rohlf [268] proposed the definition of a strict consensus tree. There are other definitions of a consensus tree, including the Adams consensus tree, the R^* consensus tree, and the greedy consensus tree. See Bryant [40] for an overview of consensus tree definitions.

Below, we will discuss four consensus tree methods, namely, strict consensus tree, majority consensus tree, greedy consensus tree, and R^* tree.

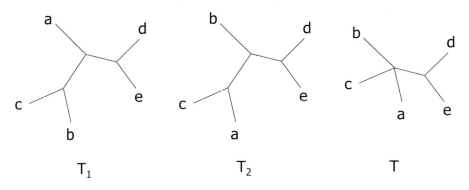

FIGURE 8.14: T is the strict consensus tree for T_1 and T_2.

8.4.1 Strict Consensus Tree

The strict consensus tree T of $\{T_1, T_2, \ldots, T_m\}$ contains exactly those splits which appear in all T_i. Precisely, let W_i be the set of splits of T_i, $i = 1, 2, \ldots, m$. The strict consensus tree is defined by the set of splits $W_1 \cap W_2 \ldots \cap W_m$. It always exists and is unique (see Exercise 7). For example, in Figure 8.14, T is the strict consensus tree of T_1 and T_2.

We first describe the algorithm to compute the strict consensus tree of two trees T_1 and T_2. Observe that every split in the strict consensus tree corresponds to a good edge (defined in Section 8.3.1). Hence, we can apply Day's algorithm to find all the splits involved in the strict consensus tree; then, the strict consensus tree can be generated. Precisely, the strict consensus tree is formed by contracting all bad edges in T_1. The time complexity of the algorithm is $O(n)$ since Day's algorithm runs in $O(n)$ time.

Now, we discuss the algorithm for constructing the strict consensus tree of m trees $\mathcal{T} = \{T_1, T_2, \ldots, T_m\}$. Let R_i be the strict consensus tree for $\{T_1, \ldots, T_i\}$. We observe that R_i equals the strict consensus tree for R_{i-1} and T_i. Hence, the strict consensus tree for $\mathcal{T}$ can be computed as follows.

1: Let $R_1 = T_1$;
2: **for** $i = 2$ to m **do**
3:    set R_i be the strict consensus tree of R_{i-1} and T_i.
4: **end for**
5: Return R_m;

For running time, the strict consensus tree of two trees R_{i-1} and T_i can be computed in $O(n)$ time as discussed earlier. Since the for-loop iterates $m - 1$ times, the time complexity is $O(mn)$.

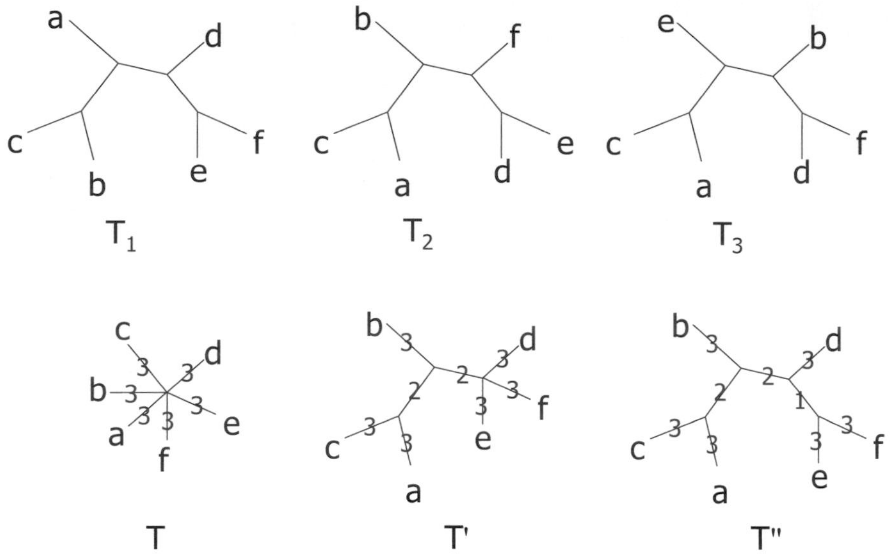

FIGURE 8.15:   Consider a set of 3 trees $\mathcal{T} = \{T_1, T_2, T_3\}$. T, T', and T'' are the strict consensus tree, the majority rule consensus tree, and the greedy consensus tree, respectively, of $\mathcal{T}$. For each edge in the consensus trees, the integer label on each edge indicates the number of occurrences of such a split.

8.4.2   Majority Rule Consensus Tree

The majority rule consensus tree contains exactly those splits that appear in more than half of the input trees. Note that the majority rule consensus tree is unique and always exists. Also, the majority rule consensus tree is a refinement of the strict consensus tree since all the splits for the strict consensus tree also appear in the majority rule consensus tree.

For example, in Figure 8.15, T and T' are the strict consensus tree and the majority rule consensus tree, respectively, of T_1, T_2, and T_3. Note that T' is a refinement of T.

Below, we describe the algorithm for constructing the majority rule consensus tree. When there are two trees T_1 and T_2, the majority rule tree of T_1 and T_2 is the same as the strict consensus tree of T_1 and T_2. We can construct it using the algorithm in Section 8.4.1. When there are more than two trees, we can compute the majority rule consensus tree using the following 3 steps.

1: Count the occurrences of each split, storing the counts in a table.
2: Select those splits with occurrences $> m/2$.
3: Create the majority rule tree using the selected splits.

For step 1, we can identify the common splits between every pair of T_i and T_j for $1 \leq i < j \leq m$ by running Day's algorithm. Such information allows us to count the occurrences of each split. We need to run $\binom{m}{2}$ instances of Day's algorithm. This step takes $O(m^2 n)$ time.

Step 2 selects the splits with counts $> m/2$. Since there are at most nm splits, this step takes $O(nm)$ time.

In step 3, we need to construct the majority rule consensus tree from the selected splits. We need the following lemma.

LEMMA 8.9

Suppose p and c are any two splits in the majority rule consensus tree. There exists a tree T_j which contains both splits p and c.

PROOF Both p and c appear in more than $m/2$ trees. By the pigeon-hole principle, there exists a tree which contains both p and c. ▯

Let $\mathcal{C} = \{c_1, \ldots, c_k\}$ be the set of splits (or edges) in the majority rule consensus tree T of $\mathcal{T} = \{T_1, T_2, \ldots, T_m\}$. Suppose we root the consensus tree T at the leaf 1, every split $c \in \mathcal{C}$ has a corresponding parent split $P[c]$ in T. Below, we describe an algorithm which compute the parent split $P[c]$ for every $c \in \mathcal{C}$. The algorithm is based on the property of the above lemma, i.e., there exists at least one tree T_i which contain both splits c and $P[c]$ for every $c \in \mathcal{C}$.

Require: A set of splits (or edges) $\mathcal{C} = \{c_1, \ldots, c_k\}$ which occur more than $m/2$ times in $\mathcal{T} = \{T_1, \ldots, T_m\}$

Ensure: A majority rule consensus tree T rooted at the leaf 1

  1: We root every tree T_i at the leaf 1.

  2: For each T_i, we get T_i' which is the tree formed by contracting all edges whose do not belong to $\mathcal{C}$.

  3: For every split $c \in \mathcal{C}$, initialize $P[c] = \emptyset$.

  4: For every split $c \in \mathcal{C}$, let B_c be the partition of c which does not contain the leaf 1.

  5: **for** $i = 1$ to m **do**

  6:     **for** every edge c in T_i' **do**

  7:        Let q be the parent edge of c in T_i'.

  8:        **if** $P[q] = \emptyset$ or $|B_c| > |B_{P[q]}|$ **then**

  9:           $P[q] = c$.

10:       **end if**

11:     **end for**

12: **end for**

13: Construct the majority rule consensus tree according to the parent-child relationship stated in $P[]$.

By the above algorithm, we can construct the majority rule consensus tree in $O(nm)$ time. In total, the majority consensus tree can be constructed in $O(nm^2)$ time.

As a final note, the majority rule consensus tree can be built in $O(nm)$ expected time [11].

8.4.3   Median Consensus Tree

Let $d(T_1, T_2)$ be the symmetric difference between T_1 and T_2, that is, the number of splits appearing in one tree but not the other. For example, for T_1 and T_2 in Figure 8.14, $\{a, d, e\}|\{b, c\}$ only appears in T_1 and $\{a, c\}|\{b, d, e\}$ only appears in T_2. Hence, $d(T_1, T_2) = 2$. T is called the median tree for $T_1, T_2, \ldots, T_m$ if it minimizes $\sum_{i=1}^{m} d(T, T_i)$.

Note that the median consensus tree is not unique. Moreover, the majority rule consensus tree is one of a median consensus tree [20]. Hence, a median consensus tree can be built using the algorithm in Section 8.4.2.

8.4.4   Greedy Consensus Tree

Given a set of trees $\mathcal{T} = \{T_1, T_2, \ldots, T_m\}$, the greedy consensus tree is created sequentially by including splits one by one. In every iteration, we include the most frequent split that is compatible with the included splits (breaking the ties randomly). The process is repeated until we cannot include any other split. Note that the greedy consensus tree always exists. However, it is not unique. Furthermore, the greedy consensus tree is a refinement of the majority rule consensus tree as the greedy consensus tree contains all the splits with frequencies at least $m/2$.

For example, in Figure 8.15, T'' is the greedy consensus tree of T_1, T_2, and T_3. Note that T_2 is another feasible greedy consensus tree.

The greedy consensus tree can be constructed as follows.

1: Count the occurrences of each split, storing the counts in a table.
2: Sort the splits to obtain $S_1, S_2, \ldots, S_t$ in decreasing order according to their number of occurrences in $\mathcal{T}$.
3: Form a tree T based on the split S_1
4: **for** $i = 2$ to t **do**
5:     **if** S_i is compatible with T **then**
6:         Modify T so that it also satisfies S_i.
7:     **end if**
8: **end for**

Below we analyze the time complexity. Step 1 takes $O(m^2 n)$ time using the Day's algorithm. Since $\mathcal{T}$ has m trees and each tree has $O(n)$ edges, $t \leq O(nm)$. Step 2 can sort the t splits in $O(nm)$ time using radix sort. Checking if a split is compatible with T takes $O(n)$ time. Hence, for Steps 4-8, the for-loop takes $O(tn) = O(n^2 m)$ time. In total, the greedy consensus tree for $\mathcal{T}$ can be constructed in $O(m^2 n + n^2 m)$ time.

$\{a,b,c\}$	$ab\|c$: 3	$bc\|a$: 0	$ac\|b$: 0
$\{a,c,d\}$	$cd\|a$: 1	$ad\|c$: 1	$ac\|d$: 1
$\{b,c,d\}$	$cd\|b$: 1	$bd\|c$: 1	$bc\|d$: 1
$\{a,b,d\}$	$ab\|d$: 3	$bd\|a$: 0	$ad\|b$: 0

FIGURE 8.16: The number of triplets occur in the three trees T_1, T_2, and T_3 in Figure 8.17.

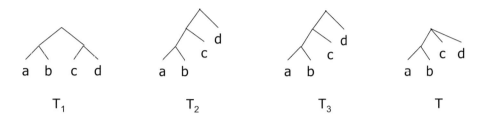

T_1 T_2 T_3 T

FIGURE 8.17: T is the R^* consensus tree for T_1, T_2, and T_3.

8.4.5 R^* **Tree**

Given a set of rooted trees $\mathcal{T} = \{T_1, \ldots, T_m\}$, this section describes the R^* tree of $\mathcal{T}$. The R^* tree is a consensus tree constructed from triplets of $\mathcal{T}$. A triplet is a restricted subtree (see Section 8.2) of any tree in $\mathcal{T}$ with respect to three taxa. (Below, we use $ab|c$ to represent a triplet where the lowest common ancestor of a and b is a descendant of the lowest common ancestor of a and c.) The R^* tree method tries to find the set of triplets which is the most commonly occurring in $\mathcal{T}$; then, the R^* tree is built from those most commonly occurring triplets. Note that the R^* consensus tree always exists and is unique. Also, the R^* consensus tree is a refinement of the majority-rule consensus tree.

As an example, consider three trees T_1, T_2, and T_3 in Figure 8.17. Figure 8.16 lists the occurrences of those triplets. For $\{a,b,c\}$, the most frequent triplet is $ab|c$. For $\{a,b,d\}$, the most frequent triplet is $ab|d$. For both $\{a,c,d\}$ and $\{b,c,d\}$, there is no frequent triplet. Hence, we set $C = \{ab|c, ab|d\}$. The R^* tree which is consistent with the set C is shown in Figure 8.17.

By the definition, the R^* tree can be constructed using three steps as follows.

1: Compute the number of occurrences of all triplets in all m trees $T_1, T_2, \ldots, T_m$.

2: For each set of three taxa, we identify the most commonly occurring triplet in the m trees. Let C be the set of the most commonly occurring triplets.

3: Construct the tree which is consistent with all triplets in the set C and is not consistent with any triplet not in C.

First, for the correctness of the algorithm, Steel [270] showed that there always exists a tree which is consistent with all triplets in the set C and is not consistent with any triplet not in C. Furthermore, such a tree is unique. Hence, the algorithm can always report a unique R^* tree.

Below, we analyze the running time of the algorithm. For step 1, since there are $\binom{n}{3}$ triplets in each tree and there are m trees, it takes $O(mn^3)$ time to count the number of occurrences of all triplets. Step 2 computes C, which takes $O(n^3)$ time. Step 3 constructs a tree which is consistent with the set C. By the triplet method, this step takes $O(\min\{O(k \log^2 n), O(k + n^2 \log n)\})$ time [161] where $k = |C| = O(n^3)$. Hence, this step takes $O(n^3)$ time. In total, the whole algorithm runs in $O(mn^3)$ time.

For the unrooted variant, a similar problem can be defined. Given a set of unrooted trees $\mathcal{T} = \{T_1, \dots, T_m\}$, we aim to construct a consensus tree, called the Q* tree, from the set of quartets which are the most commonly occurring in $\mathcal{T}$. This problem can be solved similarly [29]. The running time of the algorithm is $O(mn^4)$.

8.5   Further Reading

In this chapter and the previous two chapters, we study different methods for reconstructing phylogenetic trees and for comparing phylogenetic trees. There are many other methods in the literature. For instance, apart from character-based methods and distance-based methods, phylogenetic trees can also be reconstructed from quartets or triplets. There are also approaches based on Bayesian inference.

These two chapters did not cover some newer topics in phylogenetics. One example is the phylogenetic supertree. Currently, no method can find the phylogenetic tree for all species. To find the phylogenetic tree for all species, we have to combine a number of phylogenetic trees reconstructed for different sets of species. Figure 8.18 shows an example of combining two trees. The combined tree is called a supertree. The difficulty of this problem is how to resolve the conflicts among the trees.

Another topic is phylogenetic network. Ford Doolittle [83] said that "Molecular phylogeneticists will have failed to find the 'true tree', not because their methods are inadequate or because they have chosen the wrong genes, but because the history of life cannot properly be represented as a tree."

Evolution is in fact more than mutation. We have other types of evolutions.

- Hybridization: This process happens when two different species produce an offspring that has the genes from both parents, e.g., female tiger + male lion → liger.

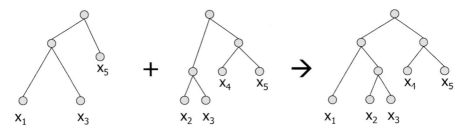

FIGURE 8.18: An example of constructing a supertree from two phylogenetic trees. Note that the two trees are leaf-labeled by a different set of species.

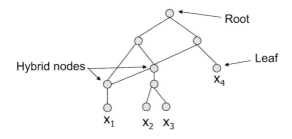

FIGURE 8.19: An example of a phylogenetic network.

- Horizontal gene transfer or lateral gene transfer. This is the process of transferring a portion of the genome from one species (donor) to another (recipient), e.g., evolution of influenza.

Note that phylogenetic trees cannot model those types of evolution. A phylogenetic network is proposed to generalize phylogenetic trees. In a phylogenetic network, each node may have more than one parent. Precisely, a phylogenetic network N is a directed acyclic graph (i.e., a direct graph with no cycle) such that

- Each node has in-degree 1 or 2 (except the root) and out-degree at most 2. (Nodes with in-degree 2 are called hybrid nodes.)

- No node has both in-degree 1 and out-degree 1.

- All nodes with out-degree 0 are distinctly labeled ("leave").

Figure 8.19 shows an example of a phylogenetic network.

8.6   Exercises

1. What is the maximum agreement subtree for T_1 and T_2? Is the maximum agreement subtree unique?

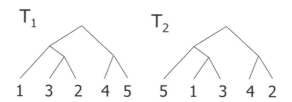

2. Given three rooted binary trees T_1, T_2, and T_3, can you give an algorithm to compute $mast(T_1, T_2, T_3)$? What is the running time?

3. Consider two rooted binary phylogenetic trees T_1 and T_2 and a particular leaf a. Give an algorithm to compute the maximum agreement subtree of T_1 and T_2 which contains the leaf a. What is the time complexity?

4. Given two unrooted phylogenetic trees U_1 and U_2, for any edge e of U_1, prove the following statement.

$$MAST(U_1, U_2) = \max\{MAST(U_1^e, U_2^f) \mid f \text{ is an edge of } U_2\}$$

5. Compute the RF distance for T_1 and T_2 using Day's algorithm. Show the steps.

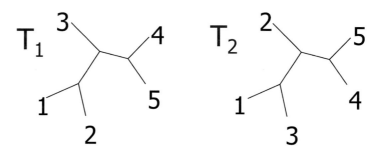

6. What is the NNI distance between the following two trees?

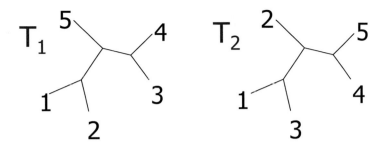

7. Given a set of trees $\mathcal{T} = \{T_1, T_2, \ldots, T_m\}$, can you prove that the strict consensus tree of $\mathcal{T}$ always exists and is unique?

8. What are the strict consensus tree and the majority-rule consensus tree of T_1, T_2, and T_3?

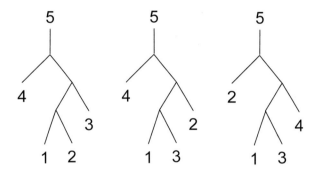

9. Can you compute the R^* tree for the following three trees?

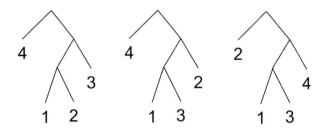

10. Given two rooted trees T_1 and T_2 leaf-labeled by the species set S, the triplet distance is the number of subsets $\{x, y, z\} \subseteq S$ such that $T_1|\{x, y, z\} \neq T_2|\{x, y, z\}$. Suppose $|S| = n$, can you give an $O(n^2)$-time algorithm to compute the triplet distance?

11. Given m unrooted degree-3 trees $T_1, T_2, \ldots, T_m$ leaf-labeled by S. Can you propose an efficient algorithm to compute the number of subsets

$\{w, x, y, z\} \subseteq S$ such that $T_1|\{w, x, y, z\} = T_2|\{w, x, y, z\} = \ldots = T_m|\{w, x, y, z\}$. What is the running time?

12. Given m unrooted trees $T_1, T_2, \ldots, T_m$ leaf-labeled by S, can you propose an efficient algorithm to compute the Q* tree? What is the running time?

13. Consider two phylogenies T_1 and T_2 leaf-labeled by L where $|L| = n$. Can you propose an $O(n^2)$-time algorithm which computes $|R_1 \cap R_2|$ for all subtree pairs (R_1, R_2) of (T_1, T_2)?

Chapter 9

Genome Rearrangement

9.1 Introduction

Before 1900, people only knew that species can evolve through mutations. Then, in 1917, people discovered evidence of genome rearrangement. Sturtevant [276] showed that strains of *Drosophila melanogaster* coming from the same or from distinct geographical locations may be different in having blocks of genes rotated by 180°. Such a rearrangement is known as reversal. In 1938, Dobzhansky and Sturtevant [81] studied genes in chromosome 3 of *Drosophila miranda* and 16 different strains of *Drosophila pseudoobscura*. They observed that the 17 strains can be arranged as vertices in a tree where every edge corresponds to one reversal. Hence, Dobzhansky and Sturtevant proposed that species can evolve through genome rearrangements. In the 1980s, Jeffrey Palmer and co-authors [227] studied the evolution of plant organelles by comparing the gene order of mitochondrial genomes. They pioneered studies of the shortest (most parsimonious) rearrangement scenarios between two genomes.

Evidents on genome rearrangements have been observed for higher organisms. Humans and mice are highly similar in DNA sequences (98% in sequence similarity). Moreover, their DNA segments are swapped. For example, chromosome X in humans can be transformed to chromosome X in mice using seven reversals.

In this chapter, we will study methods to compute the shortest series of genome rearrangements to transform one genome to another.

9.2 Types of Genome Rearrangements

In addition to reversal, which is the most common rearrangement, there are other types of genome rearrangements. Below, we list the known rearrangement operations within one chromosome (A, B, C, D are DNA segments and B^r is the reverse complement of B).

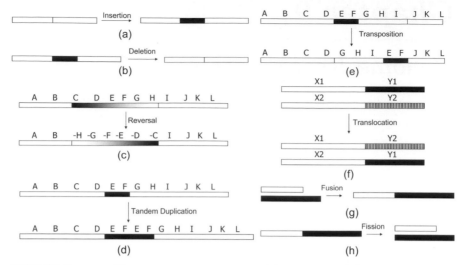

FIGURE 9.1: Various rearrangements: (a) insertion, (b) deletion, (c) reversal, (d) tandem repeat, (e) transposition, (f) translocation, (g) fusion, and (h) fission.

Reversal: Reversing a DNA segment ($ABC \rightarrow AB^rC$)

Insertion: Inserting of a DNA segment into the genome ($AC \rightarrow ABC$)

Deletion: Removal of a DNA segment from the genome ($ABC \rightarrow AC$)

Duplication: A particular DNA segment is duplicated two times in the genome ($ABC \rightarrow ABBC$, $ABCD \rightarrow ABCBD$)

Transposition: Cutting out a DNA segment and inserting it into another location ($ABCD \rightarrow ACBD$). This operation is believed to be rare since it requires 3 breakpoints.

There are also rearrangements involving more than one chromosome, which are listed as follows:

Translocation: The transfer of a segment of one chromosome to another non-homologous one.

Fusion: Two chromosomes merge.

Fission: One chromosome splits up into two chromosomes.

Figure 9.1 illustrates these rearrangement types.

$$2, \underline{4, 3}, 5, 8, 7, 6, 1$$
$$2, \underline{3, 4, 5}, 8, 7, 6, 1$$
$$2, 3, 4, 5, \underline{8, 7, 6}, 1$$
$$\underline{8, 7, 6, 5, 4, 3, 2, 1}$$
$$1, 2, 3, 4, 5, 6, 7, 8$$

FIGURE 9.2: Example of transforming an unsigned permutation $\pi = (2, 4, 3, 5, 8, 7, 6, 1)$ into $(1, 2, \ldots, 8)$ by unsigned reversal. Each step reverses the underlined region. The minimum number of reversals required is 4. Thus, $d(\pi) = 4$.

9.3 Computational Problems

Given two genomes with a set of common genes, those genes are arranged in a different order in different genomes. By comparing the ordering of those common genes, we aim to understand how one genome evolves into another through rearrangements. By the principle of parsimony, we hope to find the shortest rearrangement path. Depending on the allowed rearrangement operations, the literature studied the following problems:

- Genome rearrangement by reversals

- Genome rearrangement by translocations

- Genome rearrangement by transpositions

This chapter focuses on genome rearrangement by reversals. This problem is also called sorting by reversals.

9.4 Sorting an Unsigned Permutation by Reversals

Suppose $(1, 2, \ldots, n)$ represents the ordering of n genes in a genome. Let $\pi = (\pi_1, \pi_2, \ldots, \pi_n)$ be a permutation of $\{1, 2, \ldots, n\}$, which represents the ordering of the n genes in another genome. A reversal $\rho(i, j)$ is an operation applying on π, denoted as $\pi \cdot \rho(i, j)$, which reverses the order of the element in the interval $[i..j]$. Thus, $\pi \cdot \rho(i, j) = (\pi_1, \ldots, \pi_{i-1}, \pi_j, \ldots, \pi_i, \pi_{j+1}, \ldots, \pi_n)$. Our aim is to find the minimum number of reversals that transform π to $(1, 2, \ldots, n)$. The minimum number of reversals required to transform π to $(1, 2, \ldots, n)$ is called the unsigned reversal distance, denoted by $d(\pi)$. Figure 9.2 gives an example of sorting an unsigned permutation.

Kececioglu and Sankoff (1995) [172]	2-approximation
Bafna and Pevzner (1996) [14]	1.75-approximation
Caprara (1997) [50]	NP-hard
Christie (1998) [62]	1.5-approximation
Berman and Karpinski (1999) [27]	MAX-SNP hard
Berman, Hannenhalli, and Karpinski (2002) [28]	1.375-approximation

FIGURE 9.3:   Results related to sorting an unsigned permutation.

$$4, 5, 3, 1, 2$$
$$\overline{1, 3, 5, 4}, 2$$
$$1, 2, \overline{4, 5, 3}$$
$$1, 2, 3, \overline{5, 4}$$
$$1, 2, 3, 4, 5$$

FIGURE 9.4:   An example of transferring π to $(1, 2, \ldots, n)$ using at most $n - 1$ reversals.

The problem of sorting an unsigned permutation was introduced by Kececioglu and Sankoff [172]. It is shown to be NP-hard [50] and MAX-SNP hard [27]. This means that the problem is unlikely to be solvable in polynomial time. Even worse, it is unlikely to have a polynomial time approximation scheme. A number of approximation algorithms have been proposed, which are shown in Figure 9.3. The current best result has an approximation ratio 1.375 [28]. In the rest of this section, we will give a 4-approximation algorithm; then, we further improve it to a 2-approximation algorithm.

9.4.1   Upper and Lower Bound on an Unsigned Reversal Distance

Consider a permutation $\pi = (\pi_1, \pi_2, \ldots, \pi_n)$ of $\{1, 2, \ldots, n\}$. We first study the upper and lower bounds on an unsigned reversal distance $d(\pi)$.

We need a definition. A breakpoint is said to occur between π_i and π_{i+1} if $|\pi_i - \pi_{i+1}| > 1$. Let $b(\pi)$ be the number of breakpoints in π. As an example, when $\pi = (1, 8, 2, 3, 4, 7, 6, 5, 9)$, there are 4 breakpoints: $(1, 8), (8, 2), (4, 7)$, and $(5, 9)$. Thus, $b(\pi) = 4$.

For the lower bound, since a reversal can reduce at most two breakpoints, $d(\pi) \geq b(\pi)/2$.

For the upper bound, we present an algorithm with at most $n - 1$ reversals. The algorithm has $n - 1$ iterations. For the i-th iteration, it performs at most 1 reversal to move element i to position i. Figure 9.4 gives an example. In conclusion, we have

$$\frac{b(\pi)}{2} \leq d(\pi) \leq n - 1$$

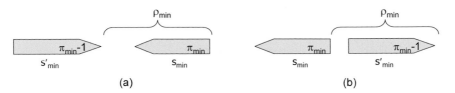

FIGURE 9.5: (a) s_{min} is to the right of s'_{min}, (b) s_{min} is to the left of s'_{min}. After the reversal ρ_{min}, the number of breakpoints is reduced by 1.

9.4.2 4-Approximation Algorithm for Sorting an Unsigned Permutation

For a permutation π of $\{1, 2, \ldots, n\}$, this section presents a 4-approximation algorithm for sorting an unsigned permutation, that is, computing a sequence of reversals $\rho_1, \ldots, \rho_\ell$ such that $\ell \leq 4d(\pi)$. Before describing the algorithm, we first prove some properties. A strip is defined to be a maximal subsequence in π without a breakpoint. A strip $(\pi_i, \pi_{i+1}, \ldots, \pi_{i+\alpha-1})$ is called decreasing if $\pi_k = \pi_i - (k - i)$ for $k = i, \ldots, i + \alpha - 1$; otherwise, it is called increasing. (Moreover, there is one exception. For the leftmost strip, if it contains only 1, it is called increasing instead of decreasing.) As an example, for $\pi = (7, 6, 5, 4, 1, 9, 8, 2, 3)$, it has three breakpoints and hence it has 4 strips: $(7, 6, 5, 4), (1), (9, 8)$, and $(2, 3)$. Among them, $(2, 3)$ is an increasing strip while the other three strips are decreasing strips.

As an observation, for any decreasing strip $(j+\alpha, \ldots, j)$ and any increasing strip $(j - \beta..j - 1)$, we can make a reversal to put the two strips next to each other such that $j-1$ and j are neighbors. Then, we say such a reversal merges the two strips. For example, by reversing the segment $(1, 9, 8, 2, 3)$, we can merge the decreasing strip $(7, 6, 5, 4)$ and the increasing strip $(2, 3)$ so that 3 and 4 are neighbors. Below lemma states the key property used by the 4-approximation algorithm.

LEMMA 9.1
If π has a decreasing strip, there exists a reversal ρ which reduces the number of breakpoints by at least one, that is, $b(\pi \cdot \rho) - b(\pi) \geq 1$.

PROOF Let s_{min} be the decreasing strip in π with the minimal element π_{min}. Let s'_{min} be the strip containing $\pi_{min} - 1$. s'_{min} must be increasing since it contains some element smaller than π_{min}. Also, let ρ_{min} be the reversal which merges s_{min} and s'_{min}.

Depending on whether s_{min} is to the right or to the left of s'_{min}, there are two cases as shown in Figure 9.5. For both cases, the reversal ρ_{min} reduces $b(\pi)$ by 1. ☐

Algorithm 4Approx-Unsigned-Reversal

1: **while** $b(\pi) > 0$ **do**
2:    **if** there exists a decreasing strip **then**
3:        we apply a reversal on π as stated by Lemma 9.1;
4:    **else**
5:        reverse an increasing strip to create a decreasing strip;
6:    **end if**
7: **end while**

FIGURE 9.6:   4-approximation algorithm for computing the unsigned reversal distance.

Figure 9.6 presents the 4-approximation algorithm, which tries to identify a sequence of reversals that reduces the number of breakpoints monotonically. By Lemma 9.1, whenever there is a decreasing strip in π, the algorithm can reduce the number of breakpoints by 1. Otherwise, we reverse an increasing strip to generate a decreasing strip. The process is repeated until there is no breakpoint. At that time, the algorithm terminates and the permutation is sorted.

Now, we prove the correctness of the algorithm. To reduce the number of breakpoints by 1, we need at most two reversals (one reversal creates a decreasing strip while another reversal reduces the number of breakpoints based on Lemma 9.1). Hence, the above algorithm will perform at most $2b(\pi)$ reversals. As shown in Section 9.4.1, the optimal solution performs at least $b(\pi)/2$ reversals. Thus, the approximation ratio is 4.

9.4.3   2-Approximation Algorithm for Sorting an Unsigned Permutation

We can further reduce the approximation ratio from 4 to 2 for sorting an unsigned permutation. The basic 2-approximation algorithm tries to maintain that every reversal reduces the number of breakpoints by 1 and retains some decreasing strip after the reversal. If such a condition cannot be maintained, it ensures reduction of two breakpoints after two reversals and retains some decreasing strip.

Without loss of generality, we assume π initially has some decreasing strip. Otherwise, we perform one reversal to convert an increasing strip to become decreasing. Let s_{min} and s_{max} be the decreasing strips in π with the minimum element π_{min} and the maximum element π_{max}, respectively. (Note that s_{min} and s_{max} may be the same strip.) Let s'_{min} and s'_{max} be the strips containing $\pi_{min} - 1$ and $\pi_{max} + 1$, respectively. Both s'_{min} and s'_{max} must be increasing strips. Let ρ_{min} be the reversal which merges s_{min} and s'_{min}. Let ρ_{max} be the reversal which merges s_{max} and s'_{max}.

Below lemma states the key property of the 2-approximation algorithm.

Algorithm 2 Approx-Unsigned-Reversal

1: If there exists no decreasing strip in π, we reverse any increasing strip in π to create a decreasing strip.

2: **while** $b(\pi) > 0$ **do**

3:     **if** $\pi \cdot \rho_{min}$ contains a decreasing strip **then**

4:         We reverse π by ρ_{min}; // this reversal reduces $b(\pi)$ by at least 1

5:     **else if** $\pi \cdot \rho_{max}$ contains a decreasing strip **then**

6:         We reverse π by ρ_{max}; // this reversal reduces $b(\pi)$ by at least 1

7:     **else**

8:         We reverse π by $\rho_{max} = \rho_{min}$; // this reversal reduces $b(\pi)$ by 2

9:         Reverse any increasing strip in π to create a decreasing strip.

10:     **end if**

11: **end while**

FIGURE 9.7: 2-approximation algorithm for computing the unsigned reversal distance.

LEMMA 9.2

Consider a permutation π that has a decreasing strip. Suppose both $\pi \cdot \rho_{min}$ and $\pi \cdot \rho_{max}$ contain no decreasing strip. Then, the reversal $\rho_{min} = \rho_{max}$ removes 2 breakpoints.

PROOF s'_{min} must be to the left of s_{min}. Otherwise, after the reversal, we still maintain a decreasing strip (see Figure 9.5). Similarly, s_{max} must be to the left of s'_{max}.

If s_{min} is to the left (or right) of both s_{max} and s'_{max}, after the reversal ρ_{max}, we still have a decreasing strip s_{min}. Similarly, if s_{max} is to the left (or right) of both s_{min} and s'_{min}, after the reversal ρ_{min}, we still have a decreasing strip s_{max}.

If s'_{max} is in between s'_{min} and s_{min}, after the reversal ρ_{min}, s'_{max} becomes a decreasing strip. Similarly, if s'_{min} is in between s_{max} and s'_{max}, after the reversal ρ_{max}, s'_{min} becomes a decreasing strip.

Hence, the only possible arrangement such that there is no decreasing strip after performing either ρ_{min} or ρ_{max} is that s'_{min}, s_{max}, s_{min}, and s'_{max} are in left to right order.

We claim that there is no element between s'_{min} and s_{max}. If there is a decreasing strip in between, after the reversal ρ_{max}, such a decreasing strip still remains. If there is an increasing strip in between, after the reversal ρ_{min}, such an increasing strip becomes decreasing. Similarly, we can show that there is no element between s_{min} and s'_{max}. Therefore, the reversals ρ_{min} and ρ_{max} are the same and reversing $\rho_{min} = \rho_{max}$ removes two breakpoints.   ☐

Figure 9.7 presents an algorithm based on Lemma 9.2. The algorithm will

$$+0, +3, +1, +6, +5, -2, +4, +7$$
$$+0, -5, -6, -1, -3, -2, +4, +7$$
$$+0, -5, -6, -1, +2, +3, +4, +7$$
$$+0, -5, -6, +1, +2, +3, +4, +7$$
$$+0, -5, -4, -3, -2, -1, +6, +7$$
$$+0, +1, +2, +3, +4, +5, +6, +7$$

FIGURE 9.8: Example of transforming a signed permutation $\pi = (+0, +3, +1, +6, +5, -2, +4, +7)$ into the identity permutation by signed reversal. The minimum number of reversals required is 5. Thus, $d(\pi) = 5$.

reduce the number of breakpoints by 2 for every 2 reversals. Hence, it will perform $b(\pi)$ reversals. The optimal solution performs at least $b(\pi)/2$ reversals (see Section 9.4.1). Thus, the approximation ratio is 2.

9.5   Sorting a Signed Permutation by Reversals

Genes have orientation. To capture the orientation of genes, the problem of sorting a signed permutation is defined. Consider a signed permutation of $\{0, 1, 2, \ldots, n\}$, that is, $\pi = (\pi_0, \pi_1, \pi_2, \ldots, \pi_n)$. We set $\pi_0 = 0$ and $\pi_n = n$ to denote the boundary of the genome. A reversal $\rho(i, j)$ is an operation applied on π, denoted as $\pi \cdot \rho(i, j)$, which reverses the order and flips the signs of the elements in the interval $i..j$. Thus, $\pi \cdot \rho(i, j) = (\pi_0, \ldots, \pi_{i-1}, -\pi_j, \ldots, -\pi_i, \pi_{j+1}, \ldots, \pi_n)$. Our aim is to find the minimum number of reversals that transforms π to an identity permutation $(+0, +1, +2, \ldots, +n)$. The minimum number of reversals required to transform π to $(+0, +1, +2, \ldots, +n)$ is called the reversal distance, denoted by $d(\pi)$. Figure 9.8 gives an example of sorting a signed permutation.

The problem of sorting a signed permutation was introduced by Sankoff et al. in 1992 [258]. Hannenhalli and Pevzner [135] gave the first polynomial time algorithm for this problem. The current best result was proposed by Tannier, Bergeron, and Sagot [283], which runs in $O(n^{3/2}\sqrt{\log n})$ time. Figure 9.9 summarizes the current known results.

9.5.1   Upper Bound on Signed Reversal Distance

A simple way to transform π to an identity permutation $(0, 1, 2, \ldots, n)$ is as follows. Disregarding the sign, we can create a correct sequence by n reversals (see Section 9.4.1). Then, the signs of the elements can be corrected by at most n sign flips (reversals of length 1). This simple algorithm for sorting the signed permutation implies $d(\pi) \leq 2n$.

Sankoff et al. (1992) [258]	Introduce the problem
Hannenhalli and Pevzner (1995) [135]	$O(n^4)$ time
Berman and Hannenhalli (1996) [26]	$O(n^2\alpha(n))$ time
Kaplan, Shamir, and Tarjan (1999) [168]	$O(n^2)$ time
Bergeron (2005) [24]	$O(n^3)$ time ($O(n^2)$ time on a vector-machine)
Tannier, Bergeron, and Sagot (2007) [283]	$O(n^{3/2}\sqrt{logn})$ time
Bader, Moret, and Yan (2001) [13]	$O(n)$ time (reversal distance only)
Bergeron, Mixtacki, and Stoye (2004) [25]	$O(n)$ time (reversal distance only)

FIGURE 9.9: Results related to sorting signed permutation. In the table, α is the inverse Ackerman's function.

Can we get a better upper bound for $d(\pi)$? Before we answer this question, we first discuss the Burnt Pancake Problem, which was introduced by Gates[1] and Papadimitriou [117]. This problem is defined as follows.

> A waiter has a stack of n pancakes. Here one side of each pancake is burnt. The waiter wants to sort the pancakes in order by size and put the burnt side down. Having only one free hand, the only available operation is to lift a top portion of the stack, invert it, and replace it. The Burnt Pancake Problem asks the number of flips needed.

Heydari and Sudborough [144] showed that the number of flips to sort the pancakes is at most $3(n + 1)/2$. Note that the stack of n pancakes can be modeled as a signed permutation. Each burnt pancake can be modeled as a signed integer and the ordering of burnt pancakes can be modeled as a permutation of signed integers. A flip can be modeled as a prefix reversal. The burnt pancake problem is thus equivalent to sorting a signed permutation by prefix reversals. Hence, the reversal distance for sorting a signed permutation by prefix reversal is at most $3(n + 1)/2$. Since prefix reversal is a kind of reversal, we conclude that, given a signed permutation π,

$$d(\pi) \le \frac{3}{2}(n + 1).$$

9.5.2 Elementary Intervals, Cycles, and Components

Consider a signed permutation $\pi = (\pi_0, \pi_1, \ldots, \pi_n)$ that begins with $\pi_0 = 0$ and ends with $\pi_n = n$. The rest of this chapter describes how to compute $d(\pi)$. First, we present some basic properties of π.

---

[1]This is the only research paper published by Bill Gates when he was studying at Harvard.

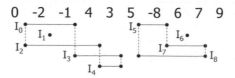

FIGURE 9.10:   Elementary intervals for $\pi = (0, -2, -1, 4, 3, 5, -8, 6, 7, 9)$

9.5.2.1   Elementary Intervals and Cycles

We introduce a point between π_i and π_{i+1} for every $0 \leq i < n$. A point is called a breakpoint if (π_i, π_{i+1}) does not equal either $(k, k+1)$ or $(-(k+1), -k)$ for some k. For example, in Figure 9.10, there are nine points. Apart from the two points between $(-2, -1)$ and $(6, 7)$, all other points are breakpoints.

For any integer $k \in \{0, 1, \ldots, n-1\}$, we define the elementary interval I_k as the interval whose two endpoints are p and p' where:

- p is the right point of k if the sign of k is positive; otherwise, p is its left point.

- p' is the left point of $k + 1$ if the sign of $k + 1$ is positive; otherwise p' is its right point.

Note that every point meets exactly two elementary intervals. Hence, the elementary intervals form disjoint cycles. For the example shown in Figure 9.10, there are four cycles where two cycles are isolated intervals containing no element. For another example, the identity permutation has n cycles where all of them are isolated intervals.

An elementary interval I_k is oriented if the signs of k and $k+1$ are different; otherwise, it is unoriented. For example, in Figure 9.10, the intervals I_0, I_2, I_7, and I_8 are oriented. Below lemma states that the reversal of an oriented interval reduces the number of breakpoints and increases the number of cycles.

LEMMA 9.3
Reversing an oriented interval reduces the number of breakpoints and increases the number of cycles by one. The new cycle is an isolated interval (interval contains no element).

PROOF   Figure 9.11 shows the effect of reversing an oriented interval I_k. After the reversal, the number of breakpoints is reduced by 1.   ☐

The above lemma states the effect of reversing an oriented interval. If we reverse any interval, below lemma states the effect.

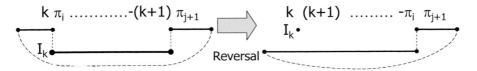

FIGURE 9.11: The effect of oriented reversal.

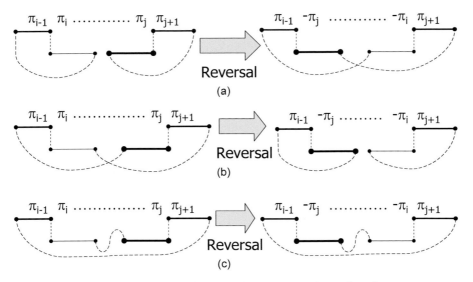

FIGURE 9.12: The effect of reversal on the number of cycles.

LEMMA 9.4

Reversing any interval in π modifies the number of cycles by $+1$, 0, or -1.

PROOF Suppose we reverse $(\pi_i, \ldots, \pi_j)$. Let v be the point between π_{i-1} and π_i and v' be the point between π_j and π_{j+1} The reversal will only affect the cycles passing through v and v'. There are two cases.

- Case 1: Two distinct cycles passing through v and v'. In this case, we will merge the two cycles. Hence, the number of cycles is reduced by 1 (see Figure 9.12(a)).

- Case 2: One cycle passing through v and v'. In this case, we will either maintain one cycle or break the cycle into two. Hence, the number of cycles is either increased by 1 (see Figure 9.12(b)) or does not change (see Figure 9.12(c)).

$\Box$

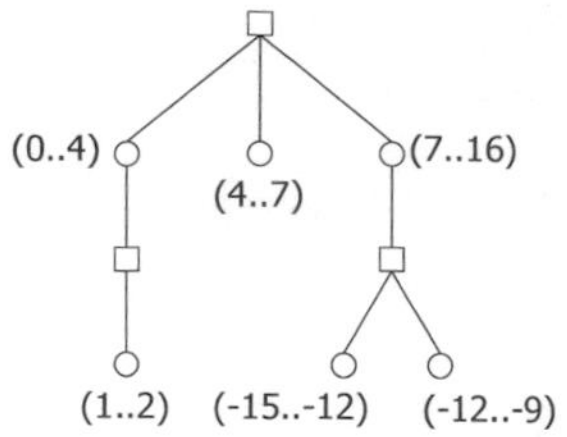

FIGURE 9.13:   Given $\pi = (0, -3, 1, 2, 4, 6, 5, 7, -15, -13, -14, -12, -10, -11, -9, 8, 16)$, the figure shows the tree T_π, where each round node represents a component and each square node represents a maximal chain.

Based on the previous lemma, we can give a lower bound for the reversal distance of π.

LEMMA 9.5
Given a signed permutation π of $\{0, 1, \ldots, n\}$, $d(\pi) \geq n - c$ where c is the number of cycles in π.

PROOF   Note that the identity permutation $(0, 1, 2, \ldots, n)$ is the only permutation which has n cycles. Since each reversal increases the number of cycles by at most 1 (see Lemma 9.4), we need at least $n - c$ reversals to generate the identity permutation.   ▯

9.5.2.2   Components

For any $i < j$, an interval in π which is called a component if

1. the interval either (1) starts from i and ends at j or (2) starts from $-j$ and ends at $-i$.

2. the interval contains all numbers between i and j; and

3. the interval is not equal to the union of two or more components.

We denote a component starting from i and ending at j as $(i..j)$ and a component starting from $-j$ and ending at $-i$ as $(-j.. - i)$. For example, the permutation π in Figure 9.10 has four components: $(0..5)$, $(5..9)$, $(-2.. - 1)$, and $(6..7)$. Another example permutation in Figure 9.13 has six components: $(0..4)$, $(4..7)$, $(7..16)$ $(1..2)$, $(-15.. - 12)$, and $(-12.. - 9)$.

Below, we describe the properties of components.

LEMMA 9.6
Two different components of a permutation are either disjointed, nested with different endpoints, or overlapping on one element.

PROOF   Consider two components A and B. They can be either disjoint, nested, or overlapping.

If A and B are overlapping but not nested, we claim that they can only overlap on one element. Assume A and B overlap on more than one element. Without loss of generality, suppose A is in the form of $(i..i+j+p)$ and B is in the form of $(i+j..i+j+p+q)$. By definition, $(i+j..i+j+p)$ consists of numbers $\{i+j, \ldots, i+j+p\}$. Hence, $(i+j..i+j+p)$ is a component. Also, $(i..i+j)$ consists of numbers $\{i, \ldots, i+j\}$. Since A is the union of $(i..i+j)$ and $(i+j..i+j+p)$, this contradicts with the fact that A is a component.

If A and B are nested, we claim that they cannot share endpoints. Assume A is nested inside B and shares one endpoint. Without loss of generality, suppose A is in the form of $(i..i+j)$ and B is in the form of $(i..i+j+k)$. By definition, both $(i..i+j)$ and $(i+j..i+j+k)$ are components. Then, B is not a valid component. We arrive at a contradiction. ☐

For example, in Figure 9.13, $(-2.. -1)$ and $(5..9)$ are disjoint, $(0..5)$ and $(5..9)$ overlap on one element, and $(6..7)$ is nested within $(5..9)$.

When two components overlap on one element, they are said to be linked. Successive linked components form a chain. A maximal chain is a chain that cannot be extended. (It may consist of a single component.) The relationship among components can be represented as a tree T_π as follows. (See Figure 9.13 for an example.)

- Each component is represented as a round node

- Each maximal chain is represented as a square node whose components are ordered children of it.

- A maximal chain is a child of the smallest component that contains this maximal chain.

Any point in π is said to belong to the component C if C is the smallest component which contains the point. For example, the point between 6 and 7 in Figure 9.10 belongs to the component $(6..7)$ but does not belong to $(5..9)$. Any component satisfies the following two properties.

LEMMA 9.7
For every elementary interval, its two endpoints belong to the same component.

PROOF   Consider any elementary interval I_k and any component C in the form either $(i..j)$ or $(-j.. -i)$ where $i < j$. If $k < i$ or $k > j$, both endpoints of I_k are outside of C. If $i \leq k < j$, by definition, both endpoints of I_k are contained in C. The lemma follows. ☐

COROLLARY 9.1

For any cycle, the endpoints of the elementary intervals on the cycle belong to the same component.

PROOF   Suppose there exist two adjacent elementary intervals I and I' on a cycle, which belong to two different components. Assume the shared endpoint of I and I' belongs to a component C. By Lemma 9.7, the other two endpoints of I and I' also belong to the component C. We arrive at a contradiction.   ☐

A component is called unoriented if it contains a breakpoint but does not contain any oriented interval; otherwise, it is oriented. For Figure 9.10, there is no unoriented component. For another example permutation in Figure 9.13, $(0..4)$, $(7..16)$, and $(1..2)$ are oriented while $(4..7)$, $(-15..-12)$, and $(-12..-9)$ are unoriented.

9.5.3   The Hannenhalli-Pevzner Theorem

Consider a signed permutation π of $\{0, 1, \ldots, n\}$ with c cycles. This section describes a polynomial time algorithm to generate the sequence of reversals which transforms π to the identity permutation.

9.5.3.1   Sorting a signed permutation when there is no unoriented component

Assume the signed permutation π has no unoriented component. This section describes the Bergeron's basic algorithm for sorting π.

We define the score $Score(I)$ of an oriented interval I in a permutation π to be the number of oriented intervals in the resulting permutation $\pi \cdot I$. The algorithm is built on the following theorem.

THEOREM 9.1

Reversing the oriented interval I of maximal score does not create new unoriented components.

PROOF   Note that once we perform the reversal for I, the signs of the elements within the interval I will change. For any elementary interval I_k, the orientation of I_k will be changed if either k or $k + 1$ (but not both) is within I. In this case, we say I_k intersects with I.

Suppose reversing the oriented interval I introduces a new unoriented component C. Then, there exists an oriented interval I', which intersects with I, belongs to C.

Let T be the total number of oriented intervals in π. We denote U and O be the number of oriented and unoriented intervals, respectively, in π

which intersect with I. We denote U' and O' be the number of oriented and unoriented intervals, respectively, in π which intersect with I'. Then, $Score(I) = T + U - O - 1$ and $Score(I') = T + U' - O' - 1$.

We claim that any unoriented interval in π, which intersects with I, also intersects with I'. If this is not the case, let J be an unoriented interval which intersects with I but does not intersect with I'. After reversing I, J becomes oriented and intersects with I'. Since J intersects with I', by Lemma 9.7, J belongs to C. This contradicts the assumption that C is unoriented. Thus, $U' \geq U$.

We also claim that any oriented interval that intersects with I' also intersects with I. Otherwise, let J be an oriented interval that intersects with I' but not with I. After reversing I, J remains oriented. Since J intersects with I', both J and I' belong to C by Lemma 9.7. This contradicts the assumption that C is unoriented. Hence, $O \geq O'$.

If $U = U'$ and $O = O'$, after reversing I, both I and I' become isolated intervals. This contradicts that C is unoriented. This means that $U' > U$ and $O > O'$. Thus, $Score(I) = T + U - O - 1 < T + U' - O' - 1 = Score(I')$. This contradicts the fact that I maximizes $Score(I)$. ☐

Based on the above theorem, Bergeron proposed the following algorithm.

> **Bergeron's algorithm**: While π has an oriented interval, we reverse π by the oriented interval I that maximizes $Score(I)$.

The corollary below shows the correctness of the algorithm.

COROLLARY 9.2
Given a signed permutation π with c cycles and no unoriented component, the reversal distance $d(\pi) = n - c$. Bergeron's algorithm can transform π to the identity permutation using $d(\pi)$ reversals.

PROOF   Theorem 9.1 guarantees that every iteration must have either some oriented intervals or only have isolated cycles. Hence, the algorithm reports a sequence of reversals which transforms π to the identity permutation. We claim that this list of reversals is optimal. Let c be the number of cycles in π. By Lemma 9.3, since every oriented reversal will generate one additional cycle, after $n - c$ oriented reversals, we get n cycles, which is an identity permutation. This implies $d(\pi) \leq n - c$. As $d(\pi) \geq n - c$ (see Lemma 9.5), the claim is proved. ☐

For time complexity, a trivial implementation of Bergeron's algorithm runs in $O(n^4)$ time. Moreover, the running time can be improved to $O(n^3)$ (see Exercise 4). Figure 9.14 illustrates the steps of Bergeron's algorithm. Note that selecting oriented interval with suboptimal score may stop unsuccessfully.

$\pi_1 = (+0, \underline{+3}, +1, +6, +5, -2, +4, +7)$
$score(+1, -2) = 2, score(+3, -2) = 4$
$\pi_2 = (+0, -5, -6, -1, \underline{-3}, -2, +4, +7)$
$score(+0, -1) = 2, score(-3, +4) = 4, score(-5, +4) = 2, score(-6, +7) = 2$
$\pi_3 = (+0, -5, -6, \underline{-1}, +2, +3, +4, +7)$
$score(+0, -1) = 0, score(-1, +2) = 2, score(-5, +4) = 2, score(-6, +7) = 2$
$\pi_4 = (+0, -5, \underline{-6, +1, +2, +3, +4}, +7)$
$score(-5, +4) = 2, score(-6, +7) = 2$
$\pi_5 = (+0, \underline{-5, -4, -3, -2, -1}, +6, +7)$
$score(+0, -1) = 0, score(-5, +6) = 0$
$\pi_6 = (+0, +1, +2, +3, +4, +5, +6, +7)$

FIGURE 9.14:   An example illustrates the steps of Bergeron's algorithm for transforming $\pi_1 = (+0, +1, +3, +1, +6, +5, -2, +4, +7)$ to the identity permutation.

For example, in π_3, we should not select $(+0, -1)$. Otherwise, we get zero oriented intervals and the algorithm stops.

9.5.3.2   Sorting a Signed Permutation When There Are Unoriented Components

This section describes the Hannenhalli-Pevzner Theorem. It also gives a polynomial time algorithm to optimally sort a signed permutation in general. The algorithm uses two particular types of reversals which transform the signed permutation π to another signed permutation which has no unoriented components. Finally, we obtain the identity permutation by applying Bergeron's algorithm stated in the previous section.

The two types of reversals are cut and merge operations. The cut operation is a reversal which reverses any elementary interval in a component. The merge operation is a reversal which reverses any interval whose two endpoints are in two different components.

LEMMA 9.8
For an unoriented component C, reversing any interval whose endpoints are within C will not split or create any cycle. Moreover, it will make C oriented. (Note that reversing an elementary interval in C, which is a cut operation, is a special case of this lemma.)

PROOF   Assume we reverse the interval $(a..b)$ where both the left point of a and the right point of b are endpoints within C. Furthermore, we assume the left point of a and the right point of b are points in the same cycle. Otherwise, the reversal must merge two cycles (see Figure 9.12(a)).

As C is unoriented, all elements in C have the same sign. Without loss of generality, suppose all elements are positive. Every elementary interval I_k within C connects the right point of k to the left point of $k+1$. Hence, there exists a path P_1 starting from the right endpoint of b, to the left endpoint of $b+1$ and finally via the right endpoint of $a-1$ and ending at the left endpoint of a. The remaining path of the cycle is denoted as P_2. Reversing $(a..b)$ is equivalent to swapping the connecting points between P_1 and P_2. Hence, we will not split or create any cycle.

Moreover, reversing the interval from $(a..b)$ to $(-b..-a)$ introduces oriented intervals, which makes C oriented. ▯

LEMMA 9.9

If a reversal has its two endpoints in different components A and B (that is a merge operation), then only the components on the path from A to B in T_π are affected. In particular,

1. *If a component C contains either A or B but not both, it will be destroyed after the reversal.*

2. *If the lowest common ancestor of A and B in T_π is a component C, if A or B is unoriented, then C becomes oriented after the reversal.*

3. *If the lowest common ancestor of A and B in T_π is a chain C, a new component D is created. If A or B is unoriented, D will be oriented.*

PROOF For (1), after the reversal, one of the bounding elements of C will change sign, but not the other. Hence, the component is destroyed.

For (2), suppose A is unoriented. The reversal changes the sign of one bounding element of A and introduces one oriented interval. This oriented interval belongs to C since (1) implies that the component A and all components in the path between A and C excluding C are destroyed. Hence, C becomes oriented.

For (3), suppose A and B are included in the components A' and B' in the chain C. Without loss of generality, assume $A' = (a..a')$ precedes $B' = (b..b')$. After the reversal, the components A' and B' will be destroyed and a new component $D = (a..b')$ will be created. Similar to (2), if A is unoriented, the reversal changes the sign of one bounding element of A, which introduces one oriented interval in D. Hence, D becomes oriented. ▯

The above lemma implies that merging two unoriented components A and B destroys or orients components on the path between A and B in T_π, without creating new unoriented components.

Before proving the Hannenhalli-Pevzner Theorem, we first define the cover of T_π. A cover $\mathcal{C}$ of T_π is a collection of paths joining all the unoriented components in π. A path is called a long path if it contains two or more unoriented components; otherwise, it is called a short path. For example, for the tree T_π in Figure 9.13, there are three unoriented components: $(4..7)$, $(-15.. - 12)$, and $(-12.. - 9)$. We can cover these unoriented components using a long path between $(4..7)$ and $(-15.. - 12)$ and a short path containing $(-12.. - 9)$.

We can orient the unoriented components on both short and long paths. By Lemma 9.8, we can orient the unoriented component on a short path by a cut operation. By Lemma 9.9, we can apply the merge operation on the two outermost unoriented components on a long path so that we can destroy or orient all the unoriented components on the long path.

The cost of a cover is defined as the sum of the costs of its paths, where the costs of a short path and a long path are 1 and 2, respectively. An optimal cover is a cover of minimal cost.

THEOREM 9.2 (Hannenhalli-Pevzner Theorem)
Given a permutation π of $\{0, 1, \ldots, n\}$. Suppose it has c cycles and the associated optimal cover of T_π has cost t. Then, $d(\pi) = n - c + t$.

PROOF　We claim that $d(\pi) \leq n - c + t$. Suppose the optimal cover of T_π has m long paths and q short paths, that is, $t = 2m + q$. By applying q cut operations to the q short paths and m merge operations to the m long paths, we remove all the unoriented components. Also, each merge operation reduces the number of cycles by one. Hence, the resulting permutation π' has $c - m$ cycles and has no unoriented component. By Corollary 9.2, $d(\pi') = n - (c - m)$ and we have $d(\pi) \leq d(\pi') + m + q = n - c + 2m + q = n - c + t$.

We also claim that $d(\pi) \geq n - c + t$. Let d be the optimal reversal distance. $d = s + m + q$ where s is the number of reversals that split cycles, m is the number of reversals that merge cycles, and q is the number of reversals that do not change the number of cycles. Since the identity permutation has n cycles, we have $c + s - m = n$. Thus, $d = n - c + 2m + q$. Some of the m reversals which merges unoriented components correspond to a merge operation on a path in T_π. Some of the q reversals which apply on some unoriented components are cut operations. Those operations should be applied on all unoriented components. Otherwise, we cannot transform π to an identity permutation. Hence $t \leq 2m + q$ and we have $d \geq n - c + t$.　□

Figure 9.15 shows the algorithm for computing the signed reversal distance. The algorithm for generating the sequence of reversals is shown in Figure 9.16. Below lemma states their time complexity.

LEMMA 9.10
Algorithms Signed_Reversal_Distance and Sort_Signed_Reversal run in $O(n)$

Algorithm Sort_Signed_Reversal

1: We identify the optimal cover $\mathcal{C}$ of the tree T_π as in Algorithm Signed_Reversal_Distance.
2: For each long path in the cover $\mathcal{C}$, identify the leftmost and the rightmost unoriented components and merge them.
3: For each short path in the cover $\mathcal{C}$, cut the unoriented component on the short path.
4: Run Bergeron_basic.

FIGURE 9.15: Algorithm Sort_Signed_Reversal.

Algorithm Signed_Reversal_Distance

1: Find all components in π.
2: Construct T_π.
3: Find the optimal cover $\mathcal{C}$ of T_π and determine t.
4: Find the number of cycles, c.
5: Report the signed reversal distance, which is $n - c + t$.

FIGURE 9.16: Algorithm Signed_Reversal_Distance.

time and $O(n^3)$ time, respectively.

PROOF   (Sketch) For the Signed_Reversal_Distance algorithm, we can find all components in π and construct T_π in $O(n)$ time (see Exercise 5). The optimal cover can also be found in $O(|T_\pi|)$ time, which is $O(n)$ (see Exercise 6).

Similar to the Signed_Reversal_Distance algorithm, the Sort_Signed_Reversal algorithm takes $O(n)$ time to construct the optimal cover of T_π. Then, each reversal corresponding to the long and short paths in the cover $\mathcal{C}$ can be performed in $O(n)$ time. In total, $O(n^2)$ time is needed to perform all these reversals. Finally, Bergeron's algorithm takes $O(n^3)$ time.   ▯

9.6   Further Reading

Since reversal is the most common rearrangement, this chapter focused on the problem of sorting by reversal in monochromosomal genomes. Biologists observed that the chance of reversing a segment depends on its length. Hence, people assign weight to each reversal and reformulate the problem to compute a reversal distance which minimizes the total weight.

Many genomes have more than one chromosome. The three most common multi-chromosomal rearrangement events are translocations, fusions, and fissions. Hannenhalli and Pevzner [134] studied the problem of sorting the permutation by these four operations. They showed that this problem is solvable within polynomial time by transforming the multi-chromosome rearrangement problem to a mono-chromosome rearrangement problem. Their algorithm takes $O(n^4)$ time to find a sequence of rearrangements which minimizes these four events.

In addition to reversal, people have also studied other rearrangement events like duplication, transposition, insertion, and deletion.

9.7 Exercises

1. Consider the permutation $\pi = (1, 8, 9, 4, 3, 2, 5, 6, 7, 11, 10, 12)$. Using the 2-approximation algorithm, what is the number of unsigned reversals required to transform it to an identity permutation? Is this optimal?

2. For the permutation $\pi = (1, 6, 5, 7, 8, 4, 2, 3, 9)$, by the 2-approximation algorithm, what is the number of unsigned reversals required to transform it to an identity permutation? Is this optimal?

3. Consider the signed permutation $\pi = (0, -4, 6, 3, -5, 1, 2, 7, 8)$.

   (a) What is the set of elementary intervals? Which intervals are oriented?

   (b) What are the cycles?

   (c) What are the components? Are the components oriented?

   (d) Can you compute the optimal sequence of reversals?

4. Present an $O(n^3)$-time implementation of Bergeron's algorithm.

5. For a signed permutation π, show that

   (a) The number of cycles in π can be found in $O(n)$ time.

   (b) All components in π can be found in $O(n)$ time.

   (c) T_π can be constructed in $O(n)$ time.

6. For a signed permutation π, give an $O(n)$-time algorithm to compute the optimal cover of T_π.

7. Consider the following signed permutation, which is an unoriented component.
$$(0, 4, 1, 3, 5, 2, 6)$$

(a) How many cycles does it have?

(b) Can you reverse $(1, 3, 5, 2)$? Does it create a new component? Is it oriented or not? How many cycles does it have?

(c) Can you reverse (4)? Does it create a new component? Is it oriented or not? How many cycles does it have?

(This question demonstrates the effect of reversing any interval in an unoriented component (see Lemma 9.8).)

8. Can you demonstrate the detailed steps of how the algorithm sorts the following signed permutation?

$$(0, 16, 1, 4, 5, 2, 3, 6, -11, -9, -10, -8, 7, 12, 14, 13, 15, 17)$$

9. Can we simulate a transposition by reversals? How? Is there any difference in the simulation process for an unsigned or a signed permutation?

10. Given an unsigned permutation π, can you give an algorithm which checks if k reversals is enough to sort π? What is the running time?

11. Consider any unsigned permutation π with b breakpoints. The transposition distance is the minimum number of transpositions which can convert π to the identity permutation. Can you give a tight lower bound of transposition distance in terms of b?

Chapter 10

Motif Finding

10.1 Introduction

Although every cell of an organism has exactly the same genome, different cells express a different set of genes to form different types of tissues. How does a cell know what genes are required and when they should express?

In the 1960s, Jacob and Monod [160]uncovered this mystery. They showed that bacteria make an inhibitor to keep β-galactosidase production turned off. Without the inhibitor, the β-galactosidase production will turn on. This is the first evidence that certain types of genes provide instruction to control the expression of other genes.

The process of controlling the expression of genes is known as gene regulation. Gene regulation dictates when, where (in what tissue(s)), and how much of a particular protein is produced. This decides the development of cells and their responses to external stimuli. The most direct control mechanism is transcription regulation, which controls whether the transcription process of a gene should be initiated. In eukaryotic cells, RNA-polymerase II is responsible for the transcription process. However, it is incapable of initiating transcription on its own. It does so with the assistance of a number of DNA-binding proteins called transcription factors (TFs). TFs bind the DNA sequence and interact to form a pre-initiation complex (PIC). RNA-polymerase II is recruited in the PIC, and then the transcription begins (see Figure 10.1). The crucial point of the regulation mechanism is the binding of TFs to DNA. Disruptions in gene regulation are often linked to a failure of TF binding, either due to mutation of the DNA binding site, or due to mutation of the TF itself.

The DNA sites that are bound by TFs are called transcription factor binding sites (TFBS). People observed that the binding sites for the same TF contain similar DNA sequences of length 5–20 bp. Such a recurring DNA pattern is known as a motif. For example, one of the earliest discovered TFs is TATA-box binding protein (TBP) and its motif is TATAAA.

Although the binding sites are sequence specific, interestingly, not all bases are found to be equally important for effective binding. While some base positions can be substituted without affecting the affinity of the binding, base substitution in other positions can completely obliterate the binding. For ex-

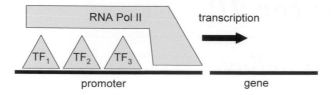

FIGURE 10.1:   The TFs bind the DNA sequence to form a pre-initiation complex (PIC), which recruits RNA-polymerase II to initiate the transcription. Missing of the binding of the TFs may prevent the recruitment of RNA-polymerase II, which stops the transcription.

ample, for TATAAA, substituting 'A' at the fifth position to 'T' may not affect the binding affinity by a lot; however, substituting 'A' at the last position to 'T' may severely affect the binding.

As identifying the binding motifs of TFs is important for understanding the regulation mechanism, this chapter focuses on methods of discovering a DNA motif. We assume a set S of regulatory sequences is given. Our task is to find a DNA motif, which is the recurring pattern in S.

10.2   Identifying Binding Regions of TFs

TFs bind specific regions in the genome, which are known as cis-regulatory sequences. In general, there are two types of cis-regulatory sequences: promoters and enhancers/repressors.

Promoters usually refer to the DNA regions immediately upstream of the transcription start sites (TSSs) of genes (see Figure 10.2 for an example). The promoter contains binding sites for TFs that directly interact with RNA polymerase II to facilitate transcriptional initiation. The rest of the promoter sequence is nonfunctional and meant to separate the binding sites at an appropriate distance.

Enhancers/repressors are also known as cis-regulatory modules (CRMs). They are cis-regulatory sequences which enhance or repress the transcription activity of genes. They can stimulate the transcriptional initiation of one or more genes located at a considerable distance (see Figure 10.2 for an example). CRMs are often involved in inducing tissue-specific or temporal expression of genes. A CRM may be 100–1000 bp in length and contain several closely arranged TF binding sites (TFBSs). Thus a CRM resembles the promoter in its TFBS composition and the mechanism by which it functions. However, a CRM typically contains a higher density of TFBSs than the promoter, has repetitive TFBSs, and involves a greater level of cooperative or composite

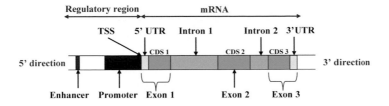

FIGURE 10.2: The locations of gene coding and noncoding regions and the promoter in a DNA strand. The promoter region is present surrounding the start of (and mostly upstream of) the transcription region. Other elements such as enhancers may be present far from the transcription start site. (See color insert after page 270.)

interactions among the TFs. The activity of a CRM is interesting as it can control gene expression from any location or strand orientation. The present understanding of its mechanism is that TFs bound at the CRM interact directly with TFs bound to the promoter sites through the coiling or looping of DNA.

The regions bound by the same TF can be identified in a number of ways. Below, we briefly describe three different ways. The three methods extract regulatory sequences from (1) promoters of homologous genes in different species, (2) promoters of co-expressed genes, and (3) chromatin immunoprecipitation (ChIP) fragments.

Promoters of homologous genes in different species: Homologous genes in different species have similar functions and properties. Hence, promoters of those homologous genes are expected to contain binding sites of the same transcription factors.

Promoters of co-expressed genes: Transcription factors initiate the expression of genes. Co-expressed genes are likely to be regulated by the same transcription factors. Hence, we expect promoter regions of the co-expressed genes to contain binding sites of the same transcription factors.

Chromatin immunoprecipitation (ChIP) fragments: Chromatin immunoprecipitation (ChIP) is a procedure which determines if a given TF binds to a specific DNA sequence in vivo (i.e., within a living cell). Figure 10.3 illustrates the ChIP process. The first step in a ChIP experiment is to cross-link the TFs with the genome. Then, sonication is applied to break the cells and to shear the genome into small fragments. Next, using a TF specific antibody, genomic regions bound by such TF are pulled down. The genomic regions extracted are expected to be bound by the targeted TF, i.e., those genomic regions are expected to contain the TF binding sites. Using microarray (ChIP-chip) or sequencing (ChIP-seq), those genomic regions are detected.

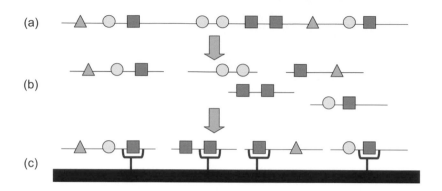

FIGURE 10.3:   The ChIP process: (a) A genome with binding of different transcription factors (TFs). (b) After shearing, we partition the genome into many small fragments. (c) Using the antibody for the square TF, we extract all DNA segments that are bound by the square TF.

Any of the above three methods extracts genomic sequences containing binding sites which allow the same TF to bind on them. Hence, those sequences are expected to contain a similar pattern which is the binding motif of the TF. The rest of this chapter presents various methods for predicting such a motif.

10.3   Motif Model

Suppose we are given a set of TF binding sites of a particular TF. The DNA sequences at the binding sites usually share some common binding pattern. Such a binding pattern is known as the TF motif.

Although every position of the TF motif has a certain nucleotide preference, the motif is not rigid. The ambiguity of TF binding appears to be intentional in nature as a way to control gene expression. Variable affinity of the TF to different DNA sites causes a kinetic equilibrium between TF concentration and occupancy (i.e., which binding sites are actually occupied by the TF in-vivo). This provides a mechanism to control the transcription rate of the genes.

This section describes how to represent the TF motif so that it can cap-

Code	Definition	Mnemonic
A	Adenine	A
C	Cytosine	C
G	Guanine	G
T	Thymine	T
R	A, G	puRine
Y	C, T	pYrimidine
K	G, T	Keto
M	A, C	aMino
S	G, C	Strong
W	A, T	Weak
B	C, G, T	Not A
D	A, G, T	Not C
H	A, C, T	Not G
V	A, C, G	Not T
N	A, G, C, T	aNy

FIGURE 10.4:   The IUB nucleotide codes.

ture variability of the TF binding sites. Two representations are proposed: consensus motif and positional weight matrix (PWM).

A consensus motif represents the common features of the effective binding site sequence. Given a set of aligned binding sites, the consensus motif represents each position by the most frequent base; if an equal number of different bases are present, the position is represented by a set of those bases. For short-hand notation, the set is represented by the IUB codes (codes of the International Union of Biochemistry, see Figure 10.4). The TF has a high affinity for sequences that match this consensus pattern, and a relatively low affinity for sequences different from it. Figure 10.5 shows an example of converting a set of aligned binding sites to a consensus sequence.

A numerical way of characterizing the binding preferences of a TF is the positional weight matrix (PWM) (also called the position-specific scoring matrix (PSSM)), which shows the degree of ambiguity in the nucleotide at each binding site position. The PWM represents each position by the frequency of the 4 bases in the aligned binding sites. Formally, a PWM is a $4 \times \ell$ matrix where ℓ is the length of the PWM. The 4 entries in every column of the PWM correspond to the frequency of A, C, G, and T. The sum of the 4 entries in each column equals 1. For the example PWM in Figure 10.5, its length is 6 and the sum of frequencies of every column equals 1.

A PWM can be visualized graphically using a sequence logo. Figure 10.6 shows the sequence logo for the PWM in Figure 10.5. The height of each character indicates the relative importance of the character in that position.

For the consensus sequence and PWM, nucleotides at different positions in the binding sites are assumed to be independent. In fact, dependencies exist. We can enrich the motif model by incorporating those dependencies.

TTGACA
TCGACA   *Consensus*
TTGACA   *Pattern*
TTGAAA   ──────────→   **TTGACA**
ATGACA
TTGACA   *Positional*
GTGACA   *Weight*
TTGACT   *Matrix (PWM)*
TTGACC
TTGACA

nucleotide	alignment position					
	1	2	3	4	5	6
A	0.1	0	0	1	0.1	0.8
C	0	0.1	0	0	0.9	0.1
G	0.1	0	1	0	0	0
T	0.8	0.9	0	0	0	0.1

FIGURE 10.5: Transforming a set of aligned binding sites into a consensus motif or a PWM.

FIGURE 10.6: The sequence logo for the PWM in Figure 10.5. (See color insert after page 270.)

Zhou and Liu [319] incorporate pairwise dependencies into the motif model. Barash et al. [19] use a Bayesian network to model a motif, allowing arbitrary dependencies between positions. Hong et al. [153] represent the motif using boosted classifiers. Xing et al. [313] represent the motif using a hidden Markov Dirichlet multinomial model.

Although incorporating position dependencies improves the motif model, it is also easier to over-fit the model. Benos et al. [23] observed that in reality, the PWM and the consensus sequence usually provide a good approximation. Hence, this chapter assumes motifs are represented as the PWM and the consensus motif.

Due to extensive research, many TF motifs are known. There exist biological databases of known TF motifs, for example, the TRANSFAC database [309] and the JASPAR database [255]. The databases usually represent the TF motifs using the PWM and the consensus motif model.

10.4   The Motif Finding Problem

The motif finding problem is formally defined as follows:

Input: A set of DNA sequences that are possibly bound by the same transcription factor. Such sequences can be obtained using experimental techniques described in Section 10.2.

Aim: We aim to search for the common binding pattern called the motif, which is represented by the consensus sequence or the PWM.

For example, Figure 10.7 shows a set of sequences. The consensus motif is *TTGACA*.

```
GCACGCGGTATCGTTAGCTTGACAATGAAGACCCCCGCTCGACAGGAAT
GCATACTTTGACACTGACTTCGCTTCTTTAATGTTTAATGAAACATGCG
CCCTCTGGAAATTAGTGCGGTCTCACAACCCGAGGAATGACCAAAGTTG
GTATTGAAAGTAAGTAGATGGTGATCCCCATGACACCAAAGATGCTAAG
CAACGCTCAGGCAACGTTGACAGGTGACACGTTGACTGCGGCCTCCTGC
GTCTCTTGACCGCTTAATCCTAAAGGGAGCTATTAGTATCCGCACATGT
GAACAGGAGCGCGAGAAAGCAATTGAAGCGAAGTTGACACCTAATAACT
```

FIGURE 10.7: Given a set of sequences, the motif finding problem is to find a prominent pattern. In the above example, the most prominent pattern *TTGACA* occurs 10 times.

There are three main computational approaches to finding a motif. The first approach is to scan the input sequences for literature known motifs. This approach is described in Section 10.5. The second approach is to discover motifs by finding over-represented patterns in the input sequences. This approach allows us to discover novel motifs. There are two main methods in this approach, namely, statistical and combinatorial methods. Those methods are described in Sections 10.6–10.10. The third approach is to find a motif with the help of additional information. Those methods are described in Section 10.11.

10.5   Scanning for Known Motifs

Given a set of input sequences and the PWM motif Θ of a transcription factor, the motif scanning problem aims to identify the occurrences of Θ in the input sequences. If Θ is over-represented in the input sequences, the TF corresponding to Θ is predicted to bind to the input sequences.

We first define the PWM score, which determines if a length-ℓ DNA pattern P looks similar to the length-ℓ PWM Θ. For $i = 1, 2, \ldots, \ell$, Pr(the position i

of the motif is $P[i]$) equals $\Theta[P[i], i]$. Hence, we define the likelihood score of Θ as $\Pi_{i=1}^{\ell}\Theta[P[i], i]$. For example, for the PWM in Figure 10.5 and a sequence $P = TTGACA$, the likelihood score is $(0.8 \times 0.9 \times 1 \times 1 \times 0.9 \times 0.8)$.

Consider a sequence S of length n, a PWM Θ of length ℓ, and a threshold τ. The motif scanning problem is to scan S to find all length-ℓ substrings P of S such that the likelihood score of P is bigger than τ.

A standard solution is to scan the sequence S and to compute the likelihood score of all length-ℓ substrings P one by one. This method takes $O(n\ell)$ time.

Here, we describe how to use the suffix tree (see Chapter 3) to speed up the scanning process [207]. Observe that if two length-ℓ substrings P and Q in S share the same length-ℓ' prefix (that is, $P[1..\ell'] = Q[1..\ell']$), then the product of the frequencies of the first ℓ' characters are the same, that is, $\Pi_{i=1}^{\ell'}\Theta[P[i], i] = \Pi_{i=1}^{\ell'}\Theta[Q[i], i]$. Suffix tree allows us to avoid the redundant computation.

Let T be the suffix tree for the sequence S. For every node v in T whose path label is $v_1, \ldots, v = v_k$, we denote $D[v] = \Pi_{i=1}^{k}\Theta[v_i, i]$. Our aim is to compute $D[v]$ for all nodes v in T of depth ℓ. Note that we have the recursive formula $D[v] = D[u] \cdot \Theta[v, k]$ where u is the parent of v and v is of depth k. Hence, by traversing T in the top-down order and applying the recursive formula for $D[v]$, we can compute $D[v]$ for every node v of depth ℓ. Finally, for every node v of depth ℓ where $D[v] > \tau$, the corresponding path label of v is the occurrence of the motif Θ.

The time complexity includes the construction time for the suffix tree T, which is $O(n)$, plus $O(c\ell)$ where c is the number of nodes in T of depth at most ℓ. Note that c is upper bounded by $\min\{n, 4^\ell\}$.

We can further speed up the process by pruning. For any node v of depth smaller than ℓ, when $D[v] < \tau$, we don't need to traverse down the tree since $D[u]$ must be smaller than the threshold τ for all descendants u of v in T.

As a final note, since the likelihood score is usually small, people use the log likelihood score instead in practice. Precisely, for the PWM motif Θ and a length-ℓ pattern P, the log likelihood score is $-\log(\Pi_{i=1}^{\ell}\Theta[P[i], i]) = -\sum_{i=1}^{\ell}\log\Theta[P[i], i]$.

10.6 Statistical Approaches

If we want to discover novel motifs, we cannot use the previous method of scanning the known motifs. Instead, we identify motifs by discovering over-represented patterns. There are two classes of methods depending on whether we would like to predict the PWM or consensus motifs. This section studies statistical approaches which predict PWM motifs. The next section describes combinatorial methods for predicting consensus motifs.

Θ	1	2	3	4	5
A	0.2	0.8	0.1	0.7	0.8
C	0	0.1	0.2	0	0.1
G	0.1	0.1	0.1	0.2	0.1
T	0.7	0	0.6	0.1	0

Θ_0	
A	0.25
C	0.25
G	0.25
T	0.25

FIGURE 10.8: An example of a motif model Θ and a background model Θ_0.

Methods using statistical approaches include Consensus [143], Gibbs motif sampler [184], AlignACE [157], ANN-Spec [311], BioProspector [193], Improbizer [12], MDScan [194], MEME [17], MotifSampler [286], NestedMICA [84], etc. Most of the methods apply statistical techniques such as Gibbs sampling and expectation maximization (EM) or their extensions. Below, we explain these two techniques by discussing Gibbs motif sampler and MEME.

10.6.1 Gibbs Motif Sampler

The Gibbs motif sampler [184] assumes that exactly one site per input sequence has a motif while the rest of the sequences are background. The motif is represented as a length-ℓ PWM Θ where $\Theta[x, i]$ is the frequency that the position i of the motif is the base x. The background is represented by a length-1 PWM Θ_0 where $\Theta_0[x]$ is the frequency of x in the background (non-motif region). Any length-ℓ pattern $X = x_1 x_2 \ldots x_\ell$ is said to be similar to the motif Θ if the likelihood score $Pr(X|\Theta) = \Pi_{i=1}^{\ell} \Theta[x_i, i]$ is high while the background score $Pr(X|\Theta_0) = \Pi_{i=1}^{\ell} \Theta_0(x_i)$ is low. In other words, X is similar to Θ if $\frac{Pr(X|\Theta)}{Pr(X|\Theta_0)}$ is bigger than 1. For example, consider Θ and Θ_0 in Figure 10.8. $Pr(TATAA|\Theta) = 0.7 \times 0.8 \times 0.6 \times 0.7 \times 0.8 = 0.18816$, $Pr(TATAA|\Theta_0) = 0.25^6 = 0.000244$, and $\frac{Pr(TATAA|\Theta)}{Pr(TATAA|\Theta_0)} = 771.15$. Since the value is much bigger than 1, $TATAA$ is higher similar to Θ.

Given a set of sequences $\mathcal{S} = \{S_1, S_2, \ldots, S_n\}$, the Gibbs sampler aims to find a length-ℓ motif model Θ and one binding site per sequence in S which maximize

$$\sum_{Z \in \mathcal{Z}} \log \frac{Pr(Z|\Theta)}{Pr(Z|\Theta_0)}$$

where $\mathcal{Z}$ is the set of binding sets and Θ_0 is the background model.

Unfortunately, the above maximization problem is NP-hard. Gibbs sampling is applied to solve this optimization problem. The Gibbs motif sampler uses a randomized approach to iteratively improve a motif. Initially, a set $\mathcal{Z}$ of length-ℓ words is extracted where $\mathcal{Z}$ contains one random length-ℓ word per sequence; then, a PWM motif Θ is generated by counting the frequency

---

Gibbs motif sampling algorithm

Require: A set of sequences $\mathcal{S} = \{S_1, S_2, \ldots, S_n\}$

Ensure: A PWM motif Θ of width ℓ.

 1: Randomly choose one random length-ℓ word per sequence and form $\mathcal{Z}$

 2: Create a PWM Θ from all patterns in $\mathcal{Z}$ and Θ_0 from all non-pattern positions

 3: **while** the motif Θ is not good enough **do**

 4:    Randomly choose a sequence S_i and delete the length-ℓ word on S_i from $\mathcal{Z}$

 5:    Create a PWM Θ from all patterns in $\mathcal{Z}$ and Θ_0 from all non-pattern positions

 6:    For every length-ℓ word X in sequence S_i, we define its weight to be $\frac{Pr(X|\Theta)}{Pr(X|\Theta_0)}$

 7:    We choose a length-ℓ word in S_i randomly according to the weight and include it into $\mathcal{Z}$. Then we generate a new motif Θ and a new background Θ_0.

 8: **end while**

FIGURE 10.9:   The Gibbs sampling algorithm.

of each base at each position in the set $\mathcal{Z}$. Similarly, the background Θ_0 is created by counting the frequencies of the bases over all non-pattern positions (i.e., all positions in $\mathcal{S}$ not belonging to $\mathcal{Z}$). Then the Gibbs sampler will randomly replace the words in $\mathcal{Z}$ to improve the motif Θ iteratively. Precisely, the algorithm randomly selects a sequence S_i and replaces the word on S_i in $\mathcal{Z}$ by another word randomly selected from S_i based on the weight $\frac{Pr(X|\Theta)}{Pr(X|\Theta_0)}$. The detail of the algorithm is described in Figure 10.9.

Now, we analyze the running time of the Gibbs motif sampling algorithm. Let m be the average length of the sequences in $\mathcal{S}$. Each iteration of the Gibbs motif sampling algorithm is dominated by Step 6, which takes $O(m\ell)$ time. The number of iterations is approximately linear to the number of sequences n and a factor T which is expected to grow with increasing m. For typical protein sequences, $T \approx 100$. Hence, the approximate running time of the Gibbs motif sampling algorithm is $O(nm\ell T)$.

As a final note, the Gibbs motif sampler gives different results for different runs. In practice, the motif is chosen from the best out of a number of runs. Also, there are various variants of the standard Gibbs sampling approach. These include AlignACE [250] and BioProspector [193].

10.6.2 MEME

Multiple EM for Motif Elicitation (MEME) [18] is a popular software for finding motifs. It is based on expectation maximization, which was originally proposed by Lawrence and Reilly [183]. MEME breaks the input sequences into a set of length-ℓ overlapping substrings $\mathcal{X} = \{X_1, \ldots, X_n\}$. It assumes $\mathcal{X}$ is a mixture of length-ℓ substrings where $100\lambda\%$ and $100(1 - \lambda)\%$ of them are generated from the motif model and the background model, respectively. Similar to Gibbs sampling, the motif model is represented by a length-ℓ PWM Θ while the background model is represented by a length-1 PWM Θ_0.

The aim of MEME is to find the motif model Θ, the background model Θ_0, and the proportion λ which best explain the input sequence. Precisely, let $Pr(X_i \mid \Theta, \Theta_0, \lambda)$ be the probability that X_i is generated by the model $(\Theta, \Theta_0, \lambda)$ and $Pr(\mathcal{X} \mid \Theta, \Theta_0, \lambda) = \Pi_{i=1}^{n} Pr(X_i \mid \Theta, \Theta_0, \lambda)$. Our aim is to find $(\Theta, \Theta_0, \lambda)$ which maximizes $Pr(\mathcal{X} \mid \Theta, \Theta_0, \lambda)$ or its log likelihood $\log Pr(\mathcal{X} \mid \Theta, \Theta_0, \lambda)$, which is,

$$\sum_{i=1}^{n} \log Pr(X_i \mid \Theta, \Theta_0, \lambda) = \sum_{i=1}^{n} (\log(\lambda Pr(X_i|\Theta) + (1 - \lambda)Pr(X_i|\Theta_0)))$$

The above problem, however, is NP-hard. MEME applies expectation maximization (EM) algorithm to refine $(\Theta, \Theta_0, \lambda)$ iteratively. The EM algorithm performs two steps: the expectation step and the maximization step.

Expectation step: Given the estimated $(\Theta, \Theta_0, \lambda)$ from the previous iteration, the expectation step determines, for every $X_i \in \mathcal{X}$, the chance that X_i is a motif, i.e., $p_i = Pr(X_i|\Theta, \Theta_0, \lambda) = \frac{\lambda Pr(X_i|\Theta)}{\lambda Pr(X_i|\Theta)+(1-\lambda)Pr(X_i|\Theta_0)}$.

Maximization step: Given p_i for all $X_i \in \mathcal{X}$, the maximization step updates $\Theta, \Theta_0, \lambda$. Precisely, $\lambda = \frac{1}{n} \sum_{i=1}^{n} p_i$. For any $x \in \{A, C, G, T\}$ and $j = 1, 2, \ldots, \ell$,

$$\Theta[x, j] = \frac{\sum\{p_i \mid X_i[j] = x, i = 1..n\}}{\sum_{i=1}^{n} p_i}$$

For any $x \in \{A, C, G, T\}$,

$$\Theta_0[x] = \frac{\sum\{(1 - p_i) \mid X_i[j] = x, i = 1..n, j = 1..\ell\}}{\ell \sum_{i=1}^{n}(1 - p_i)}$$

Figures 10.10 and 10.11 demonstrate the expectation step and the maximization step, respectively.

Figure 10.12 gives the algorithm of MEME. Steps 1 and 2 generate an initial motif and background. Then, it repeats the expectation step (Step 4) and the maximization step (Steps 5–7) to improve the motif until it cannot be improved or a certain number of iterations has been reached.

```
TGATATAACGATC
X₁=TGATA          p₁
X₂=  GATAT        p₂
X₃=   ATATA       p₃
X₄=    TATAA      p₄
X₅=     ATAAC     p₅
X₆=      TAACG    p₆
X₇=       AACGA   p₇
X₈=        ACGAT  p₈
X₉=         CGATC p₉
```

Θ	1	2	3	4	5
A	0.2	0.8	0.1	0.7	0.8
C	0	0.1	0.2	0	0.1
G	0.1	0.1	0.1	0.2	0.1
T	0.7	0	0.6	0.1	0

Θ_0	
A	0.25
C	0.25
G	0.25
T	0.25

$$p_1 = \frac{Pr(TGATA|\Theta)\lambda}{Pr(TGATA|\Theta)\lambda + Pr(TGATA|\Theta_0)(1-\lambda)}$$

$$= 0.00056\lambda \,/\, (0.00056\lambda + 0.25^5(1-\lambda))$$

FIGURE 10.10: Consider the sequence $TGATATAACGATC$ and $\ell = 5$. Then $\mathcal{X} = \{X_1 = TGATA, \ldots, X_9 = CGATC\}$. Using the estimated Θ and Θ_0 in Figure 10.8, the expectation step computes $p_i = Pr(X_i|\Theta, \Theta_0, \lambda)$. The right portion of the figure demonstrates the computation of p_1.

```
p₁ TGATA
p₂ GATAT
p₃ ATATA
p₄ TATAA
p₅ ATAAC
p₆ TAACG
p₇ AACGA
p₈ ACGAT
p₉ CGATC
```

$$\lambda = (\Sigma\, p_i)/9$$

$$\Theta[T,1] = \frac{p_1 + p_4 + p_6}{p_1 + p_2 + \ldots + p_9}$$

$$\Theta_0[C] = \frac{(1-p_5)+(1-p_6)+(1-p_7)+(1-p_8)+2(1-p_9)}{5((1-p_1)+(1-p_2)+\ldots+(1-p_9))}$$

FIGURE 10.11: Given p_i (for $i = 1, 2, \ldots, 9$) computed in Figure 10.10, this figure demonstrates the maximization step, that is, how to compute $\Theta, \Theta_0, \lambda$.

MEME algorithm

Require: A set of $\mathcal{X} = \{X_1, X_2, \ldots, X_n\}$ of length-ℓ sequences

Ensure: A PWM motif Θ of width ℓ.

1: Randomly choose a parameter $0 < \lambda < 1$ and let $\mathcal{Z}$ be a random subset of $\mathcal{X}$ with λn sequences.

2: Create a PWM Θ from all patterns in $\mathcal{Z}$ and Θ_0 from $\mathcal{X} - \mathcal{Z}$.

3: **while** the motif Θ is not good enough **do**

4: For every $X_i \in \mathcal{X}$, set

$$p_i = \frac{\lambda Pr(X_i|\Theta)}{\lambda Pr(X_i|\Theta) + (1-\lambda)Pr(X_i|\Theta_0)}$$

5: $\lambda = \frac{1}{n}\sum_{i=1}^{n} p_i$.

6: For every $j = 1..\ell$ and every $x \in \{A, C, G, T\}$,

$$\Theta[x, j] = \frac{\sum\{p_i \mid X_i[j] = x, i = 1, 2, \ldots, n\}}{\sum p_i}$$

7: For every $x \in \{A, C, G, T\}$,

$$\Theta_0[x] = \frac{\sum\{(1-p_i) \mid X_i[j] = x, i = 1, 2, \ldots, n, j = 1, 2, \ldots, \ell\}}{\ell \sum(1-p_i)}$$

8: **end while**

FIGURE 10.12: The MEME algorithm.

We now analyze the running time of MEME. For each iteration of the EM algorithm, both the expectation and the maximization steps take $O(n\ell)$ time. The number of iterations is approximately the size of $\mathcal{X}$. Hence, the running time of the EM algorithm is $O(n^2\ell)$ time.

Moreover, the EM algorithm may be trapped into a local maximal if we just execute the EM algorithm for one initial motif. To avoid this problem, we usually need to try n different initial motifs. The testing of the goodness of an initial motif takes approximately $O(n\ell)$ time. Hence, we need $O(n^2\ell)$ time to find a good initial motif. In total, the running time of MEME is $O(n^2\ell)$ time.

10.7   Combinatorial Approaches

The previous section discussed statistical approaches for motif finding. This section describes the combinatorial approaches. The combinatorial approach finds short patterns (of length ℓ) which are over-represented in the input sequences, allowing small number (says, d) of mismatches. Pevzner and Sze [237] proposed to formulate the problem as follows. Given a set $\mathcal{S} = \{S_1, S_2, \ldots, S_m\}$ of m sequences, each of length n. We also have two integer parameters ℓ and d where $d < \ell < n$. A length-ℓ pattern M is said to be an (ℓ, d)-motif in a sequence S_i if there exists a length-ℓ substring of S_i which has at most d mismatches from M. Our aim is to find a (ℓ, d)-motif M which occurs in as many sequences in $\mathcal{S}$ as possible. Figure 10.13 demonstrates an example. Note that the problem of identifying (ℓ, d)-motifs occurring in a set $\mathcal{S}$ of sequences is NP-hard. Below, we discuss various approaches to resolve this problem.

tagtact<u>aggtcggactcgcg</u>tcttgccgc
ca<u>aggtccggctctcatatt</u>caacggttcg
tacgcgccaaa<u>ggcggggctcgcatccggc</u>
acctctgtgacgtctc<u>aggtcgggctctcaa</u>

FIGURE 10.13:   An example of 4 DNA sequences. Consider a length-15 pattern $M = AGGTCGGGCTCGCAT$. This pattern minimizes $\sum_{i=1}^{4} \delta(S_i, M)$. Note that every sequence contains a $(15, 2)$-motif $M = AGGTCGGGCTCGCAT$, that is, every sequence contains a length-15 substring which has at most 2 mismatches from M.

Pattern-driven algorithm

Require: A set of m sequences $\mathcal{S} = \{S_1, S_2, \ldots, S_m\}$, each of length n
and two integer parameters ℓ and d

Ensure: An (ℓ, d)-motif M_{opt} which minimizes $\sum_{i=1}^{m} Score(S_i, M_{opt})$

1: $Score_{opt} = \infty$
2: **for** every length-ℓ pattern M from $AA\ldots A$ to $TT\ldots T$ **do**
3:      $Score = 0$
4:      **for** $i = 1$ to m **do**
5:          if $\delta(S_i, M) > d$, try another pattern
6:          $Score = Score + \delta(S_i, M)$
7:      **end for**
8:      **if** $Score < Score_{opt}$ **then**
9:          $M_{opt} = M$ and $Score_{opt} = Score$
10:     **end if**
11: **end for**
12: Report M_{opt}

FIGURE 10.14: The exhaustive pattern-driven algorithm.

10.7.1 Exhaustive Pattern-Driven Algorithm

The first practical technique for finding an unknown consensus pattern is to enumerate all possible patterns [303]. Given a set of sequences $\mathcal{S} = \{S_1, S_2, \ldots, S_m\}$, the aim is to find a length-ℓ pattern M that appears in all sequences and minimizes the number of mismatches. Precisely, we hope to find a pattern M such that $score(\mathcal{S}, M) = \sum_{i=1}^{m} \delta(S_i, M)$ is minimized where $\delta(S, M)$ is the minimum number of mismatches between M and any length-ℓ substring of S.

For the pattern $M = AGGTCGGGCTCGCAT$ and the four sequences in Figure 10.13, $\delta(S_1, M) = \delta(S_2, M) = \delta(S_3, M) = \delta(S_4, M) = 2$. Thus, $score(\mathcal{S}, M) = 2 + 2 + 2 + 2 = 8$. In fact, 8 is the minimum score for $\mathcal{S}$.

Waterman, Arratia, and Galas [303] proposed an exhaustive pattern-driven algorithm to find such a pattern. Basically, the algorithm generates all possible patterns and identifies a pattern M which minimizes $score(\mathcal{S}, M)$. The algorithm is shown in Figure 10.14.

The time and space analysis is as follows. There are 4^ℓ different patterns of length ℓ. For each pattern M, $\delta(S_i, M)$ can be computed in $O(n\ell)$ time. To scan through all m sequences, it takes $O(nm\ell)$ time. In total, it takes $O(mn\ell 4^\ell)$ time to process all patterns over $\mathcal{S}$. The space complexity is $O(mn)$, for storing the m sequences. Since the algorithm runs in exponential of ℓ, it is only feasible for short pattern.

A number of methods employ this approach, including [296], [297], [216], [292], and [266].

To improve the efficiency, newer methods like Amadeus [189] enumerate

---

Extended Sample-driven algorithm
Require: A set of m sequences $\mathcal{S} = \{S_1, S_2, \ldots, S_m\}$, each of length n
    and an integer parameter ℓ
Ensure: An (ℓ, d)-motif M_{opt} which minimizes $\sum_{i=1}^{m} Score(S_i, M_{opt})$
  1: $Score_{opt} = \infty$
  2: **for** every length-ℓ substring P occurs in $\mathcal{S}$ **do**
  3:    **for** every length-ℓ pattern M which has at most d mismatches from
      P **do**
  4:      $Score = 0$
  5:      **for** $i = 1$ to m **do**
  6:        $Score = Score + \delta(S_i, M)$
  7:      **end for**
  8:      **if** $Score < Score_{opt}$ **then**
  9:        $M_{opt} = M$ and $Score_{opt} = Score$
10:      **end if**
11:    **end for**
12: **end for**
13: Report M_{opt}

FIGURE 10.15:   The extended sample-driven algorithm.

patterns with no mismatch ($d = 0$). Then, through a series of refinements, the motifs are identified.

10.7.2   Sample-Driven Approach

The pattern-driven approach may be slow when the pattern length ℓ is long. An alternative solution is the sample-driven approach. The basic sample-driven method assumes the correct motif has an exact occurrence in $\mathcal{S}$. The method samples all length-ℓ substrings in $\mathcal{S}$ and tests if they are (ℓ, d)-motifs. As there are nm patterns in $\mathcal{S}$ and each pattern takes $O(nm\ell)$ time to test, the running time of the sample-driven method is $O((nm)^2\ell)$. This approach is more efficient than the pattern-driven approach when $nm < 4^\ell$. However, when the correct motif does not have an exact match in any of the sequences in $\mathcal{S}$, this approach will fail to find the correct motif.

To avoid missing the correct motif, one can extend the method to test, for every length-ℓ substring P in $\mathcal{S}$, all d-neighbors of P, that is, all $O(\binom{\ell}{d}3^d)$ strings that differ from P by d or less mismatches. Figure 10.15 presents the detail of the algorithm. This extension guarantees that we will not miss the correct motif in the sampling. However, the running time of the algorithm is increased to $O((nm)^2 \binom{\ell}{d}3^d\ell)$ time.

Multiprofiler [173] is a generalization of the extended sample-driven method.

10.7.3 Suffix Tree-Based Algorithm

Sagot [252] introduced the suffix tree-based approach to search for a motif. Given a length-n sequence S, we aim to find all (ℓ, d)-motifs in S which have at least q occurrences. This section describes how to utilize a suffix tree (see Chapter 3) to find all such (ℓ, d)-motifs efficiently.

First, we have some definitions. Let T be the suffix tree for the input sequence S. For any node u in T, we denote $S(u)$ as the path label of u. Conversely, any substring M in S corresponds to a node $u(M)$ in T. For any substring M of S, its number of occurrences is $occ(M)$, which equals the number of leaves in the subtree of T rooted at $u(M)$.

For any DNA pattern M, let Occ_M be the set of all substrings of S which has at most d mismatches from M. Note that the number of occurrences of M in S allowing at most d mismatches equals $\sum_{M' \in Occ_M} occ(M')$.

When $d = 0$ (i.e., no mismatch is allowed in the motif), the suffix tree can help to find motifs in linear time. Note that the set of $(\ell, 0)$-motifs is just $\{S(u) \mid u$ is at depth ℓ of T and $occ(S(u)) \geq q\}$. Since the suffix tree can be constructed in $O(n)$ time and the suffix tree can be traversed to find all length-ℓ patterns in $O(n)$ time, the algorithm runs in $O(n)$ time.

For $d > 0$, we observe that, given Occ_M and any base α,

$$Occ_{M\alpha} = \{M'\alpha \mid M' \in Occ_M, u(M'\alpha) \text{ exists in } T\} \cup \qquad (10.1)$$
$$\{M'\beta \mid M' \in Occ_M, \beta \neq \alpha, u(M'\beta) \text{ exists in } T, \text{ and}$$
$$M' \text{ has at most } d - 1 \text{ mismatches from } M\}$$

Hence, we can generate all the (ℓ, d)-motifs in S with at least q occurrences incrementally as follows. Let λ be a length-0 motif, that is, an empty string. We define $Occ_\lambda = \{\lambda\}$ and $occ(\lambda) = n$. Then, we recursively extend the motif using the observation. The extension is performed until the length of the motif is ℓ or the number of occurrences of the motif is less than q. Figure 10.16 states the detail of the algorithm.

We give an example to demonstrate the suffix tree-based motif finding algorithm. Consider three sequences $S_1 = aacgt\$$, $S_2 = acgc\#$, and $S_3 = cct\%$. Figure 10.17 shows the generalized suffix tree of S_1, S_2, and S_3. Suppose we aim to find a $(3, 1)$-motif occurring at least three times in the 3 sequences.

Initially, we have a length-0 motif λ with $Occ(\lambda) = \{\lambda\}$. In level 1, by extending λ by $\{a, c, g, t\}$, we obtain 4 $(1, 1)$-motifs a, c, g, and t with $Occ_a = \{a, c, g, t\}$, $Occ_c = \{a, c, g, t\}$, $Occ_g = \{a, c, g, t\}$, and $Occ_t = \{a, c, g, t\}$, respectively.

In level 2, we try to extend the 4 $(1, 1)$-motifs. For example, we can extend c to ca. By Equation 10.1, we have $Occ_{ca} = \{aa, cc, cg, ct\}$. As the substrings in Occ_{ca} appear 5 times in S_1, S_2, and S_3, ca is a $(2, 1)$-motif occurring at least 3 times in S. In total, this level extracts twelve $(2, 1)$-motifs, aa, ac, ag, at, ca, cc, cg, ct, ga, gc, gg, and gt.

In level 3, we try to extend the twelve $(2, 1)$-motifs. Then we get two $(3, 1)$-motifs act and cgt.

SpellModel(M, Occ_M)

1: **if** M is of length ℓ **then**
2:      Report M
3: **end if**
4: **for** $\sigma \in \{a, c, g, t\}$ **do**
5:      $\#occ = 0, Occ_{M\sigma} = \emptyset$
6:      **for** every $M' \in Occ_M$ **do**
7:          **for** the first character β in every outgoing edge of $u(M')$ in T **do**
8:              Let $Leave_\beta$ be the number of leaves under node $u(M'\beta)$
9:              **if** $\alpha = \beta$ **then**
10:                  $Occ_{M\sigma} = Occ_{M\sigma} \cup \{M'\beta\}$
11:                  $\#occ = \#occ + Leave_\beta$
12:              **else if** the number of mismatches between M and $M' < d$ **then**
13:                  $Occ_{M\sigma} = Occ_{M\sigma} \cup \{M'\beta\}$
14:                  $\#occ = \#occ + Leave_\beta$
15:              **end if**
16:          **end for**
17:      **end for**
18:      **if** $\#occ \geq q$ **then**
19:          SpellModel$(M\sigma, Occ_{M\sigma})$
20:      **end if**
21: **end for**

FIGURE 10.16: The suffix tree-based algorithm. Let T be the suffix tree of S. By calling SpellModel$(\lambda, \{\lambda\})$ where λ is an empty string, we can extract all (ℓ, d)-motifs in S with at least q occurrences.

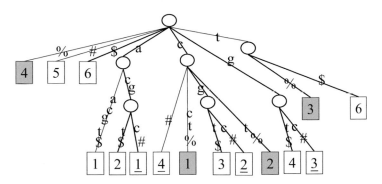

FIGURE 10.17: The figure shows the generalized suffix tree for $S_1 = aacgt\$$, $S_2 = acgc\#$, and $S_3 = cct\%$. The boxes correspond to the leaves in the suffix tree. The normal leaf boxes correspond to S_1; the leaf boxes with underlined integers correspond to S_2; and the leaf boxes with a gray background correspond to S_3.

Below, we analyze the time complexity of the algorithm in Figure 10.16. The algorithm traverses the suffix tree T up to level ℓ. Note that the number of nodes at level ℓ of T is at most n. For each level-ℓ node whose path label is M, the algorithm generates all the length-ℓ patterns in S with at most d mismatches from M. The number of such patterns is bounded above by $\binom{\ell}{d}3^d$. Hence, the running time is $O(n\binom{\ell}{d}3^d)$.

Apart from Sagot [252], other works utilize suffix tree and its variant to discover motifs including Marsan and Sagot [206], Eskin and Pevzner [95], Pavesi et al. [230], and Ettwiller et al. [98].

10.7.4 Graph-Based Method

The graph-based method was proposed by Pevzner and Sze [237]. The idea is to transform the motif finding problem to a clique finding problem in a graph. The transformation is as follows. Given a set of sequences $\mathcal{S} = \{S_1, S_2, \ldots, S_m\}$ and suppose we are looking for a (ℓ, d)-motif. We construct an m-partite graph G. (An m-partite graph is a graph where the vertices can be partitioned into m sets such that no edge appears within each set.)

- Every vertex in G corresponds to a length-ℓ word in $\mathcal{S}$.

- For two words x and y appearing in two different sequences in $\mathcal{S}$, x and y are connected by an edge if they have less than $2d$ mismatches. (If x and y are occurrences of a (ℓ, d)-motif M, $H(x, y) \leq H(x, M) + H(M, y) \leq 2d$ where $H(x, y)$ is the hamming distance between x and y. Hence, $2d$ is chosen as the threshold.)

The problem of finding an (ℓ, d)-motif appearing in q sequences corresponds to finding a clique of size q. Thus, the problem of finding motifs is reduced to the problem of finding large cliques.

However, finding cliques is an NP-complete problem. When we require that all sequences contain an (ℓ, d)-motif, WINNOWER and SP-star are proposed [237]. WINNOWER is suggested to filter edges which definitely do not belong to any large cliques. SP-star is a heuristic to iteratively find some good stars. Later, cWINNOWER [188] improved the sensitivity of WINNOWER. Fratkin et al. [112] suggested another graph-theoretic approach to find a motif by mining a maximum density subgraph instead.

10.8   Scoring Function

The algorithms discussed in the last two sections find motifs which are over-represented in the input sequences. Moreover, more than one candidate motif pattern are usually reported by those algorithms. Among them, which one is correct? A number of different scoring functions have been proposed to rank the motif patterns. They include the total distance score [303], the sum of pair score (SP-score) [237], the log likelihood score [193, 250], statistical over-representation relative to the background [266, 296], relative entropy [143], and sequence specific score [230, 311].

Consider an (ℓ, d)-motif M which appears m times in $\mathcal{S}$. Let those m occurrences of M be $w_1, w_2, \ldots, w_m$. If those m occurrences match well with M, we expect that they are similar to M. In this case, Waterman, Arratia, and Galas [303] proposed the total distance score, which is $\sum_{i=1}^{m} \delta(w_i, M)$ where $\delta(x, y)$ is the hamming distance between x and y. Pavel and Sze [237] expect that all occurrences of the motif are similar. Hence, the SP score is proposed where $SPscore(w_1, w_2, \ldots, w_m) = \sum_{i,j} \delta(w_i, w_j)$. If M fits well with $w_1, w_2, \ldots, w_m$, we expect to have a small SP score.

Both the total distance score and the SP score do not make any assumption on the non-motif regions (called background). A number of methods rank the motifs based on the discrimination of the motif from the background model.

Gibbs sampler [193, 250] uses the log likelihood score, which is the log ratio of the chance that the sequences can be explained by the motif model and the chance that the sequences can be explained by the background model.

YMF (Yeast Motif Finder) [266] and Trawler [98] suggested measuring how over-represented is the motif M using $Z - score(M) = \frac{(obs(M) - E(M))}{\sigma(M)}$ where $obs(M)$ is the observed frequency of M in the input sequences, $E(M)$ and $\sigma(M)$ are the expected frequency and the standard deviation of the occurrences of M in the background model.

Another method is the relative entropy, which measures how surprise in observing the motif relative to the background model. The relative entropy is defined as $D(M|\Theta_0) = \sum_{i=1}^{\ell} \sum_{b \in \{a,c,g,t\}} f_{b,i} \ln\left(\frac{f_{b,i}}{p_b}\right)$, where $f_{b,i}$ is the frequency of the base b at position i of the motif and p_b is the frequency of the base b in the background model.

ANN-Spec [311] and Weeder [230] use the sequence specificity score, which tries to maximize the number of sequences containing the motif. In other words, if a prediction gives multiple sites of the motif in one sequence and none in the other sequences, it is much less significant than a prediction which gives a balanced number of sites in each sequence. The score is defined as $\sum_{i=1}^{m} \log(1/E_i(M|\Theta_0))$ where $E_i(M|\Theta_0)$ is the expected number of motif M occurring in sequence S_i according to the background model.

10.9 Motif Ensemble Methods

As discussed in the previous sections, there are many different methods for discovering motifs. We observe that, for the same dataset, different motif finders may predict different motifs and different binding sites. For instance, for the benchmark dataset of Tompa et al. [293], Wijaya et al. [307] observed that an individual motif finder can only discover at most 22.2% of the true binding sites; Moreover, if we unite the output generated by the ten motif finders studied in [293], we can discover 43.9% of the true binding sites. Hence, different motif finders find different types of motifs. One immediate question is whether we can utilize the information of different motif finders to predict better motifs.

This section describes a practical yet effective approach for motif discovery, namely, the ensemble approach. The fundamental idea of the ensemble approach is to combine results of several motif finders and summarizes their results to generate new results.

Seven ensemble methods have been developed up to now. SCOPE [55] and BEST [58, 162] rerank all motifs predicted by individual finders based on a certain scoring function and report the top motif. WebMotifs [119, 248], MultiFinder [156], and RGSMiner [155] assume motifs predicted by several motif finders are likely to be the real motifs. They cluster the motifs and report the motif in the best cluster. All the above methods only work on the predicted motifs from the individual finders. If none of the reported motifs can capture the binding sites accurately, we will not be able to get a correct result. EMD [154] takes a step further and considers also the predicted binding sites. Precisely, EMD groups the motifs of the same rank from different motif finders. For each group, a new motif is derived from the predicted binding sites of the motifs in the group. However, it may not be true that different motif finders rank the same true motif at the same rank. MotifVoter [307] tries to select a group of motifs predicted by multiple motif finders such that those motifs are similar among themselves while different from the rest of the predicted motifs; then, a motif is derived through analyzing the predicted binding sites of those motifs. In this section, we detail the algorithm of MotifVoter.

10.9.1   Approach of MotifVoter

Given a set of sequences, MotifVoter executes m component motif finders, each reporting n motifs. Note that each motif also defines a set of predicted binding sites. MotifVoter has two stages.

1. Motif filtering: At this stage MotifVoter aims at removing the spurious motifs from all candidate motifs predicted by the m motif finders. The main idea of this stage is to find a cluster of motifs with high conformity based on two criteria.

   First, the *discriminative* criterion requires the selected cluster of motifs to share as many binding sites as possible while motifs outside the cluster should share none or few binding sites. Second, the *consensus* criterion requires the motifs in the selected cluster to be contributed by as many motif finders as possible.

   Section 10.9.2 first describes the details of how to formalize the discriminative and consensus criteria for the motif filtering step. Then, a heuristics algorithm is presented for finding a cluster of motifs to satisfy the two criteria.

2. Sites extraction: Based on the candidate motifs retained in Stage 1, it identifies a set of sites with high confidence that they are real. Finally, from these final sites, we generate the PWM of the motif.

   The details of the sites extraction step are given in Section 10.9.3.

MotifVoter uses 10 basic motif finders. The first group consists of 3 motif finders based on the (l, d) model, which are MITRA [95], Weeder [230], and SPACE [306]. The second group consists of 7 motif finders based on the PWM model, which includes AlignACE [157], ANN-Spec [311], BioProspector [193], Improbizer [12], MDScan [194], MEME [17], and MotifSampler [286].

10.9.2   Motif Filtering by the Discriminative and Consensus Criteria

Given two motifs x and y, we first describe how to measure their similarity based on the binding sites defined by these motifs. Let $I(x, y)$ be the set of regions covered by the binding sites defined by both x and y. Let $U(x, y)$ be the set of regions covered by binding site defined by either x or y. The similarity of x and y, denoted $sim(x, y)$, is expressed as $|I(x, y)|/|U(x, y)|$. Note that $0 \leq sim(x, y) \leq 1$ and $sim(x, x) = 1$.

Now, we define the scoring function to capture the discriminative criterion. Given m motif finders and each motif finder reports its top-n candidate motifs, there will be a set P of mn candidate motifs. Among all the candidate motifs in P, some of them will approximate the real motif while others will not. We would like to identify the subset X of P such that the candidate motifs in X are likely to approximate the real motif. The principal idea is that if the candidate motifs in X can model the real motif, motifs in X should be highly similar while motifs outside X should be distant from one another (discriminative criteria).

Let X be some subset of candidate motifs of P. The mean similarity among the candidate motifs in X, denoted as $sim(X)$, is defined as:

$$\frac{\sum\limits_{x,y \in X} sim(x,y)}{|X|^2}$$

The w score of X, denoted by $w(X)$, is defined as:

$$\frac{|X|^2 \cdot sim(X)}{\sqrt{\sum\limits_{x,y \in X} (sim(x,y) - sim(X))^2}}$$

The function $w(X)$ measures the degree of similarity among the candidate motifs in X. If many candidate motifs in X approximate the real motif, $w(X)$ is expected to be high. On the other hand, we expect that the complement of X, that is $P - X$, should have a low $w(P - X)$. Thus $w(P - X)$ constitutes the *discriminative* criterion in the clustering procedure. In other words, if X is the set of candidate motifs which approximate the real motif, we expect to have a high $A(X)$ score, where:

$$A(X) = \frac{w(X)}{w(P - X)}$$

Note that there may be multiple sets of X with the same $A(X)$ score. Among those X's, we would select the one that contains as many motif finders as possible. The latter constitutes the *consensus* criterion.

In summary, this stage aims to find $X \subseteq P$ which maximizes $A(X)$ while X contains the candidate motifs predicted by the maximum number of motif finders. The naïve method to identify X is to enumerate all possible X's and check if they satisfy the above two criteria. However, this approach is computationally infeasible. Below, we propose a heuristic to identify X.

The detail of the heuristic algorithm is shown in Figure 10.18. Let P be all motifs found by the m motif finders, where each motif finder returns n motifs. Steps 1–3 compute the pairwise similarity scores for all pairs of motifs. Based on these similarity scores, the following heuristic is applied to find X (Steps 4–9).

Algorithm MotifVoter

Require: A set P of k candidate motifs

Ensure: a PWM motif

 1: **for** each $z, y \in P$ **do**
 2:　　compute $sim(z, y)$
 3: **end for**
 4: **for** each $z \in P$ **do**
 5:　　sort $p_1, p_2, \ldots, p_k$ such that $sim(z, p_i) \geq sim(z, p_{i+1}); \forall i = 1..k$
 6:　　consider sets $X_{z,j} = \{z, p_1, \ldots, p_j\}; \forall j = 1, \ldots, k$
 7:　　compute $A(X_{z,j})$ for all such $X_{z,j}$
 8:　　set $Q(X_{z,j})$ = number of motif finders contributing to $X_{z,j}$
 9: **end for**
10: set $X = X_{z,j}$ with the maximum $A(X_{z,j})$ score, if there are two such $X_{z,j}$'s, pick the one with the largest $Q(X_{z,j})$
11: extract and align sites of motifs in X
12: construct PWM

FIGURE 10.18:　MotifVoter.

Enumerating all possible subsets in P takes exponential time. Instead, MotifVoter ranks the motifs in $P - \{z\}$, for every $z \in P$, as $p_1, \ldots, p_{|P|-1}$ such that $sim(z, p_i) > sim(z, p_{i+1})$. MotifVoter only considers subsets $X_{z,j} = \{z, p_1, \ldots, p_j\}$ for every $1 \leq j \leq |P| - 1$. The rationale behind examining only these subsets is because it is likely that $A(\{z, a, b\}) > A(\{z, b\})$ if $sim(a, z) > sim(b, z)$.

For each $X_{z,j}$, we compute the value of $A(X_{z,j})$ and the number of motif finders contributing to $X_{z,j}$, denoted as $Q(X_{z,j})$. Finally, X is assigned to be $X_{z,j}$ which maximizes $A(X_{z,j})$. If more than one subset maximizes $A()$, we pick the one with the largest $Q(X_{z,j})$. In case of a tie again, we pick one randomly.

10.9.3　Sites Extraction and Motif Generation

Given the set of motifs X, some sites defined by the motifs in X are real while some are not. This section presents a way to extract some high confidence sites. The high confidence sites are extracted using two requirements. First, we accept all sites which are occurrences of at least two motifs x and y in X where x and y are predicted by two different motif finders. The reason is that it is unlikely that several motif finders predict the same spurious binding sites.

Second, we accept all sites of the motif $x \in X$ with the highest confidence. The confidence score is defined as follows. Let $B(x)$ be the total number of nucleotides covered by the sites defined by the motif x. Let $O(x)$ be the total number of nucleotides covered by the sites defined by motif x as well as that

FIGURE 1.8 The structure of a double-stranded DNA. The two strands show a complementary base-pairing.

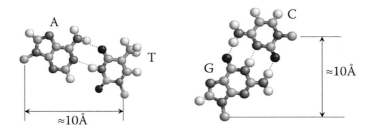

FIGURE 1.9 Watson-Crick Base-Pairing.

(a) Prokaryotes

DNA

↓ transcription

↓ translation

↓ modification

cytoplasm

(b) Eukaryotes

DNA

↓ transcription

↓ Add 5' cap and poly A tail

↓ RNA splicing

nucleus ↓ export

↓ translation

↓ modification

cytoplasm

FIGURE 1.12   Central Dogma for Prokaryotes and Eukaryotes.

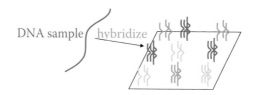

DNA sample hybridize

FIGURE 1.18   This figure shows an array with 9 probes. Each probe is the reverse complement of some short DNA fragment which forms a fingerprint (that is, unique identifier) of certain target DNA sequence. When the array is exposed to certain DNA sample, some probes
will light-up if the corresponding target DNA sequences exist in the sample.

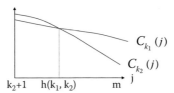

FIGURE 2.11 Graphical interpretation of Lemma 2.2.

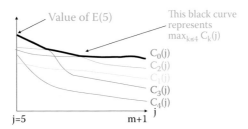

FIGURE 2.12 The figure shows 5 curves $C_k(j)$ for $0 \le k \le 4$. The thick black curve corresponds to $Env_4(j) = \max_{0 \le k \le 4} C_k(j)$.

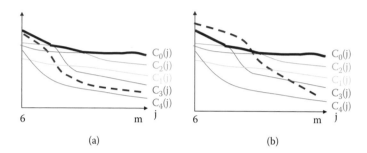

FIGURE 2.13 Possible outcome of overlaying: (a) below, and (b) above.

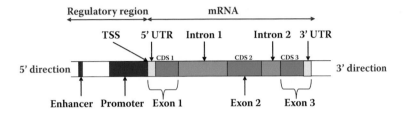

FIGURE 10.2 The locations of gene coding and noncoding regions and the promoter in a DNA strand. The promotor region is present surrounding the start of (and mostly upstream of) the transcript region. Other elements such as enhancer may be present far distant from the transcription start site.

FIGURE 10.6 The sequence logo for the PWM in Figure 10.5.

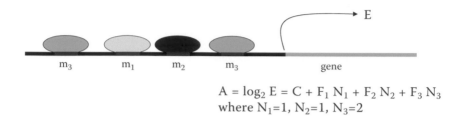

$$A = \log_2 E = C + F_1 N_1 + F_2 N_2 + F_3 N_3$$
$$\text{where } N_1=1, N_2=1, N_3=2$$

FIGURE 10.21 The linear model correlating motifs with gene expression. The ellipses represent the TFs. We have three TFs and N_1, N_2, and N_3 are the number of occurrences of the above three TFs in the promoter. $A = \log_2 E$ is the expression level of the gene. C, F_1, F_2, and F_3 are the model parameters.

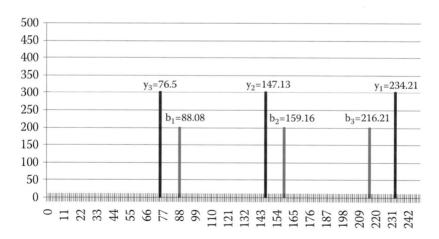

FIGURE 12.9 The figure shows the theoretical spectrum for $P = SAG$, which includes peaks for y-ions (in red) and b-ions (in green). The b-ions are drawn lower than the y-ion peaks to show their relative scarcity.

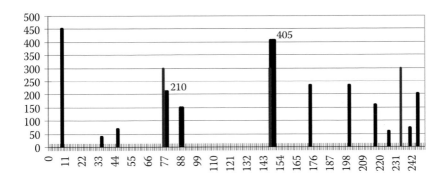

FIGURE 12.10 The figure shows an example spectrum for *P = SAG*. The tandem mass spectrum peaks are in black color while the theoretical y-ion peaks are in red color.

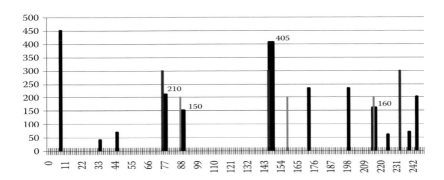

FIGURE 12.12 The figure shows the same spectrum as Figure 12.10 for *P = SAG*. The tandem mass spectrum peaks are in black color. The theoretical y-ion and b-ion peaks are in red and green colors respectively.

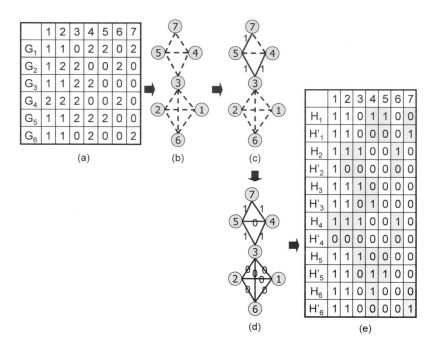

FIGURE 13.3 (a) shows a set of G of 5 genotypes. (b) shows the graph G_M. (c) shows the same graph G_M after we infer the colors of all edges. The red colored (color 0) edges are in-phase while the blue colored (color 1) edges are out-of-phase. (d) shows the set of haplotypes H that can resolve G.

defined by some motif y in X, which is predicted by some other motif finder. The confidence score of motif x is defined as $O(x)/B(x)$.

Given the sites predicted by MotifVoter, we generate a PWM to model the motif. To achieve this, we first align these sites using MUSCLE [89], then a PWM is generated from this alignment to model the motif.

10.10 Can Motif Finders Discover the Correct Motifs?

The previous sections describe a number of motif finders. Can they identify motifs in biological datasets? Which motif finder is good?

There is no universal benchmark dataset for evaluating motif finders. Below, we perform an extensive evaluation on different motif finders over 326 datasets from Tompa et al.'s benchmark, *E. coli*, and ChIP-Chip experiments.

Tompa et al.'s datasets [293] (*http://bio.cs.washington.edu/assessment/*) have been constructed based on real transcription factor binding sites drawn from four different organisms (human, fruit fly, mouse, and yeast). The background sequences are based on three categories: real, generic, and Markov. It consists of 56 datasets. The number of sequences per dataset ranges from 1 to 35 and the sequence lengths are up to 3000 base pairs. Such diverse characteristics of the benchmark makes it a good candidate for evaluating the robustness of a motif finder.

E. coli datasets (*http://regulondb.ccg.unam.mx/*) are taken from RegulonDB [254]. This species is not included in Tompa et al.'s benchmark and hence it can be used as an independent dataset to evaluate the performance of different motif finders. The sequences in the *E. coli* dataset are generated from the intergenic regions. In total, there are 62 datasets. The average number of sequences per dataset is 12 and the average sequence length is 300 bp.

Both Tompa et al.'s benchmark and *E. coli* datasets provide the actual binding site locations of each motif. Hence, we measure the performance of each motif finder using sensitivity (nSn) and positive predictive value (nPPV), which are defined as follows.

$$nSN(recall) = \frac{TP}{TP + FN}$$

$$nPPV(precision) = \frac{TP}{TP + FP}$$

where TP (true positive) is the number of overlapped bases between the real and the predicted sites, FP (false positive) is the number of bases not in the real sites but in the predicted sites, and FN is the number of bases in the real sites but not in the predicted sites. Figure 10.19 shows the comparison results.

	Tompa et al.		E. coli	
	nSN	nPPV	nSN	nPPV
QuickScore	0.017	0.030	-	-
Consensus	0.021	0.113	-	-
GLAM	0.026	0.038	-	-
MITRA	0.031	0.062	0.206	0.193
OligoDyad	0.040	0.154	-	-
MDScan	0.053	0.063	0.217	0.289
AlignACE	0.055	0.112	0.171	0.243
MotifSampler	0.060	0.107	0.110	0.339
SeSiMCMC	0.061	0.037	-	-
YMF	0.064	0.137	-	-
MEME	0.067	0.107	0.224	0.267
Improbizer	0.069	0.070	0.225	0.264
MEME3	0.078	0.091	-	-
WEEDER	0.086	0.300	0.173	0.3
ANN-Spec	0.087	0.088	0.079	0.346
BioProspector	0.111	0.215	0.158	0.304
SPACE	0.126	0.334	0.225	0.341
SCOPE*	0.098	0.149	0.229	0.423
BEST*	0.266	0.044	0.244	0.217
RGSMiner*	≤ 0.282	≤ 0.347	≤ 0.441	≤ 0.446
WebMotifs*	≤ 0.282	≤ 0.347	≤ 0.441	≤ 0.446
EMD*	≤ 0.338	≤ 0.071	0.262	0.296
MotifVoter*	0.410	0.486	0.603	0.617

FIGURE 10.19: This table is obtained from [307]. It lists the average sensitivity (nSN) and precision (nPPV) of each motif finder for Tompa et al.'s benchmark and **E. coli** datasets. For the **E. coli** dataset, the results of some stand-alone motif finders (marked with a dash) were not obtained, because either these programs are not available or the output does not give binding sites for us to evaluate the results (e.g., YMF). Motif finders marked with asterisks (*) are ensemble methods. Note that we are unable to execute some ensemble methods and we only estimate their upper bound performance.

	Yeast	Mammal
	SSD	SSD
AlignACE	1.206	1.530
MotifSampler	1.278	1.132
MEME	1.292	0.750
Weeder	1.314	1.360
ANNSpec	1.344	1.137
MDSscan	1.361	1.036
BioProspector	1.370	1.246
MITRA	1.515	1.470
SPACE	1.539	1.121
Improbizer	1.639	1.475
BEST*	1.159	1.323
WebMotifs*	1.344	-
SCOPE*	1.492	1.247
MotifVoter*	1.919	1.824

FIGURE 10.20:   The table (obtained from [307]) lists the average Sum of Squared Distance (SSD) between the true motif and the predicted motif for each motif finder, on yeast and mammalian ChIP datasets. Note that the SSD score is at most 2.

We also evaluate the motif finders using the ChIP-Chip datasets of yeast [136] and mammals. For mammalian ChIP-Chip datasets, we evaluate the datasets for nine transcription factors: CREB [316], E2F [242], HNF4/HNF6 [225], MYOD/MYOG [49], NFKB [260], NOTCH [228], and SOX [33].

For ChIP-Chip datasets, though the actual binding sites are unknown, we are given the PWM motif for each TF. Hence, we use Sum of Squared Distance (SSD) to evaluate if the predicted PWM motif by each motif finder is similar to the known PWM motif. Note that SSD is one of the best measurements for comparing PWMs [203], and is defined as:

$$SSD(X,Y) = \frac{1}{w}\sum_{i=1}^{w}(2 - \sum_{b=A}^{T}(p_x(b) - p_y(b))^2)$$

where $p_x(b)$ and $p_y(b)$ are the probabilities of base b occurring at position i in motifs X and Y, respectively. Note that $0 \leq SSD(X,Y) \leq 2$ and $SSD(X,X) = 2$. The higher the SDD score, the more similar are the two motifs. Figure 10.20 shows the comparison results.

From the results, among all the individual motif finders, SPACE gives the best sensitivity and specificity. Moreover, we also observe that motif ensemble methods can enhance the performance.

10.11   Motif Finding Utilizing Additional Information

Previous sections assume the input is only a set of regulatory sequences. The rest of this chapter studies whether we can utilize more information to improve motif finding.

10.11.1   Regulatory Element Detection Using Correlation with Expression

Section 10.2 suggested to group genes into disjoint clusters based on the similarity in their expression profiles. Those genes in the same cluster are called co-expressed genes. The promoter regions of those co-expressed genes can then be analyzed for the presence of common motifs.

However, as gene control is multifactorial, the same transcription factor may partner with other factors to co-regulate different sets of genes. Hence, genes regulated by the same motif may not be mutually disjoint. Thus, partitioning the genes into disjoint clusters will cause loss of information.

To avoid the information loss, Harmen et al. [47] suggested that we abandon gene clustering. Instead, they suggested modeling the expression level of a gene as the weighted sum of the motifs appearing in the promoter of a gene. Figure 10.21 shows the model. Precisely, the model is as follows. Let G be the set of genes and for every $g \in G$, let A_g be its expression level. Let $\mathcal{M}$ be the set of motifs and, for every $\mu \in \mathcal{M}$, let $N_{\mu g}$ be the number of occurrences of the motif μ in the promoter of the gene g. Assume additivity of the contributions from different motifs. We define

$$A_g = C + \sum_{\mu \in \mathcal{M}} F_\mu N_{\mu g},$$

where C and F_μ, for $\mu \in \mathcal{M}$, are the model parameters. C represents a baseline expression level when no motif exists in the promoter regions of a gene, whereas F_μ is the increase/decrease in expression level caused by the presence of motif μ. The sign of F_μ determines whether the putative protein factor that binds to the motif μ acts as an activator or as an inhibitor. Here positive means activator and negative means inhibitor.

For any set $\mathcal{M}$ of motifs, the fitness of the model can be measured by the sum of square error $\epsilon(\mathcal{M})$ which is defined as follows.

$$\epsilon(\mathcal{M}) = \sum_{g \in G} \left(A_g - C - \sum_{\mu \in \mathcal{M}} F_\mu N_{\mu g} \right)^2$$

where C and F_μ are the best parameters which minimize the value.

Our problem is to find a set of motifs $\mathcal{M}$ which can fit well with the expression level of the genes. Precisely, we would like to find the set $\mathcal{M}$ which

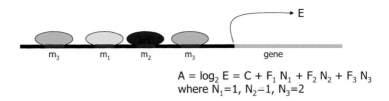

$$A = \log_2 E = C + F_1\, N_1 + F_2\, N_2 + F_3\, N_3$$
where $N_1{=}1$, $N_2{=}1$, $N_3{=}2$

FIGURE 10.21: The linear model correlating motifs with gene expression. The ellipses represent the TFs. We have three TFs and N_1, N_2, and N_3 are the number of occurrences of the three TFs in the promoter. $A = \log_2 E$ is the expression level of the gene. C, F_1, F_2, and F_3 are the model parameters. (See color insert after page 270.)

minimizes $\epsilon(\mathcal{M})$. This problem is NP-hard. This section resolves two problems. First, we present a polynomial time algorithm to compute the fitness of a fixed set of motifs $\mathcal{M}$, that is, computing the parameters C and F_μ, for all $\mu \in \mathcal{M}$, which minimizes $\epsilon(\mathcal{M})$. Then, we present the algorithm REDUCE [47] which applies a heuristic algorithm to find $\mathcal{M}$ which minimizes $\epsilon(\mathcal{M})$.

10.11.1.1 Finding the Fitness of a Set of Motifs $\mathcal{M}$

Given a set of motifs $\mathcal{M}$, this section discusses how to find the parameters C and F_μ for all $\mu \in \mathcal{M}$ which minimizes $\epsilon(\mathcal{M})$.

When $\mathcal{M}$ is an empty set, the model is $A_g = C$ for all genes g. To minimize the sum of square error $\sum_g (A_g - C)^2$, we just need to set $C = \overline{A} = \frac{(\sum_{g \in G} A_g)}{|G|}$. Thus, $\epsilon(\emptyset) = \sum_{g \in G}(A_g - \overline{A})^2$.

In general, C and F_μ for all $\mu \in \mathcal{M}$ can be computed by solving a set of linear equations based on the following lemma.

LEMMA 10.1
Given the expression A_g for a set of genes $g \in G$ and a set $\mathcal{M}$ of motifs, we can compute C and F_μ for all $\mu \in \mathcal{M}$ to minimize $\epsilon(\mathcal{M})$ as follows. F_μ for all $\mu \in \mathcal{M}$ are computed by solving a system of $|\mathcal{M}|$ linear equations, for every $\eta \in \mathcal{M}$,

$$\sum_g (A_g - \overline{A})(N_{\eta g} - \overline{N}_\eta) = \sum_\mu F_\mu \Big(\sum_g (N_{\mu g} - \overline{N}_\mu)(N_{\eta g} - \overline{N}_\eta)\Big).$$

C can be computed by

$$C = \overline{A} - \sum_{\mu \in \mathcal{M}} F_\mu \overline{N}_\mu.$$

where $\overline{A} = \frac{1}{|G|}(\sum_{g \in G} A_g)$ and $\overline{N}_\mu = \frac{1}{|G|}(\sum_{g \in G} N_{\mu g})$.

PROOF   Our aim is to find C and F_μ for all $\mu \in \mathcal{M}$ which minimizes

$$\epsilon(\mathcal{M}) = \sum_{g \in G} \left(A_g - C - \sum_{\mu \in \mathcal{M}} F_\mu N_{\mu g} \right)^2$$

By differentiating $\epsilon(\mathcal{M})$ with respect to C, we have $\sum_g (A_g - C - F_\mu N_{\mu g}) = 0$. Hence, $C = \overline{A} - \sum_{\mu \in \mathcal{M}} F_\mu \overline{N}_\mu$.

By substituting C into the formula of $\epsilon(\mathcal{M})$ and differentiating $\epsilon(\mathcal{M})$ with respect to F_η for every $\eta \in \mathcal{M}$, we get the linear equation stated in the lemma.

⬜

By applying Gaussian elimination on the system of $|\mathcal{M}|$ linear equations in Lemma 10.1, we can compute F_μ for all $\mu \in \mathcal{M}$ using $O(|\mathcal{M}|^2|G|)$ time. Then, C and $\epsilon(\mathcal{M})$ can be computed in $O(|\mathcal{M}|)$ time and $O(|\mathcal{M}||G|)$ time, respectively.

In particular, when $\mathcal{M} = \{\mu\}$, we have the following lemma.

LEMMA 10.2
Given the expression A_g for a set of genes $g \in G$ and a motif μ, the following equations compute C and F_μ which minimizes $\epsilon(\{\mu\})$.

$$F_\mu = \frac{\sum_g (A_g - \overline{A})(N_{\mu g} - \overline{N}_\mu)}{\sum_g (N_{\mu g} - \overline{N}_\mu)^2}$$

$$C = \overline{A} - F_\mu \overline{N}_\mu$$

PROOF   Follows directly from Lemma 10.1.   ⬜

10.11.1.2   REDUCE Algorithm

This section presents the algorithm REDUCE [47] which is a heuristic algorithm to find $\mathcal{M}$ which minimizes $\epsilon(\mathcal{M})$.

The idea of REDUCE is to iteratively find significant motifs one by one. In the first iteration, REDUCE enumerates all possible length-ℓ motifs and finds the motif μ which minimizes $\epsilon(\{\mu\})$. Then, it sets $\mathcal{M} = \{\mu\}$.

In the k-th iteration, we have selected $k-1$ motifs, i.e., $\mathcal{M} = \{\mu_1, \ldots, \mu_{k-1}\}$. REDUCE performs three steps to select the k-th motif. First, REDUCE computes C and F_μ for all $\mu \in \mathcal{M}$ which minimizes $\epsilon(\mathcal{M})$ by Lemma 10.1. Second, it eliminates the contribution of $\mathcal{M}$ to the expression level of each gene $g \in G$, that is, set $A'_g = A_g - C - \sum_{\mu \in \mathcal{M}} F_\mu N_{\mu g}$. Last, REDUCE finds $\eta \notin \mathcal{M}$ which minimizes $\sum_{g \in G}(A'_g - C_\eta - F_\eta N_{\eta g})^2$. Then, REDUCE includes η into $\mathcal{M}$. Figure 10.22 gives the detail of the REDUCE algorithm. Below, we analyze its running time.

Algorithm REDUCE

Require: The expression level A_g for all $g \in G$ and the promoter regions of those genes

Ensure: A set of length-ℓ motifs $\mathcal{M}$

1: Initialize $N_{\epsilon g} = 0$
2: Scan the promoter of each gene g. For each observed k-mer ϵ, increment $N_{\epsilon g}$
3: Set $\mathcal{M} = \emptyset$
4: **repeat**
5:     Using Lemma 10.1, we compute C and F_μ for all $\mu \in \mathcal{M}$ which minimizes $\epsilon(\mathcal{M})$.
6:     Set $A'_g = A_g - C - \sum_{\mu \in \mathcal{M}} F_\mu N_{\mu g}$ for all $g \in G$.
7:     For every length-ℓ motif $\eta \notin \mathcal{M}$, compute $\epsilon'(\eta) = \min_{C_\eta, F_\eta} \sum_{g \in G} (A'_g - C_\eta - F_\eta N_{\eta g})^2$ by Lemma 10.2.
8:     Select η which maximizes $\epsilon'(\eta)$.
9:     Set $\mathcal{M} = \mathcal{M} \cup \{\eta\}$
10: **until** the selected motif η is not significant

FIGURE 10.22: Iterative procedure for finding significant motifs.

Let n be the total length of all promoters of G. Suppose REDUCE finally reports q motifs. Steps 1 and 2 of REDUCE initialize $N_{\mu g}$ for all ℓ-mer μ and all $g \in G$. Step 1 takes $O(4^\ell)$ time and Step 2 takes $O(n)$ time. Steps 5–9 are iterated q times to generate q motifs. For the k-th iteration, the running time is dominated by Steps 5 and 7. Step 5 runs in $O(k^2|G|)$ time. For Step 7, there are 4^ℓ motifs η and computing $\epsilon'(\eta)$ takes $O(|G|)$ time. Thus, Step 7 takes $O(4^\ell|G|)$ time. Steps 5–9 are iterated q times to generate q motifs. Hence, all q iterations run in $O(q(q^2 + 4^\ell)|G|)$ time. In total, REDUCE runs in $O(n + q(q^2 + 4^\ell)|G|)$ time.

Based on this approach, a number of other methods were developed, including GMEP [60], MDscan [194], Motif Regressor [64].

10.11.2 Discovery of Regulatory Elements by Phylogenetic Footprinting

Section 10.2 suggested that we can mine motifs from homologous regulatory regions or promoter regions of a gene from several species. Since selective pressure causes functional elements to evolve at a slower rate than nonfunctional sequences, we expect that unusually well conserved sites among the homologous regulatory regions are possible candidates for motifs.

Moreover, the phylogenetic tree for the species may also be available. In such a case, can we utilize the phylogenetic tree to improve the motif finding accuracy? Blanchette and Tompa [31] gave a positive answer to this question by proposing the phylogenetic footprinting method.

The phylogenetic footprinting problem can be modeled as a substring parsimony problem as follows. Consider n homologous input sequences $S_1, S_2, \ldots,$ S_n and the corresponding phylogenetic tree T. The problem is to identify a set of ℓ-mers, one from each input sequence, that have a parsimony score at most d with respect to T, where ℓ is the motif length and d is the parsimony threshold. ℓ and d are parameters that can be specified by the user. Figure 10.23 shows an example of a set of five homologous sequences and the corresponding phylogenetic tree. Suppose the motif length is $\ell = 4$ and the parsimony threshold is $d = 2$, the highlighted 4-mers in Figure 10.23 satisfy the requirement.

FIGURE 10.23:   The figure shows the homologous sequences and the corresponding phylogenetic tree. The 4-mers which minimize the parsimony score are also highlighted. The parsimony score in this example is 2.

Unfortunately, the string parsimony problem is NP-hard. Below we present a dynamic programming algorithm for this problem. The algorithm proceeds from the leaves of T to its root. For each node u in T, for each ℓ-mer s, we define $W_u[s]$ to be the best parsimony score that can be achieved for the subtree of T rooted at u given that the ancestral sequence at u is s.

When u is a leaf, for every length-ℓ substring s, $W_u[s]$ is computed according to the following recurrence: When u is a leaf, for every length-ℓ substring s,

$$W_u[s] = \begin{cases} 0 & \text{if } s \text{ is a substring of the homologous sequence in } u \\ +\infty & \text{otherwise.} \end{cases}$$

When u is an internal node, for every length-ℓ substring s,

$$W_u[s] = \sum_{v:\ child\ of\ u} \min_{length-\ell\ substring\ t} (W_v[t] + d(s,t))$$

where $d(s,t)$ is the number of mutations between s and t.

Finally, from the root r of T, the best parsimony score is

$$\min_{length-\ell\ substring\ t} W_r[t]$$

By back-tracing, we can identify the set of ℓ-mers, one from each input sequence, that give the best parsimony score.

We now analyze the time complexity. For every leaf u in T, $W_u[s]$ is computed for all ℓ-mers s. Since there are 4^ℓ ℓ-mers and n leaves, processing all leaves takes $O(n4^\ell + m)$ where m is the total length of all homologous sequences.

For every internal node u in T, $W_u[s]$ is computed for all ℓ-mers s. Since there are 4^ℓ possible ℓ-mers s, we need to fill in 4^ℓ different $W_u[s]$ entries. If u has deg_u children, each entry takes $O(deg_u \ell 4^\ell)$ time. Since the total number of children for all internal nodes u in T is $O(n)$, processing all internal nodes takes $O(\sum_u deg_u \ell 4^{2\ell}) = O(n\ell 4^{2\ell})$ time. In total, the running time of the algorithm is $O(\ell 4^{2\ell} n + m)$ time.

10.12   Exercises

1. The accessment result of different motif finders can be found in

   http://bio.cs.washington.edu/assessment/assessment_result.html.

   Can you find a motif finder which is consistently better than the other motif finders?

2. Suppose the following set of sites are known to be bound by the transcription factor NFkB:

   AGGAAATG, GGAAATGCCG, AGGAAATG, GGAATTCC, GAAATG, GGAAAACC, ATGGAAATGAT, GGAAATCCCG, GGAAATTCCA, GGAAATTCCC, GGAAATTCCG.

   Can you align them by packing spaces to the beginning and the end of each sequence? Then, can you report the corresponding logo? (Sequence logo can be found in http://weblogo.berkeley.edu/logo.cgi.)

3. For each of the following cases, design a positional weight matrix that gives a high score for the motif.

   (a) The motif equals TATAAT.

   (b) The motif has five nucleotides where the first position equals A and the third position equals either C or G.

   (c) The motif is either TATAAT or CATCGT.

4. Consider a DNA sequence $S[1..n]$. A gapped motif M consists of an ℓ_1-mer and an ℓ_2-mer separated by a gap of size g. We would like to find all gapped motifs M which occur at least q times in S with at most d mismatches. Can you propose an algorithm to find all these gapped

motifs based on an exhaustive pattern-driven approach? What is the running time?

5. Referring to the above question, can you propose an algorithm to find all these gapped motifs based on a sample-driven approach? What is the running time?

6. Refer to the above two questions. Suppose we would like to find all gapped motifs M which occur at least q times in S with no mismatch. Can you propose an $O((\ell_1+\ell_2)n)$ time algorithm to find all those gapped motifs?

7. Given a sequence S of length n, the aim is to find an (ℓ, d)-motif which occurs the most frequently in S.

    (a) Propose an algorithm which finds an $(\ell, 0)$-motif in $O(n)$ time.

    (b) Propose an algorithm which finds an $(\ell, 1)$-motif in $O(|\Sigma|n\ell^2 \log n)$ time, where Σ is the alphabet size (for DNA, $|\Sigma| = 4$).

8. Given a sequence S, can you propose an efficient algorithm which reports all length-ℓ palindrome motifs which have at least τ exact occurrences in S? What is the time complexity?

9. Consider three genes g_1, g_2, and g_3 whose expression levels are $A_1 = 300$, $A_2 = 1000$, and $A_3 = 500$, respectively. Assume their promoters have two motifs μ_x and μ_y. The table below gives the number of occurrences of the two motifs in the promoter regions of the three genes.

	μ_x	μ_y
g_1	1	5
g_2	4	1
g_3	2	2

Using the model in Section 10.11.1, can you compute $\epsilon(\{x\})$ and $\epsilon(\{y\})$? Indicate if the motifs x and y are activators or inhibitors. Also, among x and y, which motif is better to explain the expression level of the genes?

10. Consider a set of aligned orthogonal sequences $S_1, \ldots, S_n$, each of length m. Suppose we also have the corresponding phylogenetic tree. Can you propose an algorithm to identify a length-ℓ region $i..i + \ell - 1$ such that the parsimony score for $S_1[i..i+\ell-1], \ldots, S_n[i..i+\ell-1]$ is the minimum parsimony score? What is the running time of your algorithm?

Chapter 11

RNA Secondary Structure Prediction

11.1 Introduction

Due to the advance in sequencing technologies, many RNA sequences have been discovered. However, only a few of their structures have been deduced. On the other hand, the chemical and biological properties of many RNAs (like tRNAs) can be determined primarily by their secondary structures. Therefore, determining the secondary structures of RNAs is becoming one of the most important topics in bioinformatics.

Below we list a number of applications of RNA secondary structures.

Function classification: Many RNAs which do not have similar sequences have the same function [177]. A possible explanation is that they have a similar secondary structure. For example, RNA viruses have a high mutation rate. Distant groups of RNA viruses show very little or no detectable sequence homology. In contrast, their secondary structures are highly conserved. Hence, people classify RNA viruses based on their secondary structures instead of their sequences.

Evolutionary studies: Ribosomal RNA is a very ancient molecule. It evolves slowly and exists in all living species. Therefore, it is used to determine the evolutionary spectrum of species [310]. One problem in the evolutionary study is to align the ribosomal RNA sequences from different species. Since the secondary structures of ribosomal RNAs are highly conserved, people use the structure as the basis for a highly accurate alignment.

Pseudogene detection: Given a DNA sequence which is highly homologous to some known tRNA gene, such a sequence may be a gene or a pseudogene. One method for detecting whether it is a pseudogene is to compute its secondary structure and to check whether its structure looks similar to the conserved tRNA secondary structure [196].

11.1.1   Base Interactions in RNA

Before studying the structures of RNAs, we need to understand the interactions between RNA bases. RNA consists of a set of nucleotides which can be either adenine (A), cytosine (C), guanine (G), or uracil (U). Each of these nucleotides is known as the base and can be bonded with another one via a hydrogen bond. If this happens, we say that the two bases form a base pair. There are two types of base pairs: canonical base pair and wobble base pair. The canonical base pair is formed by a double hydrogen bond between A and U, or a triple hydrogen bond between G and C. The wobble base pair is formed by a single hydrogen bond between G and U. Apart from these two types of base pairs, other base pairs like U-C and G-A are also feasible, though they are relatively rare. To simplify the study, we assume only canonical and wobble base pairs exist.

11.1.2   RNA Structures

Biologists normally describe RNA structure in three levels: primary structure, secondary structure, and tertiary structure. The primary structure of an RNA is just its nucleotide sequence. The secondary structure of an RNA specifies a list of canonical and wobble base pairs that occur in the RNA structure. The tertiary structure is the actual 3D structure of the RNA.

Since the secondary structure is good enough to explain most of the functionality of RNA, this chapter focuses on the prediction of the secondary structure of RNA. Consider a length-n RNA polymer $s_1 s_2 \ldots s_n$. Formally, the secondary structure of RNA is defined to be a set S of base pairs (s_i, s_j) such that

1. Each base is paired at most once.

2. Nested criteria: if $(s_i, s_j), (s_k, s_l) \in S$, we have $i < k < j$ if and only if $i < l < j$.

Actually, a secondary structure may contain base pairs which do not satisfy the above two criteria, though such cases are rare. If criteria 1 is not satisfied, a base triple may happen. If criteria 2 is not satisfied, a pseudoknot becomes feasible. Figure 11.1 shows two examples of pseudoknots. Formally speaking, a pseudoknot is composed of two interleaving base pairs (s_i, s_j) and (s_k, s_l) such that $i < k < j < l$.

When a pseudoknot and a base triple do not appear, the RNA structure can be described as a planar graph or be represented by a string of parentheses. (See Figure 11.4 for an example.) Then, the regions enclosed by the RNA backbone and the base pairs are defined as loops. Depending on the positions of the base pairs, loops can be classified into the following five types.

- **Hairpin loop**: A hairpin loop is a loop formed by closing a segment of an RNA strand by one base pair. An example is shown in Figure 11.2a.

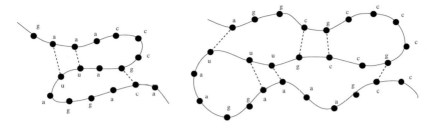

FIGURE 11.1: Example of a pseudoknot.

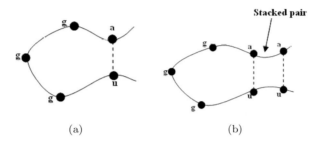

FIGURE 11.2: (a) Hairpin loop, (b) stacked pair.

- **Stacked pair**: A stacked pair is a loop closed by two base pairs that are adjacent to each other. In other words, a stacked pair is formed by two base pairs (i, j) and $(i + 1, j - 1)$. Figure 11.2b shows an example of a stacked pair.

- **Internal loop**: An internal loop is a loop closed by two base pairs like the stacked pair. Moreover, an internal loop consists of at least one unpaired base on each side of the loop between the two base pairs. Figure 11.3a shows an example of an internal loop where both sides have two unpaired bases.

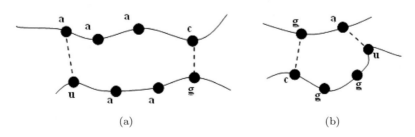

FIGURE 11.3: (a) Internal loop, (b) bulge.

- **Bulge**: A bulge consists of two base pairs like the internal loop. However, only one side of the bulge has unpaired bases. The other side must have two base pairs adjacent to each other, as shown in Figure 11.3b.

- **Multi-loop**: Any loop with three or more base pairs is a multi-loop.

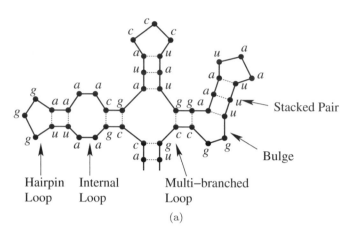

(a)

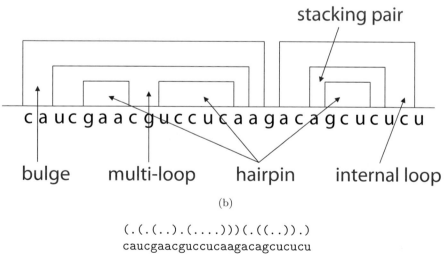

(b)

(.(.(..).(....)))(.((..)).)
caucgaacguccucaagacagcucucu

(c)

FIGURE 11.4:   Different loop types.

When there is no pseudoknot, the secondary structure can always be drawn on a plane without edge crossing. Figure 11.4(a) shows an example secondary structure with various loop types. Besides, we sometimes like to represent the RNA nucleotide sequence as a line and all the base pairs as arcs on top of the line (see Figure 11.4(b)). We also like to represent the loops by parentheses (see Figure 11.4(c)).

The rest of this chapter is organized as follows. Section 11.2 gives a brief description of how to obtain RNA secondary structures experimentally. Sections 11.3 to 11.7 describe approaches to predicting RNA secondary structures computationally.

11.2 Obtaining RNA Secondary Structure Experimentally

In the literature, several experimental methods exist for obtaining the secondary structure of an RNA, including physical methods, chemical/enzymatic methods, and mutational analysis.

- **Physical methods:**
  The basic idea behind physical methods is to infer the structure based on the distance measurements among atoms. Crystal X-ray diffraction is a physical method which gives the highest resolution. It reveals the distance information based on X-ray diffraction. For example, the structure of tRNA is obtained using this approach [176]. However, the use of this method is limited since it is difficult to obtain crystals of RNA molecules which are suitable for X-ray diffraction. Another physical method is Nuclear Magnetic Resonance (NMR), which can provide detailed local conformation based on the magnetic properties of hydrogen nuclei. However, NMR can only resolve structures of small size.

- **Chemical/enzymatic methods:**
  Enzymatic and chemical probes [93] which modify RNA under some specific constraints can be used to analyze RNA structure. By comparing the properties of the RNA before and after applying the probes, we can obtain some RNA structure information.

  Note that the RNA structure information extracted is usually limited as some segments of an RNA polymer are inaccessible to the probes. Another issue is the experimental temperatures for various chemical or enzymatic digestions. An RNA polymer may unfold due to a high experimental temperature. Caution is required in interpreting the experimental results.

- **Mutational analysis:**
  This method introduces a specific mutation to the RNA sequence. Then,

the binding abilities between the mutated sequence and some proteins are tested [280]. If the binding abilities of the mutated sequence are different from that of the original sequence, we claim that the mutated RNA sequence has structural changes. Such information helps us to deduce the secondary structure.

11.3   RNA Structure Prediction Based on Sequence Only

From laboratory experiments, it is observed that a denatured RNA renatures to the same structure spontaneously in vitro. Hence, people in general believe that the structures of RNAs are determined by their sequences. Such a belief motivates us to predict the RNA secondary structure based on the sequence only.

A lot of work has been done in this area. It can be classified into two types: (1) structure prediction based on multiple RNA sequences which are structurally similar; and (2) structure prediction based on a single RNA sequence.

For (1), the basic idea is to align the RNA sequences and to predict the structure. Sankoff [257] considered aligning two RNA sequences to infer the structure. The time complexity of his algorithm is $O(n^6)$. Corpet and Michot [65] presented a method which incrementally adds new sequences to refine an alignment by taking into account base sequence and secondary structure. Eddy and Durbin [87] built a multiple alignment of the sequences and derive the common secondary structure. They proposed the covariance models which can successfully compute the consensus secondary structure. All the above methods have the common problem that they assume the secondary structure does not have a pseudoknot. Gary and Stormo [116] proposed to solve this problem using a graph theoretical approach.

This chapter focuses on (2), that is, structure prediction based on a single RNA sequence. Section 11.4 studies the RNA structure prediction algorithm with the assumption that there is no pseudoknot. Then, some works on RNA structure prediction with a pseudoknot are introduced in Section 11.7.

11.4   Structure Prediction with the Assumption That There is No Pseudoknot

This section considers the problem of predicting an RNA secondary structure with the assumption that there is no pseudoknot. The reason for ignoring pseudoknots is to reduce the computing complexity. Although ignoring pseudoknots reduces accuracy, such an approximation still looks reasonably good as pseudoknots do not appear frequently.

Predicting an RNA secondary structure is not easy. The naïve approach relies on an exhaustive search to find the lowest free energy conformation. Such an approach fails since there are many conformations with the lowest free energy. Identifying the correct conformation is like looking for a needle in a haystack.

Over the past 30 years, people have tried to find the correct RNA conformation based on simulating the thermal motion of the RNA (e.g., CHARMM [37] and AMBER [231]). The simulation considers both the energy of the molecule and the net force experienced by every pair of atoms. In principle, the correct RNA conformation can be computed in this way. However, such an approach still fails because of the following two reasons: (1) Since we do not fully understand the chemical and physical properties of atoms, the energies and forces computed are approximated values. It is not clear whether the correct conformation can be predicted at such a level of approximation; and (2) The computation time of every simulation iteration takes seconds to minutes even for some short RNA sequences. It is impossible to compute the optimal structure in a reasonable time.

In the 1970s, scientists discovered that the stability of RNA helices can be predicted using thermodynamic data obtained from melting studies. The data implies that loops' energies are approximately independent. Based on this finding, Tinoco et al. [290, 291] proposed the nearest neighbor model to approximate the free energy of any RNA structure. Their model makes the following assumptions:

- The energy of every loop (including hairpin, stacked pair, bulge, internal loop, multi-loop) is independent of the other loops.

- The energy of a secondary structure is the sum of all its loops.

Based on this model, Nussinov and Jacobson [224] proposed the first algorithm for computing the optimal RNA structure. Their idea was to maximize the number of stacked pairs while they did not consider the destabilizing energy of various loops. Zuker and Stiegler [321] then gave an algorithm which also considers the various loops with destabilizing energies. Their algorithm takes $O(n^4)$ time, where n is the length of the RNA sequence. Lyngso, Zuker, and Pedersen [200] improve the time complexity to $O(n^3)$. Using the best known parameters proposed by Mathews et al. [209], the predicted structure on average contains more than 70% of the base pairs of the true secondary structure.

Apart from finding the optimal RNA structure, Zuker [320] also proposed an approach which can compute all the suboptimal structures whose free energies are within some fixed range from the optimal.

The next two sections present the Nussinov folding algorithm [224] and the best known RNA secondary structure prediction algorithm, which was proposed by Lyngso, Zuker, and Pedersen [200].

11.5 Nussinov Folding Algorithm

The Nussinov folding algorithm is one of the first methods to predict the RNA secondary structure. They assume that a stable secondary structure should maximize the number of base pairs. For example, Figure 11.5 considers the RNA sequence ACCAGAAACUGGU, which has at most five base pairs in the secondary structure.

ACCAGAAACUGGU

FIGURE 11.5: The secondary structure for the RNA sequence ACCAGAAACUGGU which maximizes the number of base pairs.

The Nussinov algorithm computes the maximum number of base pairs through dynamic programming. Let $S[1..n]$ be the RNA sequence. We let $V(i, j)$ be the maximum number of base pairs in $S[i..j]$.

When $i \geq j$, $S[i..j]$ either has one base or is an empty string. Hence, $V(i, j) = 0$ since we cannot form any base pair in $S[i..j]$.

When $i < j$, we have two cases depending on whether (i, j) forms a base pair.

(i, j) forms a base pair: $V(i, j) = V(i+1, j-1) + \delta(S[i], S[j])$ where $\delta(x, y) = 1$ if $(x, y) \in \{(a, u), (u, a), (c, g), (g, c), (g, u), (u, g)\}$; and 0, otherwise.

i and j do not form a base pair: In this case, there exists $i \leq k < j$ such that no base pair crosses k. Hence, $V(i, j) = \max_{i \leq k < j}\{V(i, k) + V(k+1, j)\}$.

Therefore, when $i < j$, we have

$$V(i, j) = \max \begin{cases} V(i+1, j-1) + \delta(S[i], S[j]) \\ \max_{i \leq k < j}\{V(i, k) + V(k+1, j)\} \end{cases}$$

Our aim is to compute $V(1, n)$, which is the maximum number of base pairs in $S[1..n]$. To compute $V(1, n)$, Nussinov and Jacobson applied dynamic programming to fill in the table V. Observe that the value $V(i, j)$ depends on the values $V(i, k)$, $V(k, j)$, and $V(i+1, j-1)$ for $i \leq k < j$. To ensure all dependent values are available when we fill in an entry in V, the algorithm fills

Nussinov algorithm
1: **for** $i = 1$ to n **do**
2:    $V(i, i) = V(i + 1, i) = 0$
3: **end for**
4: **for** $d = 1$ to $n - 1$ **do**
5:    **for** $i = 1$ to $n - d$ **do**
6:       $j = i + d$;
7:       $V(i, j) = \max\{V(i + 1, j - 1) + \delta(S[i], S[j]), \max_{i \leq k < j}\{V(i, k) + V(k + 1, j)\}\}$
8:    **end for**
9: **end for**
10: Report $V(1..n)$;

FIGURE 11.6: The Nussinov algorithm for computing the maximum number of base pairs for an RNA sequence S.

	1	2	3	4	5	6	7
1	0	0	0	0	1	1	2
2		0	0	0	0	1	1	2
3			0	0	0	1	1	2
4				0	0	0	1	2
5					0	0	1	1
6						0	0	0
7							0	0

FIGURE 11.7: Consider $S[1..7]$ =ACCAGCU. The figure shows the dynamic programming table V.

in the values diagonal by diagonal. Precisely, the algorithm fill in, for $d = -1$ to $n - 1$, the entries in diagonal-d where the diagonal-d consists of entries $V(i, i + d)$ for $i = 1, \ldots, n - d$. Figure 11.6 details the Nussinov algorithm.

For example, consider $S[1..7]$ =ACCAGCU. Figure 11.7 shows the corresponding dynamic programming table V. The maximum number of base pairs for S is $V(1, 7) = 2$.

For time complexity, we need to fill in $O(n^2)$'s $V(i, j)$ entries. Each $V(i, j)$ entry can be computed in $O(n)$ time. Thus, the Nussinov algorithm can be solved in $O(n^3)$ time.

11.6 ZUKER Algorithm

The Nussinov algorithm does not consider the destabilizing energy of various loops. This section presents the ZUKER algorithm which computes the optimal secondary structure considering the loop energy.

Recall that an RNA secondary structure is built upon four basic types of loops: stacked pair, hairpin loop, internal loop/bulge, and multi-loop. Mathews, Sabina, Zuker, and Turner [209] have derived the four energy functions which govern the formation of these loops. Those energy functions are:

- $eS(i, j)$: This function gives the free energy of a stacked pair consisting of base pairs (i, j) and $(i+1, j-1)$. Since no free base is included, this is the only loop which can stabilize the RNA secondary structure. Thus, its energy is negative.

- $eH(i, j)$: This function gives the free energy of the hairpin closed by the base pair (i, j). Biologically, the bigger the hairpin loop, the more unstable is the structure. Therefore, $eH(i, j)$ is more positive if $|j - i + 1|$ is big.

- $eL(i, j, i', j')$: This function gives the free energy of an internal loop or bulge enclosed by base pairs (i, j) and (i', j'), where $i < i' < j' < j$. Its free energy depends on the loop size $(|i' - i + 1| + |j' - j + 1|)$ and the asymmetry of the two sides of the loop. Normally, if the loop size is big and the two sides of the loop are asymmetric, the internal loop is more unstable and thus $eL(i, j, i', j')$ is more positive.

- $eM(i, j, i_1, j_1, \ldots, i_k, j_k)$: This function gives the free energy of a multi-loop enclosed by a base pair (i, j) and k base pairs $(i_1, j_1), \ldots, (i_k, j_k)$, where $(i_1, j_1), \ldots, (i_k, j_k)$ are non-overlapping intervals within (i, j). The multi-loop is more unstable when its loop size and the value k are big.

Based on the nearest neighbor model, the energy of a secondary structure is the sum of all its loops. By energy minimization, the secondary structure of an RNA can be predicted. To speed up the process, we take advantage of dynamic programming, which can be described using the following four recursive equations.

- $W(j)$: The energy of the optimal secondary structure for $S[1..j]$.

- $V(i, j)$: The energy of the optimal secondary structure for $S[i..j]$ given that (i, j) is a base pair.

- $VBI(i, j)$: The energy of the optimal secondary structure for $S[i..j]$ given that (i, j) closes a bulge or an internal loop.

- $VM(i, j)$: The energy of the optimal secondary structure for $S[i..j]$ given that (i, j) closes a multi-loop.

Note that the energy of the optimal secondary structure for $S[1..n]$ equals $W(n)$. To compute $W()$, it depends on $V()$. To compute $V()$, it depends on $VBI()$ and $VM()$. Below, we describe the detail of the four recursive equations.

$W(j)$ is the free energy for the optimal secondary structure for $S[1..j]$. When $j = 0$, $S[1..j]$ is a null sequence; thus, we set $W(j) = 0$. For $j > 0$, we have two cases: either (1) $S[j]$ is a free base or (2) there exists i such that $(S[i], S[j])$ forms a base pair. For case (1), $W(j) = W(j - 1)$. For case (2), $W(j) = \min_{1 \leq i < j}(V(i, j) + W(i - 1))$. Thus, we have the following recursive equations.

- $W(0) = 0$

- $W(j) = \min\{W(j - 1), \min_{1 \leq i < j}(V(i, j) + W(i - 1))\}$ for $j > 0$.

$V(i, j)$ is the free energy of the optimal secondary structure for $S[i..j]$ with (i, j) forms a base pair. When $i \geq j$, $S[i..j]$ is a null sequence and we cannot form any base pair; thus, we set $V(i, j) = +\infty$. When $i < j$, the base pair (i, j) should close one of the four loop types: hairpin, stacked pair, internal loop, and multi-loop. Thus the free energy $V(i, j)$ should be the minimum of $eH(i, j)$, $eS(i, j) + V(i + 1, j - 1)$, $VBI(i, j)$, and $VM(i, j)$. We have the following equations.

- For $i \geq j$, $V(i, j) = +\infty$

- For $i < j$,
$$V(i, j) = \min \begin{cases} eH(i, j) & \text{Hairpin loop;} \\ eS(i, j) + V(i + 1, j - 1) & \text{Stacking loop;} \\ VBI(i, j) & \text{Bulge/Internal loops;} \\ VM(i, j) & \text{Multi-loop.} \end{cases}$$

$VBI(i, j)$ is the free energy of the optimal secondary structure for $S[i..j]$ with the base pair (i, j) closes a bulge or an internal loop. The bulge or the internal loop is formed by (i, j) together with some other base pair (i', j') where $i < i' < j' < j$. The energy of this loop is $eL(i, j, i', j')$. The energy of the best secondary structure for $S[i..j]$ with (i, j) and (i', j') forms an internal loop is $eL(i, j, i', j') + V(i', j')$. By trying all possible (i', j') pairs, the optimal energy can be found as follows.

- $VBI(i, j) = \min_{i < i' < j' < j}\{eL(i, j, i', j') + V(i', j')\}$.

$VM(i, j)$ is the free energy of the optimal secondary structure for $S[i..j]$ with the base pair (i, j) closes a multi-loop. The multi-loop is formed by (i, j) together with k base pairs $(i_1, j_1), \ldots, (i_k, j_k)$ where $k > 1$ and $i < i_1 < j_1 < i_2 < j_2 < \ldots < i_k < j_k < j$. (See Figure 11.8 for an example.) Similar to the calculation of $VBI(i, j)$, we get the following equation.

- $VM(i,j) =$

  $\min_{i<i_1<j_1<...<i_k<j_k<j}\{eM(i,j,i_1,j_1,\ldots,i_k,j_k) + \sum_{h=1}^{k} V(i_h,j_h)\}$

11.6.1 Time Analysis

Based on the above discussion, computing the free energy of the optimal secondary structure for $S[1..n]$ is equivalent to finding $W(n)$. Such computation requires us to fill in four dynamic programming tables for the four recursive equations W, V, VBI, and VM. Then, by back-tracing, the optimal secondary structure can be obtained. Below, we give the time analysis for filling in the four tables.

- $W(j)$: It is an array with n entries. Each entry requires finding the minimum of n terms, $V(i,j)+W(i-1)$ for i varying from 1 to $j-1$. So, each entry needs $O(n)$ time. As a result, it costs $O(n^2)$ time in total.

- $V(i,j)$: It is an array with n^2 entries and each entry requires finding the minimum of four terms, $eH(i,j)$, $eS(i,j)+V(i+1,j-1)$, $VBI(i,j)$, and $VM(i,j)$. Since each entry can be filled in $O(1)$ time, this matrix can be computed in $O(n^2)$ time.

- $VBI(i,j)$: It is an array with n^2 entries and each entry requires finding the minimum of n^2 terms: $eL(i,j,i',j') + V(i',j')$ for $i < i' < j' < j$, in which i' and j' both vary from 1 to n at most. So, each term needs $O(n^2)$ time. As a result, it costs $O(n^4)$ time in total.

- $VM(i,j)$: It is an array with n^2 entries and each entry requires finding the minimum of exponential number of terms: $eM(i,j,i_1,j_1,\ldots,i_k,j_k)+ \sum_{h=1}^{k} V(i',j')$ where $i < i_1 < j_1 < \ldots < i_k < j_k < j$. So in total, it costs exponential time.

In summary, the execution time of the algorithm is exponential. The major bottleneck is in computing tables VM (for multi-loops) and VBI (for internal loops), which require exponential time in n and $O(n^4)$ time, respectively. To speed up, we make some assumptions on the energy function. For multi-loops, we assume that the energy of multi-loops can be approximated using an affine linear function, through which we can reduce the running time of VM from exponential to $O(n^3)$. For internal loops, we reduce the overhead of VBI to $O(n^3)$ time by the Ninio equation [229]. Therefore, we can reduce the overall complexity to $O(n^3)$ time from the original exponential time. The two speed-up methods will be discussed in detail in Sections 11.6.2 and 11.6.3.

11.6.2 Speeding up Multi-Loops

To make the problem tractable, the following simplified assumption is made. Consider a multi-loop formed by base pairs $(i,j),(i_1,j_1),\ldots,(i_k,j_k)$ as shown in Figure 11.8. The energy of the multi-loop is approximated by the sum of the background energy, the stabilizing energy due to the base pairs, and the destabilizing energy due to the unpaired bases in the multi-loop. Precisely, we define

$$eM(i,j,i_1,j_1,\ldots,i_k,j_k) = a+bk+c((i_1-i-1)+(j-j_k-1)+\sum_{h=1}^{k-1}(i_h+1-j_h-1))$$

where a, b, and c are constant. k is the number of base pairs in the loop and $((i_1-i-1)+(j-j_k-1)+\sum_{h=1}^{k-1}(i_h+1-j_h-1))$ is the number of unpaired bases in the loop.

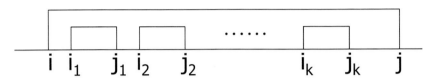

FIGURE 11.8: Structure for a multi-loop.

Given the above assumption, the RNA structure prediction algorithm can be speeded up by introducing a new recursive equation $WM(i,j)$. $WM(i,j)$ is defined to be the energy of the optimal secondary structure of $S[i..j]$ that constitutes the substructure of a multi-loop structure. In other words, inside the multi-loop substructure, every free base is given a score c while every base pair belonging to the multi-loop substructure is given a score b. Thus, we have the following equation.

$$WM(i,j) = \min \begin{cases} WM(i,j-1)+c, & j \text{ is free base;} \\ WM(i+1,j)+c, & i \text{ is free base;} \\ V(i,j)+b, & (i,j) \text{ is a pair;} \\ \min_{i<r\le j}\{WM(i,r-1)+WM(r,j)\}, & i \text{ and } j \text{ are not free} \\ & \text{and } (i,j) \text{ is not a pair;} \end{cases}$$

Given $WM(i,j)$, $VM(i,j)$ equals the sum of the multi-loop penalty, a, and the energies of the two parts $WM(i+1,r-1)$ and $WM(r,j-1)$ for some $i+1 < r \le j-1$. Thus, we have

$$VM(i,j) = WM(i+1,j-1)+a$$

After the changes, the optimal secondary structure can be computed by filling in five dynamic programming tables: $W(i)$, $V(i,j)$, $VBI(i,j)$, $VM(i,j)$, and $WM(i,j)$.

Now, we analyze the time complexity. The time complexity for filling tables $W(i)$, $V(i,j)$, and $VBI(i,j)$ is the same as the analysis in Section 11.6.1. They run in $O(n^2)$, $O(n^2)$, and $O(n^4)$ time, respectively.

For the table $WM(i, j)$, it has n^2 entries and each entry can be computed by finding the minimum of four terms: $WM(i, j - 1) + c$, $WM(i + 1, j) + c$, $V(i, j) + b$, and the minimum of $WM(i, k - 1) + WM(k, j)$ for $i < k \leq j$. The first three terms can be computed in $O(1)$ time while the last term takes $O(n)$ time. In total, filling table $WM(i, j)$ takes $O(n^3)$ time.

For the table $VM(i, j)$, it also has n^2 entries and each entry can be evaluated by finding the minimum of the n terms $WM(i + 1, k - 1) + WM(k, j - 1) + a$ for $i + 1 < k \leq j - 1$. Thus, filling table $VM(i, j)$ also takes $O(n^3)$ time.

In conclusion, the modified algorithm runs in $O(n^4)$ time.

11.6.3   Speeding up Internal Loops

Consider an internal loop/bulge formed by two base pairs (i, j) and (i', j') with $i < i' < j' < j$. Let $n_1 = i' - i - 1$ and $n_2 = j - j' - 1$ be the number of unpaired bases on both sides of the internal loop, respectively. We assume the free energy $eL(i, j, i', j')$ of the internal loop/bulge can be computed as the sum of three parts where

- $size(n_1 + n_2)$: the energy penalty function depending on the number unpaired bases;

- $stacking(i, j)$ and $stacking(i', j')$: the energy penalties for the mismatched base pairs adjacent to the two base pairs (i, j) and (i', j'), respectively; and

- $asymmetry(n_1, n_2)$: the energy penalty for the asymmetry of the two sides of the internal loop.

Precisely, we have

$$eL(i, j, i', j') = \\ size(n_1 + n_2) + stacking(i, j) + stacking(i', j') + asymmetry(n_1, n_2)$$

In practice, $asymmetry(n_1, n_2)$ is approximated by the Ninio equation [229], i.e.,

$$asymmetry(n_1, n_2) = \min\{K, |n_1 - n_2| f(m)\}$$

where $m = \min\{n_1, n_2, e\}$ with K and e are constants. The constant e is proposed to be 1 or 5 in two works [229, 235].

Note that the Ninio equation satisfies $asymmetry(n_1, n_2) = asymmetry(n_1 - 1, n_2 - 1)$. Hence, we have the following lemma which is useful for devising an efficient algorithm for computing internal loop energy.

LEMMA 11.1
Consider $i < i' < j' < j$. Let $n_1 = i' - i - 1$, $n_2 = j - j' - 1$ and $l = n_1 + n_2$. For $n_1 > e$ and $n_2 > e$, we have $eL(i, j, i', j') - eL(i + 1, j - 1, i', j') =$

$$size(l) - size(l - 2) + stacking(i, j) - stacking(i + 1, j - 1)$$

PROOF $eL(i, j, i', j') - eL(i + 1, j - 1, i', j')$
$= [size(l) + stacking(i, j) + stacking(i', j') + asymmetry(n_1, n_2)] -$
$\quad [size(l - 2) + stacking(i + 1, j - 1) + stacking(i', j') +$
$\quad asymmetry(n_1 - 1, n_2 - 1)]$
$= size(l) - size(l - 2) + stacking(i, j) - stacking(i + 1, j - 1)$ ⬜

Based on the assumptions, $VBI(i, j)$ for all $i < j$ can be found in $O(n^3)$ time as follows.

We define new recursive equations VBI' and VBI''. $VBI'(i, j, l)$ equals the minimal energy of an internal loop of size l closed by a pair (i, j). $VBI''(i, j, l)$ also equals the minimal energy of an internal loop of size l closed by a pair (i, j). Moreover, VBI'' requires the number of the bases between i and i' and the number of the bases between j and j', excluding i, i', j and j', are both more than a constant e. Formally, VBI' and VBI'' are defined as follows.

$$VBI'(i, j, l) = \min_{i < i' < j' < j \wedge i' - i - 1 + j - j' - 1 = l} \{eL(i, j, i', j') + V(i', j')\}$$

$$VBI''(i, j, l) = \min_{\substack{i < i' < j' < j \wedge i' - i - 1 + j - j' - 1 = l \wedge \\ i' - i - 1, j - j' - 1 > c}} \{eL(i, j, i', j') + V(i', j')\}$$

Together with Lemma 11.1, we have

$$VBI''(i, j, l) - VBI''(i + 1, j - 1, l) =$$
$$size(l) - size(l - 2) + stacking(i, j) + stacking(i + 1, j - 1)$$

$VBI'(i, j, l) =$

$$\min \begin{cases} VBI''(i + 1, j - 1, l) + size(l) - size(j - 2) + stacking(i, j) \\ \quad -stacking(i + 1, j - 1) \\ \min_{1 \le d \le e} V(i + d, j - l - d) + eL(i, j, i + d, j - l + d \quad 2) \\ \min_{1 \le d \le e} V(i + l + d, j - d) + eL(i, j, i + l - d + 2, j - d) \end{cases}$$

The last two entries of the above equation handle the cases where this minimum is obtained by an internal loop, in which d is less than a constant e.

Now, we analyze the time complexity. The dynamic programming tables for VBI' and VBI'' have $O(n^3)$ entries. Each entry in VBI' and VBI'' can be computed in $O(e)$ and $O(1)$ time, respectively. Thus, both tables can be filled in using $O(en^3)$ time. Given VBI', the table VBI can be filled in using $O(n^2)$ time.

Together with filling in the tables W, V, VM, WM, the time required to predict a secondary structure without a pseudoknot is $O(n^3)$.

11.7   Structure Prediction with Pseudoknots

The previous two sections discuss methods for predicting an RNA secondary structure assuming there is no pseudoknot. However, a pseudoknot really exists in RNA (Pleij and his collaborators [239] gave the first experimental evidence of pseudoknot). Although pseudoknots are not frequent, they are very important in many RNA molecules [69]. For example, pseudoknots form a core reaction center of many enzymatic RNAs, likes RNase P[195] and ribosomal RNA[178]. They also appear at the 5′-end of mRNAs, which acts as a control of translation. Therefore, discovering pseudoknots in RNA molecules is very important.

Up to now, there is no good way to predict an RNA secondary structure with pseudoknots. In fact, this problem is an NP-hard problem [6, 159, 199]. Different approaches have been attempted to tackle this problem. Heuristic search procedures are adopted in most RNA folding methods that are capable of folding pseudoknots. For example, quasi-Monte Carlo searches by Abrahams et al. [1], genetic algorithms by Gultyaev et al. [125], the Akiyama and Kanehisa method [5] based on the Hopfield network, and the Brown and Wilson method [38] based on the stochastic context-free grammar. These approaches cannot guarantee the best structure is found and are unable to say how far a given prediction is from the optimal. Other approaches are based on the maximum weighted matching [116, 278]; they report some success in predicting pseudoknots and base triples.

Based on dynamic programming, Rivas and Eddy [245], Lyngso and Pedersen [199] and Akutsu [6] proposed three polynomial time algorithms which can find an optimal secondary structure for certain kinds of pseudoknots. Their time complexities are $O(n^6)$, $O(n^5)$, and $O(n^4)$, respectively. This section introduces a simple pseudoknot and presents Akutsu's $O(n^4)$-time algorithm for predicting an RNA secondary structure with pseudoknots.

11.7.1   Definition of a Simple Pseudoknot

Consider a substring $S[i_0..k_0]$ of an RNA sequence $S[1..n]$ where i_0 and k_0 are arbitrarily chosen positions. A set of base pairs M_{i_0,k_0} is a simple pseudoknot [6] if there exist j_0, j_0' such that

- Each endpoint i appears in M_{i_0,k_0} once. Each base pair (i, j) in M_{i_0,k_0} satisfies either $i_0 \leq i < j_0' < j \leq j_0$ or $j_0' \leq i < j_0 < j \leq k_0$.

- If the pairs (i, j) and (i', j') in M_{i_0,k_0} satisfy $i < i' < j_0'$ or $j_0' \leq i < i'$, then $j > j'$.

The first part of the definition divides the sequence $S[i_0..k_0]$ into three segments: $S[i_0..j_0], S[j_0..j_0']$, and $S[j_0'..k_0]$. For each base pair in M_{i_0,k_0}, one of its

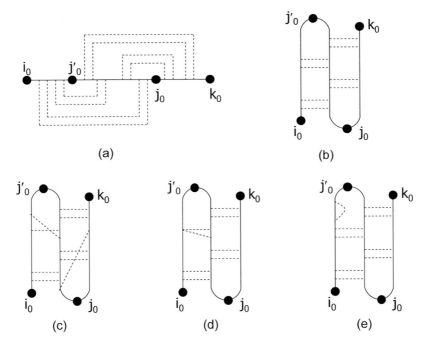

FIGURE 11.9: An illustration of a simple pseudoknot. Figures (a) and (b) are simple pseudoknots while the rest are not.

ends must be in $S[j_0..j_0']$ while the other end is either in $S[i_0..j_0]$ or $S[j_0'..k_0]$. The second part of the definition implies that the base pairs cannot intersect with each other. Figure 11.9(a) shows an example of a simple pseudoknot while Figure 11.9(b) shows an alternative view of the same pseudoknot. Figure 11.9(c–e) are some examples that are not simple pseudoknots.

With the definition of a simple pseudoknot, an RNA secondary structure with simple pseudoknots is defined as below. A set of base pairs M is called an RNA secondary structure with simple pseudoknots if $M = M' \cup M_1 \cup \ldots \cup M_t$ for some non-negative integer t such that

- For $h = 1, 2, \ldots, t$, M_h is a simple pseudoknot for $S[i_h..k_h]$ where $1 \leq i_1 < k_1 < i_2 < k_2 < \ldots < i_t < k_t \leq n$.

- M' is a secondary structure without pseudoknots for the string S' where S' is obtained by removing segments $S[i_h..k_h]$ for all $h = 1, 2, \ldots, t$.

11.7.2 Akutsu's Algorithm for Predicting an RNA Secondary Structure with Simple Pseudoknots

Given an input RNA sequence $S[1..n]$, this section presents a dynamic programming algorithm to report an RNA secondary structure with simple pseu-

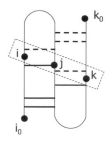

FIGURE 11.10: (i, j, k) is a triplet in the simple pseudoknot. Note that, for all the base pairs in solid lines, they are below the triplet (i, j, k).

doknots that maximizes the number of base pairs. The algorithm requires two recursive equations where $V(i, j)$ is defined to be the optimal score of an RNA secondary structure with simple pseudoknots for the sequence $S[i..j]$ and $V_{pseudo}(i, j)$ is defined to be the optimal score for $S[i..j]$ with the assumption that $S[i..j]$ forms a simple pseudoknot.

For $V(i, j)$, there are three cases: (1) $S[i..j]$ is a simple pseudoknot, (2) (i, j) forms a base pair, or (3) $S[i..j]$ can be decomposed into two compounds. Therefore, we have the following recursive equation.

$$V(i, j) = \max \begin{cases} V_{pseudo}(i, j) \\ V(i+1, j-1) + \delta(S[i], S[j]) \\ \max_{i<k\le j}\{V(i, k-1) + V(k, j)\} \end{cases}$$

where $V(i, i) = 0$ for all i. Also, $\delta(S[i], S[j]) = 1$ if $\{S[i], S[j]\} = \{a, u\}$ or $\{c, g\}$; otherwise, $\delta(S[i], S[j]) = -\infty$.

For $V_{pseudo}(i_0, k_0)$, its value can also be computed using a dynamic programming algorithm. Prior to describe the algorithm, we first give some notations. Recall that, in a simple pseudoknot of $S[i_0..k_0]$, the sequence is partitioned into three segments $S[i_0..j_0']$, $S[j_0'..j_0]$, and $S[j_0..k_0]$ for some unknown positions j_0 and j_0'. We denote the three segments as left, middle, and right segments, respectively (see Figure 11.9(a,b) for an illustration). Any tuple (i, j, k) is called a triplet if $i_0 \le i \le j_0'$, $j_0' < j \le j_0$, and $j_0 < k \le k_0$. A base pair (x, y) is said to be below the triplet (i, j, k) if $(x \le i$ and $y \ge j)$ or $(x \ge j$ and $y \le k)$. A base pair (x, y) is said to be on the triplet (i, j, k) if $(x, y) = (i, j)$ or $(x, y) = (j, k)$. Figure 11.10 is an example illustrating the concept 'below'. All the base pairs in solid lines are below the triplet (i, j, k).

For a triplet (i, j, k), it should satisfy either one of the following relations: (1) (i, j) is a base pair, (2) (j, k) is a base pair, or (3) both (i, j) and (j, k) are not base pairs. Below, we define three variables, based on the above three relationships, which are useful for computing $V_{pseudo}(i_0, k_0)$.

- $V_L(i, j, k)$ is the maximum number of base pairs below the triplet (i, j, k) in a pseudoknot for $S[i_0..k_0]$ given that (i, j) is a base pair;

- $V_R(i, j, k)$ is the maximum number of base pairs below the triplet (i, j, k) in a pseudoknot for $S[i_0..k_0]$ given that (j, k) is a base pair; and

- $V_M(i, j, k)$ is the maximum number of base pairs below the triplet (i, j, k) in a pseudoknot for $S[i_0..k_0]$ given that both (i, j) and (j, k) are not base pairs.

Note that $\max\{V_L(i, j, k), V_R(i, j, k), V_M(i, j, k)\}$ is the maximum number of base pairs below the triplet (i, j, k) in a pseudoknot for $S[i_0..k_0]$. Then $V_{pseudo}(i_0, j_0)$ can be calculated as follows.

$$V_{pseudo}(i_0, k_0) = \max_{i_0 \le i < j < k \le k_0} \{V_L(i, j, k), V_M(i, j, k), V_R(i, j, k)\}$$

11.7.2.1 $V_L(i, j, k)$, $V_R(i, j, k)$, $V_M(i, j, k)$

We define the recursive formulae for $V_L(i, j, k)$, $V_R(i, j, k)$, and $V_M(i, j, k)$.

$$V_L(i, j, k) = \delta(S[i], S[j]) + \max \left\{ \begin{array}{l} V_L(i - 1, j + 1, k), \\ V_M(i - 1, j + 1, k), \\ V_R(i - 1, j + 1, k) \end{array} \right\}$$

$$V_R(i, j, k) = \delta(S[i], S[j]) + \max \left\{ \begin{array}{l} V_L(i, j + 1, k - 1), \\ V_M(i, j + 1, k - 1), \\ V_R(i, j + 1, k - 1) \end{array} \right\}$$

$$V_M(i, j, k) = \max \left\{ \begin{array}{l} V_L(i - 1, j, k), \ V_M(i - 1, j, k), \\ V_L(i, j + 1, k), \ V_M(i, j + 1, k), \ V_R(i, j + 1, k), \\ V_M(i, j, k - 1), \ V_R(i, j, k - 1) \end{array} \right\}$$

Here we provide an intuitive explanation for the formula above. For both $V_L(i, j, k)$ and $V_R(i, j, k)$, the first term represents the base pair on the triplet (i, j, k) while the second term represents the number of base pairs below (i, j, k). For $V_M(i, j, k)$, since there is no base pair on the triplet (i, j, k), the formula only consists of the number of base pairs below the triplet. Note that the two variables, $V_R(i - 1, j, k)$ and $V_L(i, j, k - 1)$, do not appear in the formula for $V_M(i, j, k)$. This is because $V_M(i, j, k)$ indicates that both (i, j) and (j, k) are not base pairs.

To compute $V_L(i, j, k), V_R(i, j, k), V_M(i, j, k)$, some base values are required.

$$\left\{ \begin{array}{ll} V_R(i_0 - 1, j, j + 1) = \delta(S[j], S[j + 1]) & \forall j \quad (1) \\ V_R(i_0 - 1, j, j) = 0 & \forall j \quad (2) \end{array} \right.$$

$$\left\{ \begin{array}{ll} V_L(i, j, j) = \delta(S[i], S[j]) & \forall i_0 \le i < j \quad (3) \\ V_L(i_0 - 1, j, k) = 0 & \forall k = j + 1 \lor k = j \quad (4) \end{array} \right.$$

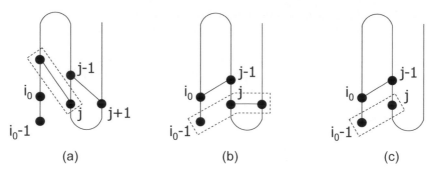

FIGURE 11.11:   Basis for the recursive equations V_L, V_M, and V_R.

$$V_M(i_0 - 1, j, k) = 0 \quad \forall k = j + 1 \vee k = j \quad (5)$$

The base case (3) can be explained by Figure 11.11(a). The base cases (1) and (2) can be explained by Figures 11.11(b) and (c), respectively. The base cases (4) and (5) are trivial since they are out of range.

11.7.2.2   Time Analysis

This section gives the time analysis. First, we analyze the time complexity for computing V_{pseudo}. Observe that the base cases for V_L, V_M, and V_R only depend on i_0. Thus, for a fixed i_0, the values for the base cases of V_L, V_M, and V_R can be computed in $O(n^2)$ time. For a fixed i_0, the values of table V_L, V_R, V_M can be computed in $O(n^3)$ time since each table has n^3 entries and each entry can be computed in $O(1)$ time.

By the definition of V_{pseudo}, we have the following recursive equation: $V_{pseudo}(i_0, k_0) = \max\{V_{pseudo}(i_0, k_0-1), \max_{i_0 < i < j < k_0}\{V_L(i, j, k_0), V_R(i, j, k_0), V_M(i, j, k_0)\}\}$. Thus, for a fixed i_0, $V_{pseudo}(i_0, k_0)$ for all k_0 can be computed in $O(n^3)$ time.

Since there are n choices for i_0, $V_{pseudo}(i_0, k_0)$ for all $1 \leq i_0 < k_0 \leq n$ can be computed in $O(n^4)$ time.

Next, we analyze the time required to fill in the table V. There are n^2 entries in the table V and each entry can be computed in $O(n)$ time. Hence, the table V can be filled in using $O(n^3)$ time.

In summary, given a sequence $S[1..n]$, we can predict the RNA secondary structure of S with simple pseudoknots in $O(n^4)$ time.

11.8   Exercises

1. Consider the following phenylalanyl-tRNA sequence:

GCGGAUUUAGCUCAGUUGGGAGAGCGCCAGACUGAAGAUCUGGAGGUCCUGUGUUCGAUCCACAGAAUUCGCACCA

Can you predict the secondary structure of this sequence using the Zuker MFOLD algorithm (http://mfold.bioinfo.rpi.edu/, with default setting)? How many structures does MFOLD predict? Is there any structure which looks similar to the cloverleaf structure?

2. Given an RNA sequence $R[1..n]$, suppose R forms a secondary structure by maximizing the number of base pairs with the assumption that there are no pseudoknots and no multi-loop. Propose the most efficient algorithm to predict the RNA secondary structure. What is the time complexity?

3. Consider an RNA sequence $S[1..n]$. The Nussinov algorithm computes an RNA secondary structure of S without pseudoknots by maximizing the number of base pairs. However, it does not consider the fact that all hairpin loops in an RNA secondary structure have minimum length ℓ (that is, if (i, j) is a base pair of a hairpin loop, $j - i - 1 \geq \ell$).

   (a) Can you give an efficient algorithm to compute the maximum number of base pairs with this additional requirement? What is the running time?

   (b) Can you describe an efficient algorithm to recover the corresponding secondary structure? (Store the optimal structure using parenthesis representation in an array $A[1..n]$. For example, if $\ell = 3$ and $S =$ACGUUUUACGU, then the output is $A =$((((...)))).)

4. Consider an RNA sequence $S[1..n]$. Suppose the only allowed base pairs are (A, U), (U, A), (C, G), and (G, C). A secondary structure of S is called a thin secondary structure if it satisfies the following criteria:

   • There exist t disjoint intervals $[i_1..j_1], [i_2..j_2], \ldots, [i_t..j_t]$ such that, for every interval $[i_k..j_k]$, there is no base pair connecting a base in the interval and a base outside the interval.

   • Every interval $[i_k..j_k]$ forms a pseudoknot. Moreover, the pseudoknot can be broken by removing one base pair in the interval.

   • Finally, the base pairs outside all the intervals do not form any pseudoknots.

   Our aim is to find a thin secondary structure maximizing the total number of base pairs. Can you propose an efficient algorithm to compute such an RNA secondary structure of S? What is the time complexity?

5. Given an RNA sequence $S[1..n]$, we would like to form a secondary structure which maximizes the number of tri-stacking pairs. A tri-stacking pair is a stacking pair of length three, that is, it is formed by 3 consecutive base pairs (i, j), $(i + 1, j - 1)$, and $(i + 2, j - 2)$ for some i, j. For

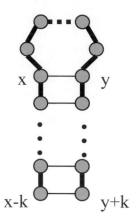

FIGURE 11.12:   Hairpin stem

example, for an RNA sequence ACCCCAGGGGACCCAGGG, the best secondary structure is .((((.)))) .(((.))), which contains three tri-stacking pairs. Let $t(S)$ be the maximum number of tri-stacking pairs formed in the RNA sequence S. Propose an $O(n^3)$-time algorithm to compute $t(S)$. (Assume $C-G$ and $A-T$ are the only allowed interacting base pairs.)

6. Consider a chromosome $S[1..n]$. Figure 11.12 illustrates a hairpin stem. Formally, a hairpin stem consists of (1) a hairpin whose size is in between two constants c_1 and c_2, that is, $c_1 \leq y - x - 1 \leq c_2$ and (2) a stem such that $S[x - i] = c(S[y + i])$ for $i = 0, 1, \ldots, k$ where $c(z)$ returns the complementary base of z (the complement of A is T and the complement of C is G). In this case, we say $S[x - k..y + k]$ is a hairpin stem.

   (a) Give an efficient algorithm that reports the longest hairpin stem $S[i..j]$ in S (that is, the hairpin stem $S[i..j]$ which maximizes $(j - i + 1)$). State the time complexity of the algorithm.

   (b) A region $S[i..j]$ is called a d-mismatch hairpin stem if the corresponding stem contains at most d base pairs which are not complementary. Give an efficient algorithm that reports the longest d-mismatch hairpin stem $S[i..j]$ in S. State the time complexity of the algorithm.

7. Given an RNA sequence $R[1..n]$. Let S be the set of base pairs. Note that S may contain pseudoknots and every base can only pair with exactly one other base. A subset $Q \subseteq S$ is said to be a chain if, for any $(p_1, q_1), (p_2, q_2) \in Q$, we have either $p_1 < q_1 < p_2 < q_2$ or $p_2 < q_2 < p_1 < q_1$. Can you propose an efficient algorithm to identify the size of

the biggest subset $Q \subseteq S$ such that Q is a chain? What is the time complexity?

8. Given an RNA sequence $R[1..n]$. Let S be the set of base pairs. Note that S may contain pseudoknots and every base can only pair with exactly one other base. A subset $Q \subseteq S$ is said to be with pseudoknot if, for some $(p_1, q_1), (p_2, q_2) \in Q$, we have $p_1 < p_2 < q_1 < q_2$. Can you propose an efficient algorithm to identify the size of the biggest subset $Q \subseteq S$ such that Q has no pseudoknot? What is the time complexity?

9. Given an RNA sequence $S[1..n]$. Suppose the only allowed base pairs are a-u, c-g, and g-u. Can you propose an efficient algorithm to compute the maximum number of stacking pairs appearing in a secondary structure of S? What is the time complexity of your algorithm?

Chapter 12

Peptide Sequencing

12.1 Introduction

A protein is a long chain of amino acids, known as a polypeptide. The sequence of amino acids in each protein is unique and plays a major role in determining its 3D conformation, its charge distribution, and ultimately its function. Thus, knowing this sequence is pertinent to further investigation of a protein's functions and its involvement in biological processes. However, sequencing the complete protein is still difficult. It is currently only possible to sequence peptide fragments of length ≈ 10 in a high-throughput manner. Nevertheless, peptide sequencing is already very useful and finds many applications.

High-throughput peptide sequencing is based on the observation that amino acids have different masses. The sequence of a peptide can be deduced through measuring the masses. The enabling technology for measuring the mass of peptides is the mass spectrometer. Mass spectrometry (MS) is the technique used to separate the components of a sample according to their masses,[1] commonly expressed in Daltons (*Da*).[2] Figure 12.1 shows a simple hypothetical MS spectrum derived from a sample with three components (assuming no noise). Each component produces a peak in the MS spectrum at its mass value. The height of a peak indicates the relative abundance of its component. Depending on the equipment used, the error of the mass measurement ranges from ± 0.01 to ± 0.5 Da.

In this chapter, we first discuss how an experimental mass spectrum of a peptide is obtained. Then, we detail the computational methods for obtaining the peptide sequences from the mass spectrum de novo or through database search.

---

[1]In reality, the mass spectrometer separates the components according to their mass charge ratios instead of their masses. Moreover, we assume the majority of the components have unit charge. We use the term "mass" for simplicity.

[2]1 Dalton is the mass of a hydrogen atom.

FIGURE 12.1:   Given a sample with three components where component 1 has mass=100 Da and volume = 10 mol, component 2 has mass=50 Da and volume = 50 mol, and component 3 has mass=33 Da and volume = 30 mol. The figure shows the corresponding mass spectrum.

12.2   Obtaining the Mass Spectrum of a Peptide

This section discusses how to obtain the mass spectrum of a peptide. Figure 12.2 summarizes the entire process of sequencing a peptide.

Since a protein is too long for direct sequencing, the first step cuts the protein into short peptide fragments through digestion of enzyme protease.[3] The resultant peptides are of length $\approx$ 10 amino acids.

For example, if the protease *trypsin* is used, the protein will be cut at K or R provided they are not followed by P. After digestion, we will get a set of peptides mostly ending with K or R. For example, the digestion of the sequence below by trypsin gives three peptides:

$$\text{ACCHC}\underline{\mathbf{K}}\text{CCVR}\text{PPC}\underline{\mathbf{R}}\text{CA} \rightarrow \text{ACCHC}\underline{\mathbf{K}}, \text{CCVRPPC}\underline{\mathbf{R}}, \text{CA}$$

Note that the protein is not cut at the first **R**, because the next residue is P.

Given the mixture of peptides, the second step separates those peptides. This step first uses High Performance Liquid Chromatography (HPLC) to separate the mixture by the polarity of the compounds. Then, the mixture is further separated by mass spectrometry according to the masses of the compounds. Each peptide of a specific mass corresponds to a peak in the spectrum (see Figure 12.3).

The peptide of a particular mass is selected and further fragmented at random positions along the peptide backbone. Usually, fragmentation is performed by Collision Induced Dissociation (CID). The peptide is passed into a collision chamber containing a pressurized inert gas (e.g., argon).[4] The high

---

[3]Protease is an enzyme which breaks peptide bonds between amino acids of proteins.
[4]An inert gas avoids any chemical reaction.

Input: A Protein Sample

A. Biology/Experimental Part:

1. Digest the protein into a set of peptides;
2. By high performance liquid chromatography (HPLC) and mass spectrometry, separate the peptides;
3. Select a particular peptide;
4. Fragment the selected peptide;
5. Obtain the tandem mass spectrum of the selected peptide.

B. Computational Part:

1. De novo sequencing, or
2. Protein database search.

FIGURE 12.2: Protein identification process using the liquid chromatography tandem mass spectrometry (LC-MS/MS) method.

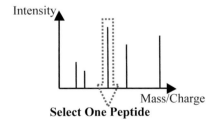

FIGURE 12.3: Spectrum of peaks after HPLC and MS. Each represents one peptide.

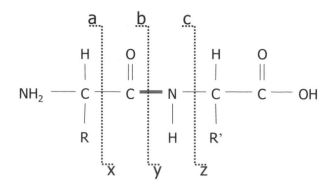

FIGURE 12.4:   Ions resulting from a peptide with 2 amino acids.

pressure causes the peptide ions to collide with the argon atoms, thus breaking the bonds along the peptide backbone.

Since most of the bonds are broken along the peptide backbone, the peptide can be broken at three locations: C–C, C–N, and N–C bonds. Figure 12.4 shows how a peptide with two amino acids is fragmented. The resulting fragment ions with the N-terminus are the a-ions, b-ions, and c-ions, while the ions with the C-terminus are the x-ions, y-ions, and z-ions.

Since the peptide C–N bond is easier to be broken during fragmentation, the most abundant ions are y-ions and b-ions, with the abundance of y-ions a bit bigger than that of b-ions. The other ions are less abundant.

After fragmentation in the collision chamber, the fragment mixture is passed to the mass spectrometer to obtain the tandem mass spectrum[5] (also called the MS/MS spectrum) of the peptide. In the mass spectrum, the fragment ions are separated according to their mass (represented horizontally) and their relative abundance (or intensity) are indicated by the heights (see Figure 12.5). We let $\mu_M(r)$ be the intensity of the peak of mass r in the spectrum M.

Ideally, the tandem mass spectrum contains only peaks corresponding to the various ions of the peptide. In real life, the spectrum contains many noise peaks while it may miss peaks of some ions. Figure 12.6 shows an example of a real tandem mass spectrum. The spectrum contains both signal peaks and noise peaks.

---

[5]The mass spectrum is called the tandem mass spectrum because it is obtained through the repeated applications of mass spectrometry.

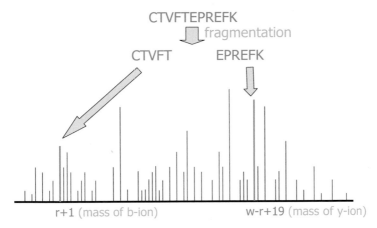

FIGURE 12.5: Consider a peptide $CTVFTEPREFK$ of mass $w = 1228.55$. The fragmentation between amino acid residues T and E generates two fragments $CTVFT$ and $EPREFK$ of mass $r = 551.65$ and $w - r = 676.9$, respectively. The mass spectrum contains the corresponding b-ion and y-ion.

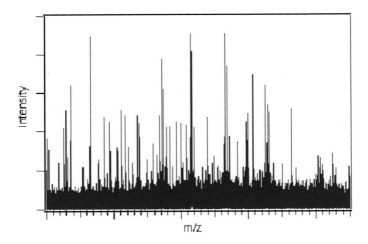

FIGURE 12.6: An example of an experimental tandem mass spectrum (MS/MS spectrum).

a	$w(a)$	a	$w(a)$
A	71.08	M	131.19
C	103.14	N	114.1
D	115.09	P	97.12
E	129.12	Q	128.13
F	147.18	R	156.19
G	57.05	S	87.08
H	137.14	T	101.1
I	113.16	V	99.13
K	128.17	W	186.21
L	113.16	Y	163.18

FIGURE 12.7:   Average mass $w(a)$ of amino acid residues.

12.3   Modeling the Mass Spectrum of a Fragmented Peptide

In order to identify a peptide sequence from its tandem mass spectrum, we need to know how to model it first. The modeling problem is: Given a peptide sequence, we want to predict its mass spectrum.

12.3.1   Amino Acid Residue Mass

As each peptide is composed of amino acid residues,[6] their masses must be known in order to derive the theoretical mass spectrum of a peptide. Let A be the set of amino acids and $w(a)$ be the mass of the amino acid residue $a \in A$ (in Daltons). Figure 12.7 shows the value of $w(a)$ for all 20 amino acids.

Note that $w(I) = w(L)$. Hence, these two amino acids are indistinguishable by any method based on mass and are treated as the same amino acid. The residue with the smallest mass is G (glycine) of mass 57.05 Da and the one with the largest mass is W (tryptophan) of mass 186.21 Da.

12.3.2   Fragment Ion Mass

Given a peptide $P = a_1 a_2 \ldots a_k$ with k amino acid residues, P is fragmented and ionized as explained in Section 12.2. Then, the tandem mass spectrum is formed from those fragmented ions of P. Figure 12.8 shows the most common fragmentation mode and its resultant ions (b-ions and y-ions). In the tandem mass spectrum, each ion generates a peak.

---

[6]An amino acid residue is an amino acid after the removal of a water molecule.

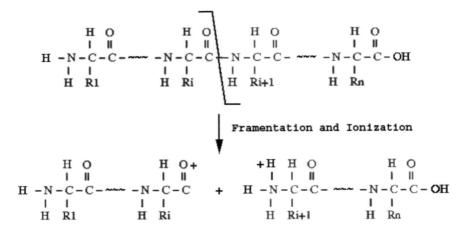

FIGURE 12.8: Fragmentation of a peptide into a b-ion (bottom left) and a y-ion (bottom right).

Below we describe the theoretical spectrum of the peptide P. For the sake of simplicity and illustration, the following assumptions are made: (1) fragmentation only occurs at a C–N bond (hence resulting in b- and y-ions), (2) each peptide molecule is fragmented at most one time, (3) each fragment has unit charge, and (4) the probability of fragmentation at each position on the peptide is uniform.

For $i = 1, 2, \ldots, k - 1$, the breaking of the C–N bond between the i-th and the $(i + 1)$-th residues results in the C-terminal ion and the N-terminal ion. The N-terminal ion is a b-ion for $a_1 \ldots a_i$. Because a b-ion has an extra H, the mass of the b-ion is $b_i = w(a_1 \ldots a_i) + 1$. The C-terminal ion is a y-ion for $a_{i+1} \ldots a_k$. Because a y-ion has three extra H ions and one extra O ion, the mass of the y-ion is $y_{i+1} = w(a_{i+1} \ldots a_n) + 19$. Note that $b_i + y_{i+1} = w(P) + 20$.

For example, for a peptide $P = SAG$, $w(P) = w(S) + w(A) + w(G) = 87.08 + 71.08 + 57.05 = 215.21$. Its b-ions and y-ions are as follows.

$$w(P) = w(S) + w(A) + w(G) = 87.08 + 71.08 + 57.05 = 215.21$$
$$y_1 = w(SAG) + 19 = 234.21$$
$$y_2 = w(AG) + 19 = 147.13$$
$$y_3 = w(G) + 19 = 76.05$$
$$b_1 = w(S) + 1 = 88.08$$
$$b_2 = w(SA) + 1 = 159.16$$
$$b_3 = w(SAG) + 1 = 216.21$$

Figure 12.9 shows the theoretical spectrum of $P = SAG$ containing 6 peaks, each corresponding to either a b-ion or a y-ion.

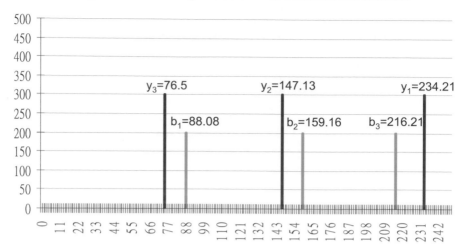

FIGURE 12.9:   The figure shows the theoretical spectrum for $P = SAG$, which includes peaks for y-ions (in red) and b-ions (in green). The b-ions are drawn lower than the y-ion peaks to show their relative scarcity. (See color insert after page 270.)

The above mass spectrum model is a simplified model. We can enrich the spectrum with peaks of other types of N-terminal and C-terminal ions. For an N-terminal peptide, in addition to b-ions, the other abundant ions are $a, a - NH_3, a - H_2O, b - H_2O, b - NH_3$ ions. For a C-terminal peptide, in addition to y-ions, the other abundant ions are $y - H_2O, y - NH_3$ ions. We can enrich the theoretical spectrum of P with peaks of those abundant ions (see Exercise 2).

Furthermore, we can enrich the model by peaks of the following ions: (a) the isotopic ions, (b) multiply charged ions, and (c) ions generated by multiple fragmentations.

12.4   De Novo Peptide Sequencing Using Dynamic Programming

Suppose a raw tandem mass spectrum M is generated from a peptide of mass W (with measurement error bounded within $\pm\delta$). The de novo peptide sequencing problem asks for a peptide sequence P which best explains the spectrum M. Precisely, we want to compute a peptide sequence P whose theoretical spectrum is similar to M (where the similarity is defined by some scoring function) and whose mass is between $W - \delta$ and $W + \delta$.

In this section, we describe a dynamic programming solution for solving this problem. The next section discusses another method based on the graph-based approach.

12.4.1 Scoring by Considering y-Ions

Since fragmentation of a peptide produces relatively high abundance of y-ions, the tandem mass spectrum is expected to contain many y-ion peaks. This section discusses a method to sequence the peptide using the peaks from y-ions.

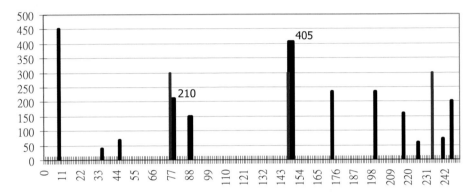

FIGURE 12.10: The figure shows an example spectrum for $P = SAG$. The tandem mass spectrum peaks are black while the theoretical y-ion peaks are red. (See color insert after page 270.)

Consider a peptide $P = a_1 \ldots a_k$. For any j, the mass of the y-ion of the C-terminal peptide $a_j \ldots a_k$ of P is $w(a_j \ldots a_k) + 19$. A simple score function $score_Y(M, P)$ is defined which measures the sum of intensity of the y-ion peaks of all C-terminal peptides of P:

$$score_Y(M, P) = \sum \{\mu_M(r) \mid |r - w(a_j \ldots a_k) - 19| \leq \delta \text{ for } j = 1, 2, \ldots, k\}$$
$$(12.1)$$

For example, consider the peptide $P = SAG$ and the spectrum M shown in Figure 12.10. $score_Y(M, P)$ is $210 + 405 + 0 = 615$.

If the value of $score_Y(M, P)$ is big, we expect P fits the spectrum M. Hence, the peptide sequencing problem can be restated as finding a peptide P which maximizes $score_Y(M, P)$ among all peptides such that $|w(P) - W| \leq \delta$.

A straightforward way to maximize $score_Y(M, P)$ is to iterate through every possible peptide P with $|w(P) - W| \leq \delta$. Among all such peptides P, we report the P which maximizes $score_Y(M, P)$. However, this is an exponential time algorithm.

Algorithm Max_Y_Ion
Require: The mass spectrum M and a weight W
Ensure: A peptide P of mass between $W - \delta$ and $W + \delta$ which maximizes $score_Y(M, P)$.
 1: Set $V(r) = 0$ for $r < 0$
 2: **for** $r = 0$ to $W + \delta$ **do**
 3:    $V(r) = \max_{a \in A}\{V(r - w(a)) + f_M(r + 19)\}$
 4: **end for**
 5: $r' = \arg\max_{W - \delta \le r \le W + \delta} V(r)$
 6: Through back-tracing, we find the peptide P of mass r' which maximizes $score_Y(M, P)$

FIGURE 12.11:   Algorithm for computing the peptide P which maximizes $score_Y(M, P)$.

This section presents a polynomial time solution using dynamic programming. For every mass r, we let $V(r)$ be the maximum $score_Y(M, P)$ among all peptides P of weight r. Our aim is to compute the maximum score $score_Y(M, P)$ among all peptides P of weight between $W - \delta$ and $W + \delta$, i.e., to compute $\max_{W - \delta \le r \le W + \delta} V(r)$.

Below, we define the recursive formula for $V(r)$. For $r \le 0$, $V(r) = 0$ since there is no peak of mass less than 0. For $r > 0$, we have the following recursive formula.

$$V(r) = \max_{a \in A}\{V(r - w(a)) + f_M(r + 19)\}$$

where $f_M(r)$ is the sum of intensity of peaks whose mass is between $r - \delta$ and $r + \delta$, that is, $\sum\{\mu_M(x) \mid |x - r| \le \delta\}$.

Figure 12.11 shows the algorithm which computes $V(r)$ based on the recursive formula. Then, the peptide sequence P can be recovered by back-tracing. Below is the time analysis. For each $0 \le r \le W + \delta$, $V(r)$ can be computed in $O(|A|)$ time. Hence, the V table can be filled in using $O(|A|W)$ time. Back-tracing takes $O(W)$ time. In total, the algorithm in Figure 12.11 runs in $O(|A|W)$ time.

12.4.2   Scoring by Considering y-Ions and b-Ions

The dynamic programming (DP) algorithm described in Section 12.4.1 considers only y-ions. Moreover, b-ions are also abundant after a peptide is fragmented. Considering both y-ions and b-ions will improve the theoretical model of the tandem mass spectrum.

Consider a peptide $P = a_1 a_2 \ldots a_k$. If M is the mass spectrum for the peptide P, M is likely to contain peaks for the y-ions with masses $y_j = w(a_j \ldots a_k) + 19$ for $j = 1, 2, \ldots, k$ and b-ions with masses $b_i = w(a_1 \ldots a_i) + 1$

for $i = 1, 2, \ldots, k$. We redefine the score function as the sum of intensities of both y-ion and b-ion peaks of P:

$$\text{score}(M, P) = \sum \{\mu_M(r) \mid |r - y_i| \leq \delta \text{ or } |r - b_i| \leq \delta \text{ for } i = 1, \ldots, k\} \quad (12.2)$$

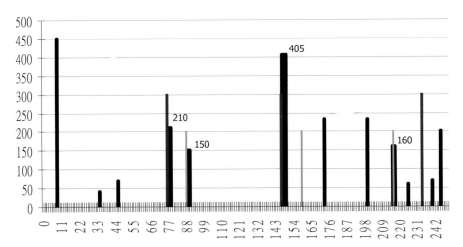

FIGURE 12.12: The figure shows the same spectrum as Figure 12.10 for $P = SAG$. The tandem mass spectrum peaks are black. The theoretical y-ion and b-ion peaks are red and green, respectively. (See color insert after page 270.)

As shown in Figure 12.12, by matching the hypothetical peaks in Figure 12.9 for $P = SAG$ to that of the tandem mass spectrum M, we have a similarity score $score(M, P) = f_M(y_1) + f_M(y_2) + f_M(b_1) + f_M(b_3) = 210 + 405 + 150 + 160 = 925$. If this similarity score is the highest among all possible peptides, we deduce that $P = SAG$ is the optimal peptide sequence for M.

At first glance, this problem can be solved by generalizing the previous dynamic programming solution to add the sum of intensities of both b- and y-ions. However, the previous dynamic programming algorithm cannot be used because whenever a peak in M is matched by a b-ion and a y-ion of approximately equal mass, the height of this peak is summed twice. Such an algorithm tends to match the high intensity peaks more than once, rather than match more peaks. Hence, we need a modified DP algorithm that avoids double counting [202].

Recall that breaking a peptide P into $a_1 \ldots a_i$ and $a_{i+1} \ldots a_k$ will generate a b-ion and a y-ion of weights b_i and y_{i+1}, respectively. Note that $b_i + y_{i+1} = W + 20$. Let m be $(W + 20)/2$. Note that $b_i < m < y_{i+1}$. If we consider (b_i, y_{i+1}) as an interval on a line, the intervals (b_i, y_{i+1}), for $i = 1, \ldots, k - 1$,

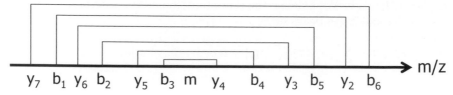

FIGURE 12.13: (b_i, y_{i+1}), for all $i = 1, \ldots, k - 1$, forms a set of nested intervals. Note that all nested intervals cover the mass $m = \frac{1}{2}(W + 20)$ where W is the mass of the peptide. Also, when some b-ions and y-ions have the same mass, it is possible that the two intervals have the same endpoints.

are nested. Figure 12.13 shows an example of the nested intervals which are separated by the same mid-point m.

For the b-ion and y-ion peaks corresponding to the outermost ℓ intervals, they are formed by breaking the prefix $a_1 \ldots a_i$ and the suffix $a_j \ldots a_k$ of the peptide P where $i + (k - j + 1) = \ell$. Let $score'(M, a_1 \ldots a_i, a_j \ldots a_k)$ be the sum of the intensities of all b-ion and y-ion peaks formed by breaking the peptide P between a_x and a_{x+1} for $x \in \{1, \ldots, i\} \cup \{j - 1, \ldots, k - 1\}$. We have the following recursive equation.

$$score'(M, a_1 \ldots a_i, a_j \ldots a_k)$$
$$= \begin{cases} score'(M, a_1 \ldots a_{i-1}, a_j \ldots a_k) + f_M(b_i, y_j) \text{ if } b_i \geq y_j \\ score'(M, a_1 \ldots a_i, a_{j+1} \ldots a_k) + f_M(y_j, b_i) \text{ otherwise} \end{cases}$$

where $b_i = w(a_1 \ldots a_i) + 1$, $y_j = w(a_j \ldots a_k) + 19$, and $f_M(r, s) =$ the sum of intensities of peaks in M near masses r and $W - r$, but not near masses s and $W - s$.

Based on the observation, we derive a modified dynamic programming solution. The dynamic programming solution computes the sum of the intensities of the peaks of the outermost interval first. Then, the sum is incremented by including the intensities of the peaks of the inner intervals one by one. We define $V(r, s)$ to be the maximum $score'(M, P_1, P_2)$, among all peptides P_1 and P_2 satisfying $w(P_1) = r$ and $w(P_2) = s$. V satisfies the recursive relation in the following lemma.

LEMMA 12.1

$$V(r, s) = \max \begin{cases} max_{a \in A}\{V(r - w(a), s) + f_M(r + 1, s + 19)\}, \ r \geq s \\ max_{a \in A}\{V(r, s - w(a)) + f_M(s + 19, r + 1)\}, \ r < s \end{cases}$$

with base case $V(r, s) = 0 (r \leq 0, s \leq 0)$.

Algorithm Sandwich

Require: A mass spectrum M, a weight W, and an error bound δ

Ensure: A peptide P such that $score(M, P)$ is maximized and $|w(P) - W| \leq \delta$

1: Let $\hat{a} = \max_{a \in A} w(a)$
2: Initialize all $V(r, s) = -\infty$; Let $V(0, 0) = 0$
3: **for** $r = 1$ to $(W/2 + \hat{a})$ **do**
4:     **for** $s = r - \hat{a}$ to $\min\{r + \hat{a}, W - r\}$ **do**
5:         **for** $a \in A$ such that $r + s + w(a) < W$ **do**
6:             **if** $r < s$ **then**
7:                 $V(r, s) = \max\{V(r, s), V(r - w(a), s) + f_M(r + 1, s + 19)\}$
8:             **else**
9:                 $V(r, s) = \max\{V(r, s), V(r, s - w(a)) + f_M(s + 19, r + 1)\}$
10:             **end if**
11:         **end for**
12:     **end for**
13: **end for**
14: Identify the best $V(r, s)$ among all r, s, a satisfying $|r - s| < \hat{a}$ and $|r + s + w(a) - W| < \delta$. Through back-tracing, we can recover a peptide $P = P'aP''$ where $w(P') = r$ and $w(P'') = s$.

FIGURE 12.14: Sandwich algorithm.

Utilizing the recursive formula for V in Lemma 12.1, Figure 12.14 shows the algorithm for computing the peptide P such that $score(M, P)$ is maximized and $|w(P) - W| \leq \delta$. Note that the algorithm does not fill in $V(r, s)$ for all $0 \leq r, s \leq W/2$. We only need to fill in $V(r, s)$ for $|r - s| \leq \hat{a}$ where $\hat{a} = \max_{a \in A} w(a)$ (why? See Exercise 5). Each entry can be computed in $O(|A|)$ time. The time complexity of this algorithm is $O(|A| \cdot W \cdot \hat{a})$.

This algorithm was implemented in Peaks [202]. Although the algorithm we described only handles b-ions and y-ions, the algorithm can be generalized to handle other abundant N-terminal and C-terminal ions.

12.5 De Novo Sequencing Using Graph-Based Approach

Another approach to solve the de novo sequencing problem is to reformulate it as a graph problem. This section discusses the method Sherenga [70], one of the first graph-based methods proposed.

Consider an experimental tandem mass spectrum M of some peptide P of mass W. Recall that a peak of mass r in M corresponds to a b-ion if

$r = w(P') + 1$ for some N-terminal fragment P' of P. Similarly, a peak of mass r corresponds to a y-ion if $r = W - w(P') + 19$ for some N-terminal fragment P' of P. A spectrum graph $G(M)$ is a directed graph defined as follows. For every peak of mass r, we generate two vertices of masses $r - 1$ and $W - r + 19$. In addition, we include two vertices v_{start} and v_{end} of masses 0 and W, respectively. A directed edge from u to v exists if $v - u$ is the mass of some amino acid $a \in A$ and the edge is labeled by a. Figure 12.15 shows an example of the spectrum graph.

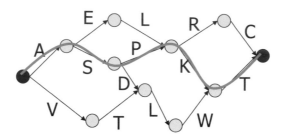

FIGURE 12.15:   An example spectrum graph which consists of a start vertex and an end vertex. There are multiple paths from the start vertex to the end vertex. The highlighted one (ASPKT) is the correct path.

Assume (1) the mass spectrometer is infinitely accurate and (2) the fragmentation process breaks the fragment at every possible position. Then, $G(M)$ contains a path from v_{start} to v_{end} which corresponds to the entire peptide. However, in reality, neither of the two assumptions holds. First, the accuracy of the measurement of the mass spectrometer has an error $\pm \delta$. To resolve it, $G(M)$ contains an edge between two vertices u and v of label a if $|v - u - w(a)| < \delta$. Second, the fragmentation process may be incomplete. To account for it, we include an edge in $G(M)$ between two vertices u and v if $|v - u|$ equals the mass of a dipeptide or a tripeptide.

There may be multiple paths from v_{start} to v_{end} in the spectrum graph $G(M)$. Sherenga assigns a score to every path such that the score of each path measures how well the path (and hence, one corresponding peptide sequence) matches with the spectrum M. Then, Sherenga reports the path with the highest score.

Below, we describe the scoring function of Sherenga. Assume that the spectrometer produces noise uniformly and randomly with probability q_R. Furthermore, we assume q_b is the probability that the b-ion peak exists in M given the b-ion appears in the theoretical spectrum. Similarly, we assume q_y is the probability that the y-ion peak exists in M given the y-ion appears in the theoretical spectrum. The probabilities q_R, q_b, q_y can be determined empirically.

For every vertex of mass v in the spectrum graph $G(M)$, the score of v, $score(v)$, is defined as the sum of two log-odd ratio $score_b(v)$ and $score_y(v)$ where

$$score_b(v) = \begin{cases} \log \frac{q_b}{q_R} & \text{if } v + 1 \text{ exists in } M \\ \log \frac{1-q_b}{1-q_R} & \text{otherwise} \end{cases}$$

$$score_y(v) = \begin{cases} \log \frac{q_y}{q_R} & \text{if } W - v + 19 \text{ exists in } M \\ \log \frac{1-q_y}{1-q_R} & \text{otherwise} \end{cases}$$

A more accurate estimation of $score(v)$ is to include other types of ions. Also, we can account for the intensity of the peak in the spectrum M.

The score of a path in $G(M)$ equals the sum of the log-odd ratio scores of all its vertices. With the scoring function, the peptide sequencing problem can be cast as a longest path problem in a weighted directed acyclic graph. Standard methods can then be used to solve this problem.

12.6   Peptide Sequencing via Database Search

For many cases, de novo peptide sequencing may not be able to recover the peptide sequence. Then, we try to match the spectrum to the protein database to identify the peptide. The problem of peptide sequencing via database search is formulated as follows: The input is a database D of protein sequences and a raw tandem mass spectrum M. Suppose the spectrum M is generated from a protein peptide in D of mass within $W - \delta$ and $W + \delta$. Our aim is to find a protein peptide from D which can best explain the tandem mass spectrum M.

This section presents a solution called SEQUEST [94] to solve the problem. SEQUEST scans the database and computes the theoretical spectrum of every peptide of mass within $W - \delta$ and $W + \delta$. Then, it computes the matching score of the theoretical spectrum of every peptide with the experimental spectrum. The peptides are ranked according to the matching score.

Precisely, SEQUEST consists of four steps. Step 1 removes the noise in the tandem mass spectrum. It avoids noise by retaining only the 200 most abundant peaks of the raw spectrum. Also, the total sum of signals of the 200 peaks is renormalized to 100.

Step 2 searches the protein database D to find all peptides such that the mass of each peptide P is within $W - \delta$ and $W + \delta$.

In Step 3, each peptide P identified in Step 2 is assigned a preliminary score S_P, which is defined as follows. For each peptide P, let Π be the set of predicted fragment ions of P. Suppose Π' is the subset of Π whose peaks appear in the experimental spectrum M. Then, S_P is defined to be

$$\frac{|\Pi'|}{|\Pi|}\left(\sum_{r\in\Pi'}\mu_M(r)\right).$$

Step 4 selects the top 500 fit sequences with the highest preliminary scores and reranks them by the matching score which is defined as follows. For each of the top 500 fit sequences P, SEQUEST creates a theoretical spectrum M_{th} containing the following peaks:

- peaks for all y-ions and b-ions of P. They are assigned intensity 50.

- peaks for ± 1 of y-ions and b-ions of P. They are assigned intensity 25.

- peaks for a-ions and natural loss of water and ammonia of P. They are assigned intensity 10.

A matching score is generated by comparing the theoretical spectrum with the experimental spectrum M using cross-correlation analysis. The cross-correlation R_τ equals $\sum_{i=1}^{W+\delta} \mu_{M_{th}}(i)\mu_M(i+\tau)$. The matching score equals $R_0 - \frac{1}{150}\sum_{\tau=-75}^{75} R_\tau$. The top 500 fit sequences are ranked by the matching score and are reported by SEQUEST.

12.7   Further Reading

For de novo peptide sequencing, in addition to Peaks [202] and Sherenga [70], other methods are Lutefish [284, 285], SeqMS [105], Compute-Q [59], NovoHMM [107], and PepNovo [111]. Apart from NovoHMM and Peaks, all methods are based on spectrum graph approaches. They introduce different scoring functions to score the predicted peptides. NovoHMM employs the Hidden Markov Model.

For database searching, in addition to SEQUEST [94], other popular software includes Mascot [233], Tandem [66], and OMSSA [118]. Recently, researchers tried to utilize de novo sequencing to help to improve the efficiency and the accuracy of database searches. Methods include SPIDER [133], GutenTag [279], MS-BLAST [263], OpenSea [262], DeNovoID [131], and InsPect [281].

Another topic is identifying post-translational modifications (PTMs). When some amino acid residues in a protein peptide are modified, the experimental mass spectrum will look different from the theoretical mass spectrum of the protein peptide. By studying the shift of the peaks in the mass spectrum, the post-translational modifications can be identified. The first approach to PTM identification was proposed by Yates et al. [314]. Early methods required the users to specify a list of modifications. Recently, new methods have been proposed which can identify arbitrary modifications (up to some size limit). Those works include SPIDER [133], OpenSea [262], MS-Alignment [294], and PTMFinder [282].

12.8 Exercises

1. Can you take an experimental spectrum and try PepNovo, SEQUEST?

2. Consider the sequence $SAGTPV$. Can you show its theoretical spectrum? In the spectrum, show the peaks for b-ion, a ion, $(b\text{-}H_2O)$-ion, y-ion, and $(y\text{-}NH_3)$-ion.

3. Consider a protein sequence $STACKPCHGCKCCTVRPTCRGS$. After the digestion of trypsin, what is the set of peptides?

4. Given a spectrum M, Section 12.4.1 describes a dynamic programming algorithm which finds a peptide P maximizing the sum of intensities of the peaks corresponding to the y-ions of P. Can you modify the algorithm so that it reports a peptide P which maximizes the sum of intensities of the peaks corresponding to the y-ions, $(y\text{-}H_2O)$-ions, and $(y\text{-}NH_3)$-ions of P.

5. For the sandwich algorithm in Figure 12.14, can you explain why we only need to fill in entries $V(r, s)$ for $|r - s| \leq \hat{a}$?

6. Consider a spectrum M with a set of peaks $\{88.08, 130.12, 175.19, 201.2, 232.24, 258.25, 303.32\}$. Suppose $q_R = 0.1$, $q_b = 0.8$, and $q_y = 0.9$. Also, the weight of the peptide is 413.44. Can you construct the spectrum graph $G(M)$? Can you predict the peptide sequence from $G(M)$?

Chapter 13

Population Genetics

13.1 Introduction

As stated in Chapter 1, although our genomes are highly similar (approximately 99.9% identical), genetic variations (polymorphism) exist in different individuals. Those variations give different skin colors, different outlooks, and also different genetic diseases. This chapter discusses various strategies to study the differences in the human population.

13.1.1 Locus, Genotype, Allele, and SNP

Humans are diploid. We have two copies of each chromosome. One is inherited from the father while another one is inherited from the mother. Any location in a particular pair of chromosomes in our genome is called a locus. For a locus of each individual, it is occupied by two nucleotides, one in each chromosome. Such a pair of nucleotides is called the genotype at the locus.

An allele is a viable nucleotide that occupies a given locus of some individuals in the human population. In general, a locus may have four possible alleles $\{A, C, G, T\}$. Moreover, since mutation is rare, most of the loci have only one possible allele. Locus with more than one possible allele is called Single-Nucleotide Polymorphism (SNP). Precisely, SNP is the locus where there is a single nucleotide variation among different individuals. It is the most common type of polymorphism among different individuals. Each SNP normally has two alleles and is called diallelic.

Consider a particular locus which is an SNP. Suppose A and a represent the two possible alleles of this locus. Then, this locus has three possible genotypes: AA, aa, and Aa. As AA and aa are pairs of the same allele, they are called homozygous; otherwise, Aa is called heterozygous.

For example, Figure 13.1 considers a particular pair of chromosomes in four individuals. For locus i, the alleles are T and C. For locus j, the allele is A. So, locus i is an SNP and locus j is not an SNP since the allele at locus j for all individuals is A. The genotypes for locus i in the four individuals are TC, CC, TT, and TT. So, locus i is homozygous for the individuals 2, 3, and 4 while it is heterozygous for the individual 1.

```
                       i j
     Individual 1:  ...ACGTCATG...
                    ...ACGCCATG...
     Individual 2:  ...ACGCCATG...
                    ...ACGCCATG...
     Individual 3:  ...ACGTCATG...
                    ...ACGTCATG...
     Individual 4:  ...ACGTCATG...
                    ...ACGTCATG...
```

FIGURE 13.1:   This figure shows an example of a pair of chromosomes for four individuals.

Given a set of individuals, we can compute the frequencies of the two alleles of the SNP. The allele with the lower frequency is called the minor allele. For the example in Figure 13.1, loci i is an SNP with two alleles T and C. The allele frequency of T is $5/8$ while the allele frequency of C is $3/8$. In this case, C is the minor allele and the minor allele frequency is $3/8$.

SNPs are expected to make up 90% of all human genetic variations. SNPs with a minor allele frequency of $\geq 1\%$ occur every 100 to 300 bases along the human genome, on average. Two thirds of the SNPs substitute cytosine (C) with thymine (T). Through the collaborative effort of many countries, we already have identified the set of common SNPs in the human population. (Please visit the webpage of the HapMap project *http://www.hapmap.org/* for more details.)

13.1.2   Genotype Frequency and Allele Frequency

Given a population, the genotype frequency is the relative frequency of a genotype on a locus. Allele frequency is the relative frequency of an allele on a locus.

For example, consider a population of 10 individuals. Let A and a be the two possible alleles of a given locus. Suppose their genotypes are $\{AA, Aa, aa, AA, AA, Aa, aa, Aa, AA, Aa\}$. Then, the numbers of AA, aa, and Aa are 4, 2, and 4, respectively. Hence, we have the following genotype frequencies: $f(AA) = 4/10$, $f(aa) = 2/10$, and $f(Aa) = 4/10$. The allele frequencies of the two alleles A and a are $p_A = (2+1+0+2+2+1+0+1+2+1)/20 = 0.6$ and $p_a = (0+1+2+0+0+1+2+1+0+1)/20 = 0.4$, respectively.

There is a simple relationship between genotype frequency and allele frequency.

$$p_A = f(AA) + 0.5f(Aa), p_a = f(aa) + 0.5f(Aa)$$

For the above example of 10 individuals, it can easily be verified that they satisfy this simple relationship.

13.1.3 Haplotype and Phenotype

Haplotype is a combination of alleles at different loci on the same chromosome. Below shows an example.

```
                      i       j         k
Copy1 of the chr  --------A------T--------G-----------
Copy2 of the chr  --------C------A--------C-----------
```

The genotypes at the three loci i, j, and k are AC, AT, and CG. They form two haplotypes: ATG and CAC.

Phenotype is the observable structure, function, or behavior of a living organism. Hair color, eyebrow color, etc. are some examples of phenotypes. It has been shown that the variation of SNPs has an effect on the phenotypes. For example, the gene apolipoprotein E (ApoE) contains two SNPs (or two loci) which result in three possible haplotypes: E2, E3, and E4. Each individual who inherits at least one haplotype E4 from his parents has a higher risk of developing Alzheimer's disease.

The haplotype in the SNPs which determines the appearance or phenotype of an individual is called the dominant haplotype; otherwise, it is called a recessive haplotype.

Not all SNPs show phenotype. The SNPs which do not affect the phenotype are called natural SNPs; otherwise, they are called causal SNPs. The prediction of phenotypes from SNPs has applications in disease diagnostics, personalized medicine, and forensics.

13.1.4 Technologies for Studying the Human Population

There are many different genotyping technologies. Nowadays, we can perform whole genome genotyping for all the common SNPs found in HapMap (US$0.1–US$0.01 per genotype)!

Note that genotyping does not tell us the haplotypes appearing in the chromosomes. For example, suppose the genotypes of two loci are AC and CT. Then, there are two possible cases:

```
Case 1:                              Case 2:


--------A-------C--------            --------A-------T--------
--------C-------T--------            --------C-------C--------
```

13.1.5 Bioinformatics Problems

Based on the above discussion, there are a number of bioinformatics problems related to SNPs.

- Genotype phasing: Since high-throughput haplotype sequencing is still difficult, we need computational methods to reconstruct the haplotypes from the genotypes.

- Tag SNP selection: Genotyping all SNPs is expensive and sometimes impossible. So, the usual practice is to select and genotype a subset of SNPs, called tag SNPs. The tag SNPs are expected to capture a majority of the variations.

- Association study: Finding the relationship between phenotypes and genetic variation.

Below, we first discuss two important concepts: Hardy-Weinberg equilibrium and Linkage disequilibrium. Then, we study algorithms for the above three informatics problems.

13.2 Hardy-Weinberg Equilibrium

In Section 13.1.2, we showed that the allele frequency can be determined from the genotype frequency. On the other hand, given the allele frequency, can we determine the genotype frequency?

In the absence of migration, mutation, and natural selection, genotypes are formed by random mating. Then, the genotype frequency at a locus is a simple function of the allele frequency. This phenomenon is termed "Hardy-Weinberg equilibrium" (HWE).

HWE is popularly used to check data quality. If the genotype frequencies at any locus deviate from HWE, it may be due to miscalls during the genotyping process.

HWE makes two assumptions:

1. Random mating: We assume the mating between male and female individuals is a random process.

2. No natural selection: We assume the process of reproduction is not in favor of certain phenotypes.

Suppose the locus i is diallelic and has two possible alleles A and a. Let p_A and p_a be the major and minor allele frequencies. Under the HWE assumptions, the two alleles at locus i are independent. Hence, the expected frequencies of the three different genotypes are: $e(AA) = p_A * p_A$, $e(aa) = p_a * p_a$, and $e(Aa) = 2p_A * p_a$. If the genotype frequencies for AA, aa, and Aa are similar to the expected frequencies $e(AA)$, $e(aa)$, and $e(Aa)$, respectively, we say the genotype frequencies satisfy HWE.

Let n be the size of the population. Let n_A and n_a be the number of occurrences of the alleles A and a, respectively, at locus i. Let n_{AA}, n_{aa}, and n_{Aa} be the number of occurrences of the genotypes AA, aa, and Aa, respectively, at locus i. We have the following table.

	AA	aa	Aa	total
allele A	$2n_{AA}$	0	n_{Aa}	n_A
allele a	0	$2n_{aa}$	n_{Aa}	n_a
	$2n_{AA}$	$2n_{aa}$	$2n_{Aa}$	n

For a fixed n_A and n_a, we observe that the number of heterozygous genotypes n_{Aa} determines the numbers of homozygous genotypes n_{aa} and n_{AA}.
Precisely, $n_{AA} = (n_A - n_{Aa})/2$ and $n_{aa} = (n_a - n_{Aa})/2$.

To determine if the genotype frequencies satisfy HWE, we ask the following
question: For a fixed number of alleles n_A and n_a, what is the probability that
the number of heterozygous genotypes equal n_{Aa}? Below, we apply the Fisher
exact test to determine such probability, which is denoted as $Pr(n_{Aa} \mid n_A, n_a)$.

Note that the number of combinations where there are n_A's A in the popu-
lation is $\binom{2n}{n_A} = \frac{(2n)!}{n_A! n_a!}$ and the number of combinations where there are n_{Aa}'s
heterozygous genotypes is $\frac{2^{n_{Aa}} n!}{n_{Aa}! n_{AA}! n_{aa}!}$. Hence, we have

$$Pr(n_{Aa} \mid n_A, n_a) = \frac{n_A! n_a!}{(2n)!} \times \frac{2^{n_{Aa}} n!}{n_{Aa}! \left(\frac{n_A - n_{Aa}}{2}\right)! \left(\frac{n_a - n_{Aa}}{2}\right)!}$$

If $Pr(n_{Aa} \mid n_A, n_a)$ is small, we say the SNP deviates from HWE and it
may be a miscall during the genotyping process. Usually, we discard SNPs
which deviate from HWE at a significance level of 10^{-3} or 10^{-4}.

For example, suppose the genotypes of the population are $\{AA, Aa, aa, AA,$
$AA, Aa, aa, Aa, AA, Aa\}$. This implies $n_A = 12$ and $n_a = 8$. The genotype
frequencies are $n_{AA} = 4$, $n_{aa} = 2$, and $n_{Aa} = 4$. By the Fisher exact test, the
probability $Pr(n_{Aa} = 4 \mid n_A = 12, n_a = 8) = \frac{2^4 10!}{4!4!2!} \times \frac{12!8!}{20!} = 0.40095$. Since
the value is bigger, this SNP satisfies HWE.

When n is big, the Fisher exact test is computationally infeasible; the χ^2
test is an approximation to the Fisher exact test. For the χ^2 test, we have
$score = \frac{(f(AA)-e(AA))^2}{e(AA)} + \frac{(f(aa)-e(aa))^2}{e(aa)} + \frac{(f(Aa)-e(Aa))^2}{e(Aa)}$ which follows the χ^2
distribution with degree of freedom $= 1$.

As a final note, though this approach can avoid miscall SNPs, it may miss
some causal SNPs. In real life, there exist different forces to change the frequencies. The forces include selection, drift, mutation, and migration. Those
forces make the causal SNPs deviate from HWE.

13.3 Linkage Disequilibrium

Linkage disequilibrium (LD) refers to the non-random association between
alleles at two different loci. In other words, it measures if two particular
alleles can co-occur more often than expected by chance. There are three LD
measurements: D, D', and r^2.

13.3.1 D and D'

Consider two loci where locus 1 has two possible alleles A and a $(p_a+p_A = 1)$ while locus 2 has two possible alleles B and b $(p_b+p_B = 1)$. If loci 1 and 2 are independent, we have $p_{AB} = p_A p_B$, $p_{Ab} = p_A p_b$, $p_{aB} = p_a p_B$, and $p_{ab} = p_a p_b$.

We define the linkage disequilibrium coefficient $D = p_{AB} - p_A p_B$. D is in the range -0.25 to 0.25. Also, it is easily shown that $D = p_{ab} - p_a p_b = p_A p_b - p_{Ab} = p_a p_B - p_{aB}$ (see Exercise 9). If linkage disequilibrium (LD) is present (i.e., A and B have some dependence), then D is non-zero.

D is highly dependent on the allele frequency and is not good for measuring the strength of LD. Therefore, a normalized measurement is defined: $D' = D/D_{\max}$ where $D_{\max}$ is the maximum possible value for D given p_A and p_B. Precisely, $D_{\max} = \max\{p_{AB} - p_A p_B\} = \min\{p_A, p_B\} - p_A p_B$. When $|D'| = 0$, the two loci are independent. When $|D'| = 1$, we call it the complete LD.

For example, consider a set of allele pairs $AB, Ab, aB, Ab, ab, ab, ab$. We have $p_{AB} = \frac{1}{7}, p_A = \frac{3}{7}$, and $p_B = \frac{2}{7}$. Hence, $D = \frac{1}{7} - \frac{3}{7}\frac{2}{7} = \frac{1}{49}$. $D_{\max} = \frac{2}{7} - \frac{3}{7}\frac{2}{7} = \frac{8}{49}$. Hence, $D' = D/D_{\max} = \frac{1}{8}$.

13.3.2 r^2

r^2 measures the correlation of two loci. We define $r^2 = \frac{D^2}{p_A p_a p_B p_b}$. When $r^2 = 1$, the two loci are completely dependent, that is, we can deduce the allele on loci 2 given the allele on loci 1, and vice versa. We call it the perfect LD.

For example, consider the set of allele pairs $AB, Ab, aB, Ab, ab, ab, ab$ again. $p_{AB} = 1/7, p_A = 3/7$, $p_B = 2/7$, and $D = 1/49$. Hence, $r^2 = (1/49)^2/(3/7 * 4/7 * 2/7 * 5/7) = 1/120$.

13.4 Genotype Phasing

High-throughput genotyping technology provides a cost effective method to generate genotypes for millions of loci of an individual. However, it cannot tell us the haplotypes of the individual.

To generate the haplotypes, wet-lab experiments were suggested. Newton et al. [219] and Wu et al. [312] introduced asymmetric PCR amplification to preferentially amplify one strand of the DNA. Ruano et al. [251] suggested isolating a single chromosome by limited dilution followed by PCR amplification. Both methods allow us to sequence the haplotype directly. Perlin et al. [234] proposed inferring haplotype information by using genealogical information in families. The above methods, however, are still low-throughput, costly, and complicated.

Since it is expensive or technically difficult to examine the two copies of a chromosome separately, the practical way is to first generate the individual's genotype using high-throughput technologies first. Then, we try to computationally infer the haplotypes from the genotypes. This problem is known as the genotype phasing problem. For instance, the HAPMAP project applies computational methods to recover the genome-wide haplotypes of individuals. Before we formally define the problem, we first give some definitions. For each locus, we denote 0 and 1 as the minor allele a and the major allele A, respectively. The genotype of the locus can be either aa, AA, and Aa and we denote them as 0, 1, and 2, respectively.

Below, we illustrate how to infer haplotypes from genotypes. For example, suppose the genotypes of an individual for four loci are 2101. Then, the corresponding pair of haplotypes is $(1101, 0101)$. In this case, we say 1101 and 0101 resolve 2101. However, in general, there may be multiple possible pairs of haplotypes for a given genotype. For example, consider the genotypes of another individual, which are 2120. Both $(1110, 0100)$ and $(1100, 0110)$ can resolve 2120. In such a case, which pair of haplotypes is real?

This problem is called the genotype phasing problem, which is formally defined as follows. Consider the set of genotypes $\mathcal{G} = \{G_1, G_2, \ldots, G_n\}$ for n individuals. The problem aims to find a set of haplotypes $\mathcal{H}$ which can best explain $\mathcal{G}$.

In this section, we discuss four methods to solve this problem: (1) Clark's algorithm, (2) Haplotyping by Perfect Phylogeny, (3) Maximum likelihood, and (4) Phase.

13.4.1 Clark's Algorithm

Clark proposed one of the first algorithms to resolve the genotype phasing problem in 1990 [63] using the parsimony criteria. His algorithm aims to find the minimum number of haplotypes which can resolve all the genotypes in $\mathcal{G}$.

However, the problem of finding the minimum number of haplotypes which can resolve all genotypes in $\mathcal{G}$ is NP-hard [129]. Thus, it is unlikely to have a polynomial time algorithm. Clark proposed a heuristics algorithm, which is as follows. For every homozygous genotype or single-site heterozygous genotype $G_i \in \mathcal{G}$, the algorithm generates a set of haplotypes unambiguously and includes them in $\mathcal{H}$. Then, for each known haplotype $H \in \mathcal{H}$, the algorithm checks if there exists some unresolved genotype $G_j \in \mathcal{G}$ which can be resolved by H and a new haplotype $H' \notin \mathcal{H}$. If the algorithm can resolve G_j by H and H', it includes H' into $\mathcal{H}$ and marks G_j as resolved. The procedure is repeated until all genotypes are resolved.

To demonstrate the execution of Clark's algorithm, consider an input set of three genotypes $\mathcal{G} = \{G_1 = 10121101, G_2 = 10201121, G_3 = 20001211\}$. The first step of Clark's algorithm is to resolve all homozygous genotypes or single-site heterozygous genotypes in $\mathcal{G}$. Note that G_1 is a single-site heterozygous genotype. It can be resolved by $H_1 = 10101101$ and $H_2 = 10111101$. We set $\mathcal{H} = \{H_1, H_2\}$.

The second step of Clark's algorithm iteratively checks if we can resolve some genotype in $\mathcal{G}$ using some $H \in \mathcal{H}$ and some other haplotype $H' \notin \mathcal{H}$. We can resolve G_2 by H_1 and a new haplotype $H_3 = 10001111$. Then, we can resolve G_3 by H_3 and a new haplotype $H_4 = 00001011$. Finally, Clark's algorithm reports that $\mathcal{G}$ can be resolved by the set of haplotypes $\mathcal{H} = \{H_1 = 10101101, H_2 = 10111101, H_3 = 10001111, H_4 = 00001011\}$.

Although Clark's algorithm is simple and popular, it has a few problems. First, Clark's algorithm might not get started since there is no homozygous genotype or single-site heterozygous genotype in $\mathcal{G}$. Second, Clark's algorithm might not resolve all individuals in $\mathcal{G}$. Third, the output returned by Clark's algorithm is order dependent. In other words, if we reorder the individuals in $\mathcal{G}$, we may get a different set of haplotypes to resolve $\mathcal{G}$.

13.4.2 Perfect Phylogeny Haplotyping Problem

The perfect phylogeny haplotyping (PPH) problem was introduced by Gusfield [130]. It makes two assumptions on the haplotypes that resolve the set of genotypes $\mathcal{G}$: (1) There is only mutation and it does not have recombination; and (2) There exists at most one mutation per allele in the history of evolution. The assumption implies that the haplotypes can be explained by a perfect phylogeny (see Section 7.2.2 for the definition of perfect phylogeny). Precisely, the perfect phylogeny haplotyping problem tries to find a set of haplotypes $\mathcal{H}$ which can resolve all genotypes in $\mathcal{G}$ given that $\mathcal{H}$ satisfies a perfect phylogeny.

For example, consider $\mathcal{G} = \{a = 220, b = 012, c = 222\}$, the set of haplotypes which resolves $\mathcal{G}$ and satisfies a perfect phylogeny is $\mathcal{H} = \{100, 010, 011\}$. Note that a is resolved by $(100, 010)$, b is resolved by $(010, 011)$, and c is resolved by $(100, 011)$. Figure 13.2 shows the corresponding perfect phylogeny for $\mathcal{H}$.

A number of different methods have been proposed for this problem. Gusfield [130] gave the first solution to this problem, which runs in $O(nm\alpha(nm))$ time, where $\alpha()$ is the inverse Ackerman function. This solution is based on the reduction to the graph realization problem. Eskin et al. [96] and Bafna et al. [16] independently gave two simple solutions to this problem, which run in $O(nm^2)$ time. In 2005, Ding et al. [78] gave an optimal $O(nm)$ time algorithm for this problem. This section discusses the solution of Bafna et al. [16]. For simplicity, we represent the genotypes $\mathcal{G} = \{G_1, \ldots, G_n\}$ as an $n \times m$ matrix G where the entry $G(i, c)$ at row i and column c equals the c-th genotype of G_i.

First, we describe the condition when a set of haplotypes fits a perfect phylogeny. This condition is known as the 4-Gamete test.

	1	2	3
G_1	2	2	0
G_2	0	1	2
G_3	2	2	2

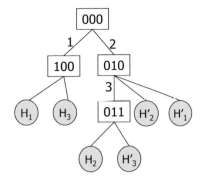

FIGURE 13.2: The perfect phylogeny for the set of haplotypes $\mathcal{H} = \{100, 010, 011\}$ that resolves $\mathcal{G} = \{a = 220, b = 012, c = 222\}$. The edge label on every edge of the tree indicates the character which is mutated.

LEMMA 13.1

A set of haplotypes $\mathcal{H}$ admits a perfect phylogeny if and only if there are no two columns i and j containing all four pairs 00, 01, 10, and 11.

PROOF   This is implied by Lemma 7.3. $\Box$

Given an $n \times m$ ternary matrix G, our aim is to create a $2n \times m$ binary matrix H satisfying the 4-gamete test where

- For every $1 \leq r \leq n$, rows $2r$ and $2r - 1$ of H form a pair of haplotypes which resolves the genotype in row r of G. Precisely, for any column c, if $G(r, c) \neq 2$, $H(2r, c) = H(2r - 1, c) = G(r, c)$; otherwise, $\{H(2r, c), H(2r - 1, c)\} = \{0, 1\}$.

We first have some definitions.

- Suppose there exist two columns c and c' in G which contain (1) either 00, 20, or 02 and (2) either 11, 21, or 12. (1) implies that columns c and c' in H must contain 00 while (2) implies that columns c and c' in H must contain 11. In this case, the pair of columns c and c' is called in-phase.

- Suppose there exist two columns c and c' in G which contain (1) either 10, 12, or 20 and (2) either 01, 21, or 02. Then, (1) and (2) imply that columns c and c' in H must contain both 10 and 01. In this case, the pair of columns c and c' is called out-of-phase.

For example, columns 2 and 4 in Figure 13.3(a) are in-phase while columns 3 and 4 in Figure 13.3(a) are out-of-phase. Columns 4 and 5 in Figure 13.3(a) are neither in-phase nor out-of-phase.

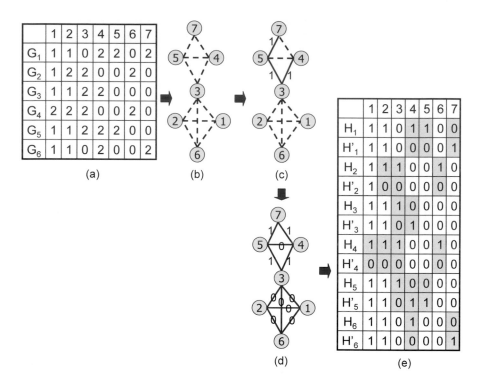

(a)   (b)   (c)   (d)   (e)

FIGURE 13.3:   (a) A set G of six genotypes. (b) The graph G_M. (c) The graph G_M with all the in-phase and out-of-phase edges colored. The red colored (color 0) edges are in-phase while the blue colored (color 1) edges are out-of-phase. Note that, for this example, there are only out-of-phase edges. (d) The same graph G_M after we infer the colors of those uncolored edges in (c). Note that the colors of every triangle satisfy Lemma 13.2. (e) The set of haplotypes H that can resolve G. (See color insert after page 270.)

Note that if there exists a pair of columns c and c' which is both in-phase and out-of-phase, the pair of columns contains 00, 01, 10, and 11. This violates the 4-gamete test and thus, M has no solution to the PPH problem. Below, we assume none of the column pairs in M is both in-phase and out-of-phase.

We now study how to determine if G has a solution to the PPH problem. We first define a graph G_M whose nodes represent the columns in G and a pair of columns (c, c') forms an edge if 22 appears in the pair of columns (c, c') in G. (See Figure 13.3(b).) For every edge (c, c') in G_M, we color (c, c') red (or 0) and blue (or 1) if (c, c') are in-phase and out-of-phase, respectively; otherwise, the edge has no color. (See Figure 13.3(c).) We observe that when (c, c') is colored, there is only one way to resolve 22 appearing in the pair of columns (c, c') as follows.

- If the pair of columns c and c' is in-phase (red color), then for every row r in G which is 22, the haplotypes in rows $2r$ and $2r - 1$ of H must be either 11 or 00; otherwise, it violates the 4-gamete test.

- If the pair of columns c and c' is out-of-phase (blue color), then for every row r in G which is 22, the haplotypes in rows $2r$ and $2r - 1$ of H must be either 10 or 01; otherwise, it violates the 4-gamete test.

For the uncolored edges, we can infer their colors. If the inferred colors have no conflict with the colored edges, Bafna et al. showed that G has a solution to the PPH problem. The lemma below formally states the necessary and sufficient condition to determine if G has a solution to the PPH problem.

LEMMA 13.2
Consider a matrix M where none of the column pairs is both in-phase and out-of-phase. There is a solution to the PPH problem for M if and only if we can infer the colors of all edges in G_M such that for any triangle (i, j, k) in G_M where $G(r, i) = G(r, j) = G(r, k) = 2$ for some row r, either 0 or 2 edges of the triangle are colored blue.

PROOF ($\rightarrow$) Suppose there is a solution to the PPH problem for M. We can infer the colors of every pair of columns by checking if the pair of columns is in-phase or out-of-phase. Then, consider any triangle (i, j, k) in G_M where there is one row r with $G(r, i) = G(r, j) = G(r, k) = 2$. We claim that there is either 0 or 2 edges of the triangle blue. Otherwise, we fail to find a pair of haplotypes to resolve row r.
($\leftarrow$) We leave it as an exercise.   ⬚

Figure 13.3(c,d,e) demonstrate how we infer the colors of the uncolored edges and compute H which resolves G.

Based on the above discussion, Bafna et al. [16] proposed an $O(nm^2)$-time algorithm to solve the PPH problem. Figure 13.4 states the framework of

---

Algorithm PPH

Require: The $n \times m$ ternary matrix M representing the genotypes. We require that none of the character pairs in M is both in-phase and out-of-phase.

Ensure: A $2n \times m$ binary matrix H representing the haplotypes, if it exists.

1: Create a graph G_M whose nodes represent the columns in G and a pair of columns (c, c') forming an edge if 22 appears in the pair of columns (c, c') in G.

2: For every edge (c, c') in G_M, (c, c') is colored red and blue if (c, c') are in-phase and out-of-phase, respectively; otherwise, the edge has no color.

3: The colors for all the uncolored edges in G_M are inferred based on Lemma 13.2.

4: We compute the matrix H so that every column pair satisfies the in-phase (red) and the out-of-phase (blue) constraints in G_M.

FIGURE 13.4: Algorithm by Bafna et al. [16] for solving the PPH problem.

the algorithm. Each edge in G_M can be generated and colored in $O(n)$ time. Since there are m^2 edges in G_M, Steps 1 and 2 take $O(nm^2)$ time. Step 3 can be solved in $O(m^2)$ time. Step 4 takes $O(nm)$ time.

13.4.3 Maximum Likelihood Approach

Excoffier and Slatkin [99] proposed the maximum likelihood approach to infer the haplotype with the most realistic haplotype frequencies under the assumption of Hardy-Weinberg equilibrium.

We first give an example to demonstrate the idea behind the method. Consider two genotypes $G_1 = 0111$ and $G_2 = 0221$. There are two possible ways to resolve them.

1. G_1 is resolved by $(0111, 0111)$ and G_2 is resolved by $(0111, 0001)$.

2. G_1 is resolved by $(0111, 0111)$ and G_2 is resolved by $(0101, 0011)$.

Which solution is better? For solution 1, there are two haplotypes 0111 and 0001. Their frequencies are $\frac{3}{4}$ and $\frac{1}{4}$. The chance of getting $G_2 = 0221$ is $\frac{3}{4} \times \frac{1}{4}$. For solution 2, there are three haplotypes 0111, 0101, and 0011. Their frequencies are $\frac{1}{2}$, $\frac{1}{4}$, and $\frac{1}{4}$. The chance of getting $G_2 = 0221$ is $\frac{1}{4} \times \frac{1}{4}$.

Solution 1 is more feasible since the probability of getting G_2 is higher. Excoffier and Slatkin [99] infer haplotypes based on this idea.

To describe the method, we need some definitions. Let $G = \{G_1, G_2, \ldots, G_n\}$ be the set of n genotypes. Given a genotype G_i, let S_i be the set of all haplotype pairs that can resolve G_i and let H_i be the set of all possible haplotypes.

For example, consider the genotype $G_2 = 0221$. Since there are two heterozygous loci, we have $2^2 = 4$ possible haplotypes: $H_2 = \{h_1 = 0001, h_2 = 0011, h_3 = 0101, h_4 = 0111\}$. The set of all haplotype pairs that resolve 0221 is $S_2 = \{h_1 h_4, h_2 h_3\}$.

We denote $H = \{h_1, h_2, \ldots, h_m\} = \cup_{i=1}^n H_i$ as the set of all possible haplotypes that can resolve G. Our aim is to estimate the frequencies F_j of h_j for $h_j \in H$. Note that $F_1 + F_2 + \ldots + F_m = 1$.

Given the frequency $F = \{F_1, F_2, \ldots, F_m\}$, the probability of generating genotype G_i, for $i = 1, 2, \ldots, m$, can be computed using the following formula.

$$Pr(G_i|F) = \sum_{h_x h_y \text{ resolve } G_i} F_x F_y$$

Excoffier and Slatkin [99] proposed finding the frequencies F which maximize the overall probability product of all $P(G_i)$, that is, the following function

$$L(F) = Pr(G|F) = \Pi_{i=1}^n Pr(G_i|F)$$

In principle, F can be identified by solving the equation for $L(F)$. However, there is no close form. Instead, the expectation maximization (EM) method is used. The EM algorithm tries to iteratively improve the estimation of the haplotype frequency $F^{(t)} = \{F_1^{(t)}, \ldots, F_m^{(t)}\}$ for $t = 0, 1, \ldots$ by performing two steps: the expectation step and the maximization step.

Assume we have an initial estimation of the haplotype frequency $F^{(0)} = \{F_1^{(0)}, \ldots, F_m^{(0)}\}$. For the t-th iteration where $t = 0, 1, \ldots$, the expectation step (E-step) tries to estimate, for every G_i and for every pair of haplotypes $h_x h_y$ that resolves G_i, the haplotype pair frequency $P(h_x h_y | G_i, F^{(t)})$. Given $F^{(t)}$, we have

$$Pr(h_x h_y | G_i, F^{(t)}) = \frac{Pr(h_x h_y | F^{(t)})}{\sum_{h_a h_b \text{ resolves } G_i} Pr(h_a h_b | F^{(t)})}$$

$$= \frac{F_x^{(t)} F_y^{(t)}}{\sum_{h_a h_b \text{ resolves } G_i} F_a^{(t)} F_b^{(t)}}$$

The maximization step (M-step) tries to estimate, for every haplotype $h_x \in H$, the haplotype frequency $F_x^{(t+1)}$. Let $\delta(h, h_x h_y)$ be the number of occurrences of h in $\{h_x, h_y\}$. $F_x^{(t+1)}$ can be estimated as

$$F_x^{(t+1)} = \frac{1}{2n} \sum_{i=1}^n \sum_{h_x h_y \text{ resolves } G_i} \delta(h_x, h_x h_y) Pr(h_x h_y | G_i, F^{(t)})$$

Below, we illustrate the EM-algorithm using an example. Consider a set of genotypes $G = \{G_1 = 11, G_2 = 12, G_3 = 22\}$. The set of possible haplotypes of G is $H = \{h_1 = 11, h_2 = 00, h_3 = 10, h_4 = 01\}$. For $x = 1, 2, 3, 4$, let $F_x^{(0)}$

be the initial haplotype frequency of h_x. (For instance, assuming uniform distribution, we can set $F_x^{(0)} = 0.25$ for $x = 1, 2, 3, 4$.)

In the expectation step, we estimate the haplotype pair frequency.

- For G_1, $h_1 h_1$ is the only possible haplotype pair which resolves G_1; hence, $Pr(h_1 h_1 | G_1, F^{(0)}) = 1$

- For G_2, $h_1 h_3$ is the only possible haplotype pair which resolves G_2; hence, $Pr(h_1 h_3 | G_2, F^{(0)}) = 1$.

- For G_3, $h_1 h_2$ and $h_3 h_4$ are the two possible haplotype pairs which resolve G_3; hence, $Pr(h_1 h_2 | G_3, F^{(0)}) = \dfrac{F_1^{(0)} F_2^{(0)}}{(F_1^{(0)} F_2^{(0)} + F_3^{(0)} F_4^{(0)})} = \dfrac{1}{2}$ and
  $Pr(h_3 h_4 | G_3, F^{(0)}) = \dfrac{F_3^{(0)} F_4^{(0)}}{(F_1^{(0)} F_2^{(0)} + F_3^{(0)} F_4^{(0)})} = \dfrac{1}{2}$.

In the maximization step, we estimate the haplotype frequency $F^{(1)}$.

- $F_1^{(1)} = \frac{1}{2n}[2Pr(h_1 h_1 | G_1, F^{(0)}) + Pr(h_1 h_3 | G_2, F^{(0)}) + P(h_1 h_2 | G_3, F^{(0)})] = \frac{7}{12}$

- $F_2^{(1)} = \frac{1}{2n} Pr(h_1 h_2 | G_3, F^{(0)}) = \frac{1}{12}$

- $F_3^{(1)} = \frac{1}{2n}[Pr(h_1 h_3 | G_2, F^{(0)}) + Pr(h_3 h_4 | G_3, F^{(0)})] = \frac{3}{12}$

- $F_4^{(1)} = \frac{1}{2n} Pr(h_3 h_4 | G_3, F^{(0)}) = \frac{1}{12}$

Through iteratively performing the expectation step and the maximization step, we can improve the estimation of F.

13.4.4 Phase Algorithm

In 2001, Stephens, Smith, and Donnelly [273] proposed Phase. Phase resolves the genotypes using Gibbs sampling. We first give an example to explain the basic idea of Phase. Consider a set of known haplotypes $\{H_1 = 10001, H_2 = 11110, H_3 = 00101\}$. Suppose the haplotype frequencies of H_1, H_2, and H_3 are $F_1 = 0.2$, $F_2 = 0.5$, and $F_3 = 0.3$, respectively. Now, let us consider a genotype 20112. There are two possible ways to resolve it.

- Solution 1: 20112 is resolved by $(00110, 10111)$.

- Solution 2: 20112 is resolved by $(00111, 10110)$.

Which solution is better? Although none of the four haplotypes is a known haplotype, we observe that 00111 and 10110 are similar to some known haplotypes (with one mutation). On the other hand, 00110 and 10111 are not similar to the known haplotypes (with at least two mutations). Hence, solution 1 seems to be correct.

The above demonstration means that a genotype is likely to use some haplotypes h which are the same or similar to some frequently observed haplotypes H. Formally, Stephens and Donnelly [272] derived an equation to approximate the probability $Pr(h|H)$. For any haplotype $\alpha \in H$, let F_α be the haplotype frequency of α in the population. Note that $\sum_{\alpha \in H} F_\alpha = 1$. Suppose the population size is n. Let θ be the scaled mutation rate and P be the substitution matrix (see Section 2.6). We have

$$Pr(h|H) = \sum_{\alpha \in H} \sum_{s=0}^{\infty} F_\alpha \left(\frac{\theta}{n+\theta}\right)^s \frac{n}{n+\theta}(P^s)_{\alpha h} \tag{13.1}$$

Using the above equation, Phase applied Gibbs sampling to reconstruct the haplotype pairs for a set of genotypes $G = \{G_1, \ldots, G_n\}$. Our aim is to determine $S = \{S_1, \ldots, S_n\}$ where S_i is the unknown haplotype pair (h_{i1}, h_{i2}) for G_i. Phase first makes an initial guess of $S^{(0)} = \{S_1^{(0)}, \ldots, S_n^{(0)}\}$. Then, for $t = 0, 1, \ldots$, Phase tries to iteratively modify $S^{(t)}$ to $S^{(t+1)}$ and hope that $S^{(t+1)}$ is a better prediction.

To generate $S^{(t+1)}$ from $S^{(t)}$, Phase performs four steps.

1: We uniformly randomly choose an ambiguous G_i. Let $H_{-i} = \cup\{h_1, h_2 \mid S_j^{(t)} = (h_1, h_2), j = 1, \ldots, i-1, i+1, \ldots, n\}$.

2: For every $(h_{i,1}, h_{i,2})$ that resolves G_i, $Pr(h_{i,1}|H_{-i})$ and $Pr(h_{i,2}|H_{-i} \cup \{h_{i,1}\})$ are computed using Equation 13.1; then, we compute

$$Pr(h_{i,1}h_{i,2}|H_{-i}) = Pr(h_{i,1}|H_{-i})Pr(h_{i,2}|H_{-i} \cup \{h_{i,1}\}\})$$

3: Randomly sample $(h_{i,1}, h_{i,2})$ according to the probability distribution $Pr(h_{i,1}h_{i,2}|H_{-i})$.

4: Let $S^{(t+1)} = S^{(t)} \cup \{(h_{i,1}, h_{i,2})\} - \{S_i^{(t)}\}$.

In [273], Stephens et al. showed that Phase is better than Clark's algorithm [63] and the EM algorithm proposed by Excoffier and Slatkin [99].

13.5 Tag SNP Selection

There are about 10 million common SNPs (SNPs with allele frequency $> 1\%$). It accounts for $\approx 90\%$ of the human genetic variation. Hence, we can study the genetic variation of an individual by getting his/her profile for the common SNPs. Even though the cost of genotyping is rapidly decreasing, it is still impractical to genotype every SNP or even a large proportion of them.

Fortunately, nearby SNPs show strong correlation to each other (that is, strong LD (see Section 13.3)). Therefore, instead of sequencing all the SNPs, a practical solution is to sequence a subset of SNPs (called tag SNPs) to

$$(0,1,0,1,0,1,0,0,0,1,0)$$
$$(0,1,0,0,1,0,1,0,0,1,0)$$
$$(0,1,0,1,0,1,0,0,1,1,1)$$
$$(1,0,1,1,0,1,0,1,1,0,1)$$
$$(1,0,1,0,1,0,0,1,1,0,1)$$
$$(1,0,0,0,1,0,1,0,1,1,1)$$

FIGURE 13.5:   An example of six haplotypes, each described by 11 SNPs.

represent the rest of the SNPs. This section studies two methods for tag SNP selection.

13.5.1   Zhang et al.'s Algorithm

Zhang et al. [315] proposed a method to select tag SNPs using dynamic programming. Their method partitions the genome into blocks to minimize the number of tag SNPs required to account for most of the common haplotypes in each block. Precisely, the input is a set of n haplotypes, each described by m SNPs. We denote $r_i(k)$ as the allele of the i-th SNP in the k-th haplotype. where $r_i(k) = 0$ or 1. The aim is to find a partition of blocks that minimizes the number of tag SNPs which can distinguish at least $\alpha\%$ of the haplotypes in each block.

For example, Figure 13.5 shows a set of six haplotypes, each described by 11 SNPs. Suppose we want to distinguish at least 80% of these haplotypes. We can partition the 11 SNPs into 3 blocks: $r_1..r_3$, $r_4..r_7$, and $r_8..r_{11}$. In total, we select four tag SNPs. For the first block, we select r_1 as the tag SNP. For the second block, we select r_4 as the tag SNP. For the third block, we select r_8 and r_{11} as the tag SNPs.

- For $r_1..r_3$, r_1 is selected as the tag SNP. If $r_1 = 0$, we predict $r_1..r_3 = 010$; otherwise, if $r_1 = 1$, we predict $r_1..r_3 = 101$. Such a tag SNP can predict correctly five out of six haplotypes, which is higher than 80%.

- For $r_4..r_7$, r_4 is selected as the tag SNP. If $r_4 = 0$, we predict $r_4..r_7 = 0101$; otherwise, if $r_4 = 1$, we predict $r_4..r_7 = 1010$. Such a tag SNP also can predict correctly five out of six haplotypes, which is higher than 80%.

- For $r_8..r_{11}$, r_8 and r_{11} are selected as the tag SNPs. If $r_8 = 0$ and $r_{11} = 0$, we predict $r_8..r_{11} = 0010$. If $r_8 = 0$ and $r_{11} = 1$, we predict $r_8..r_{11} = 0111$. If $r_8 = 1$ and $r_{11} = 1$, we predict $r_8..r_{11} = 1101$. Such a pair of tag SNPs can predict correctly all six haplotypes.

Below, we describe a dynamic programming solution to the tag SNP selection problem. We denote $f(r_i..r_j)$ as the minimum number of tag SNPs that can uniquely distinguish at least $\alpha\%$ of the haplotypes in the block $r_i..r_j$. For example, in the block $r_1..r_3$ of Figure 13.5, we have the following haplotypes: three 010, two 101 and one 100. To distinguish 80% of these haplotypes, we need one tag SNP, that is, r_1. To distinguish 100% of these haplotypes, we need at least two tag SNPs, that is, r_1 and r_3. If $\alpha = 80\%$, $f(r_1..r_3) = 1$. If $\alpha = 100\%$, $f(r_1..r_3) = 2$. Computing $f(r_i..r_j)$ is NP-complete as it is equivalent to the minimum test set problem [115]. Moreover, since the program generally will compute $f(r_i..r_j)$ when $f(r_i..r_j)$ is small, we can compute $f(r_i..r_j)$ by enumerating all possible tag SNP sets.

We use dynamic programming to compute the minimum number of tag SNPs. We denote $S(i)$ as the minimum number of tag SNPs which can distinguish at least $\alpha\%$ of the haplotypes in $r_1..r_i$. Note that $S(i)$ satisfies the following recursive equation.

$$S(0) = 0$$
$$S(i) = \min\{S(j-1) + f(r_j..r_i) \mid 1 \le j \le i\}$$

In practice, there may exist several block partitions that give the minimum number of tag SNPs. We also want to minimize the number of blocks. Let $C(i)$ be the minimum number of blocks so that the number of tag SNPs is $S(i)$. We have

$$C(0) = 0$$
$$C(i) = \min\{C(j-1) + 1 \mid 1 \le j \le i, S(i) = S(j-1) + f(r_j..r_i)\}$$

By filling in the dynamic programming tables for S and C and through back-tracing, we can compute a set of tag SNPs which can distinguish at least $\alpha\%$ of the haplotypes.

13.5.2 IdSelect

Carlson et al. [51] suggested another method, IdSelect, for selecting tag SNPs. Among a set S of all SNPs exceeding a specified minor allele frequency (MAF) threshold, the objective is to select a subset T of SNPs S such that for any SNP i in S, there exists an SNP j in T so that $r^2(i, j) > th$ for some threshold th. IdSelect is a greedy algorithm. It is shown in Figure 13.6.

A disadvantage of IdSelect is that it tends to include many rare SNPs as the tag SNPs. Since rare SNPs are harder to link with other SNPs, it is not very nature.

13.6 Association Study

Association study is an experiment designed to identify the association between SNPs and observable traits like blood pressure or disease status. An

---

Algorithm IdSelect

Require: A set S of SNPs that are above the MAF threshold.

Ensure: A subset $T \subseteq S$ such that for every SNP i in S, there exists an SNP j in T such that $r^2(i, j) > th$.

1: Let $T = \emptyset$
2: **while** S is not empty **do**
3:   Select $s \in S$ which maximizes the size of the set $\{s' \in S \mid r^2(s, s') > th\}$.
4:   $T = T \cup \{s\}$.
5:   $S = S - \{s\} - \{s' \in S \mid r^2(s, s') > th\}$.
6: **end while**
7: Report T

FIGURE 13.6:   Algorithm IdSelect for selecting tag SNPs.

	Allele a	Allele A
Lung cancer samples	60	20
Control samples	10	70

FIGURE 13.7:   The table shows an experiment for 80 lung cancer patients and 80 normal patients. From the relative risk and the odds ratio, patients with allele a have higher risk of suffering from lung cancer.

association study involves two groups of people: case samples (individuals with disease) and control samples (individuals without disease). If a certain genetic variation (including a single SNP or a set of SNPs) is associated more frequently with individuals with disease, the variation is said to be associated with disease. For example, consider the SNP in Figure 13.7. If it equals a, the patient has higher chance of having lung cancer. Hence, we say this SNP is lung cancer associated.

The information generated in an association study is useful. It helps to identify drug targets or disease markers. It also helps us understand how genetic variation affects the response to pathogens or drugs. Last, it helps us understand the differences among different races. For example, why do Asian people have a higher chance of getting hepatitis B?

This section describes two methods for single SNP association.

13.6.1   Categorical Data Analysis

Consider a case dataset and a control dataset. For an SNP at a particular locus, let a be the minor allele and A be the major allele. Let 0, 1, and 2 be the three genotypes where 0 is aa, 1 is AA, and 2 is Aa.

	0	1	2
case	$n_{0,case}$	$n_{1,case}$	$n_{2,case}$
control	$n_{0,ctl}$	$n_{1,ctl}$	$n_{2,ctl}$

(a)

	Allele a	Allele A
case	$p = 2n_{0,case} + n_{2,case}$	$q = 2n_{1,case} + n_{2,case}$
control	$r = 2n_{0,ctl} + n_{2,ctl}$	$s = 2n_{1,ctl} + n_{2,ctl}$

(b)

FIGURE 13.8: (a) Genotype frequencies for case and control samples. (b) Allele frequencies for case and control samples.

13.6.1.1 Pearson's χ^2 Test and Fisher Exact Test

Given a case and control dataset and an SNP at a particular locus, we hope to check if there is any correlation between the allele of the SNP with the case and control samples.

Suppose we are given the genotype frequencies for both case and control samples as shown in Figure 13.8(a). Note that, for $\delta \in \{case, ctl\}$, the frequency of allele a is $2n_{0,\delta} + n_{2,\delta}$ and the frequency of allele A is $2n_{1,\delta} + n_{2,\delta}$. Hence, we can construct the 2-by-2 count table in Figure 13.8(b), which shows the number of occurrences of alleles a and A in case and control. Similarly to Section 13.2, we can test if allele a is associated with the case sample by checking if the number of allele a in case samples deviates from expectation.

13.6.2 Relative Risk and Odds Ratio

We can also use the relative risk and the odds ratio to check if case samples are associated with allele a.

The relative risk (RR) is defined as the ratio of the probability of allele a occurring in the case samples versus the control samples. Precisely, we have

$$RR = \frac{p/(p+q)}{r/(r+s)}$$

The relative risk is always greater than or equal to zero. If $RR < 1$, allele a has a higher chance of occurring in the control samples. If $RR = 1$, allele a has an equal chance of occurring in the case and the control samples. If $RR > 1$, allele a has a higher chance of occurring in the case samples.

The odds ratio (OR) is defined as the ratio of the odds of the allele a occurring in the case samples versus the odds of the allele a occurring in the control samples. Precisely, we have

$$OR = \frac{p/r}{q/s} = \frac{ps}{qr}$$

The odds ratio is always greater than or equal to zero. If $OR = 1$, allele a is equally likely to occur in both case and control samples. If $OR > 1$, allele a is likely to occur in case samples. If $OR < 1$, allele a is likely to occur in control samples. Furthermore, the 95% confidence interval of OR is $[ORe^{-1.96\sigma}, ORe^{-1.96\sigma}]$, where $\sigma = \sqrt{\frac{1}{p} + \frac{1}{q} + \frac{1}{r} + \frac{1}{s}}$.

Consider the eighty lung cancer patients and the eighty control patients. For this particular SNP, Figure 13.7 shows the 2-by-2 count table. The relative risk and odds ratio for having allele a for cancer patients are $RR = (60/70)/(20/90) = 3.86$ and $OR = (60 * 70)/(20 * 10) = 21$. This means that patients whose SNP has allele a have a higher risk of suffering from lung cancer.

Given a set of SNPs for the case and control samples, we can rank those SNPs based on their relative risk and odds ratio. The most significant SNP has a higher chance of being related to the phenotype.

13.6.3   Linear Regression

Another method for correlating SNP and phenotype is by linear regression. Suppose the input is a set of pairs $\{(x_i, y_i) \mid i = 1, 2, \ldots, n\}$ where x_i is the allele of the SNP and y_i is the phenotypic score. This approach identifies a linear relationship between x_i and y_i. Precisely, our aim is to compute β_0 and β_1 such that

$$y_i = \beta_0 + \beta_1 x_i + \epsilon_i$$

where $\sum_{i=1..n} \epsilon_i^2 = \sum_{i=1..n} (y_i - \beta_0 - \beta_1 x_i)^2$ is minimized. $\sum_{i=1..n} \epsilon_i^2$ is called the sum of squares error (SSE).

We can determine β_0 and β_1 which minimize $\sum_{i=1...n} \epsilon_i^2$ by partial differentiation.

By differentiation with respect to β_0, we have

$$\frac{\partial(\sum_{i=1..n} \epsilon_i^2)}{\partial \beta_0} = \sum_{i=1..n} 2(y_i - \beta_0 - \beta_1 x_i) = 0$$

Hence, we have

$$\left(\sum_{i=1}^{n} y_i\right) - n\beta_0 - \beta_1 \left(\sum_{i=1}^{n} x_i\right) = 0 \tag{13.2}$$

By differentiation with respect to β_1, we have

$$\frac{\partial(\sum_{i=1..n} \epsilon_i^2)}{\partial \beta_1} = \sum_{i=1..n} 2(y_i - \beta_0 - \beta_1 x_i)(-x_i) = 0$$

Hence, we have

$$\sum_{i=1..n} x_i y_i - \beta_0 \sum_{i=1..n} x_i - \beta_1 \sum_{i=1..n} x_i^2 = 0 \qquad (13.3)$$

By solving Equations 13.2 and 13.3, we have

$$\beta_1 = \frac{\sum_{i=1..n} (x_i - \mu_x)(y_i - \mu_y)}{\sum_{i=1..n} (x_i - \mu_x)^2} \qquad (13.4)$$

$$\beta_0 = \mu_y - \beta_1 \mu_x \qquad (13.5)$$

where $\mu_x = \frac{1}{n} \sum_{i=1..n} x_i$ and $\mu_y = \frac{1}{n} \sum_{i=1..n} y_i$.

Note that $\beta_1 \neq 0$ means that the SNP is associated with the phenotype. To test if $\beta_1 \neq 0$, we apply an F-test. We let $\hat{y}_i = \beta_0 + \beta_1 x_i$. Let the mean square error (MSE) be $\frac{1}{n-2} \sum_{i=1..n} (y_i - \hat{y}_i)^2 = \frac{1}{n-2} \sum_{i=1..n} \epsilon_i^2$ and let the mean square error of regression (MSR) be $\sum_{i=1..n} (\hat{y}_i - \mu_y)^2$. Then, $\frac{MSR}{MSE}$ follows the F-distribution. If $\frac{MSR}{MSE} > F_{1,n-2,0.95}$, we reject the hypothesis that $\beta_1 = 0$.

For example, consider a set of $n = 16$ samples. Figure 13.9 shows the genotypes and the phenotypic scores of those 16 samples. We have $\mu_x = \sum_{i=1..n} x_i = 0.8125$ and $\mu_y = \sum_{i=1..n} y_i = 2.55625$. By Equations 13.4 and 13.5, we have $\beta_0 = 2.2415$ and $\beta_1 = 0.3874$. Hence, the best fit linear equation is $y = 2.2415 + 0.3874x + \epsilon$. Note that $MSR = 1.266338$ and $MSE = 0.040931$. We have $\frac{MSR}{MSE} = 30.03819$, which is bigger than $F_{1,14,0.95} = 4.6$. Hence, we accept $\beta_1 \neq 0$ and we expect the phenotypic score is associated with the genotype.

13.6.4 Logistic Regression

The previous section assumed the phenotype of each individual is a real value. For a case and control study, the phenotype is binary (that is, either case or control). We can determine the correlation between the SNP and the phenotype by fitting the linear relationship between Pr_{case} and the genotype X using the following linear function.

$$Pr_{case} = \beta_0 + \beta_1 X + \epsilon \qquad (13.6)$$

However, linear regression is not ideal because of the following reasons:

1. For a given X, the predicted Pr_{case} by Equation 13.6 can be bigger than 1 or less than 0.

2. Linear regression assumes the variance of the phenotype is independent of the genotype X. However, it is not true when the phenotype is binary.

This problem can be resolved by applying logistic regression. We first define the logistic function (also called the sigmoid function), which is

$$F(t) = \frac{1}{1 + e^{-t}}$$

i	Genotype (x_i)	phenotypic score (y_i)
1	0	2
2	0	2.1
3	0	2.4
4	0	2.3
5	0	2.2
6	0	2.5
7	1	2.4
8	1	2.5
9	1	2.6
10	1	3
11	1	2.7
12	1	2.8
13	1	2.3
14	2	2.9
15	2	3.2
16	2	3

FIGURE 13.9:   An example of the genotypes and phenotypic scores of 16 individuals.

Figure 13.10 shows the logistic function.

Instead of fitting the linear function, logistic regression tries to fit

$$Pr_{case} = F(\beta_0 + \beta_1 X) = \frac{1}{1 + e^{-\beta_0 - \beta_1 X}}.$$

In other words, it is fitting

$$\frac{Pr_{case}}{1 - Pr_{case}} = \beta_0 + \beta_1 X.$$

Similar to linear regression, when $\beta_1 \neq 0$, the SNP is associated with the phenotype.

13.7   Exercises

1. Can you prove Lemma 13.2?

2. Given the following genotype population, compute the allele frequencies. Based on the allele frequencies and assuming HWE, can you compute the expected genotype frequencies? Can you perform a χ^2 test for HWE? Do you accept the hypothesis that the population is in Hardy-Weinberg equilibrium?

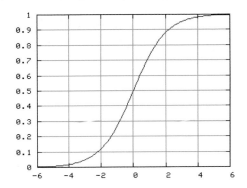

FIGURE 13.10: The logistic function.

Genotype	AA	Aa	aa
Actual	100	50	50

3. This chapter described the Hardy-Weinberg equilibrium for two alleles. Can you generalize the Hardy-Weinberg equilibrium to handle three alleles?

4. Can you deduce the haplotypes for the following four genotypes using Clark's algorithm?

   - $G_1 = 122101$
   - $G_2 = 210101$
   - $G_3 = 022121$
   - $G_4 = 201121$

5. Consider the following set of genotypes.

$$G_1 : 1012$$
$$G_2 : 2220$$
$$G_3 : 1020$$
$$G_4 : 2210$$

   (a) Every iteration in Clark's algorithm resolves at least one G_i by including at most 1 new haplotype. Suppose we always try to resolve G_i with the smallest index i first. What is the set of haplotypes generated?

   (b) Is the answer in (a) optimal? If not, can we resolve G_i in a different order to obtain the optimal solution?

6. Can you deduce the haplotypes for the following three genotypes using the EM algorithm? We assume that the initial haplotype frequencies are uniform and we only execute the EM algorithm for one round.

   - $G_1 = 122$
   - $G_2 = 210$
   - $G_3 = 212$

7. Consider the following genotype matrix M. Can you detail the steps of Bafna et al.'s method to solve the perfect phylogeny haplotyping problem?

M	1	2	3	4	5	6
G_1	1	1	0	2	2	0
G_2	1	2	2	0	0	2
G_3	1	1	2	2	0	0
G_4	2	2	2	0	0	2
G_5	1	1	2	2	2	0

8. Consider the following genotype matrix M. Can you detail the steps of Bafna et al.'s method to solve the perfect phylogeny haplotyping problem?

M	1	2	3	4
G_1	0	2	1	1
G_2	2	0	1	1
G_3	2	2	1	1
G_4	0	0	1	2
G_5	0	0	2	1
G_6	0	0	2	2

   Suppose there is an additional row $G_7 = (2, 2, 2, 1)$. Can you detail the steps of Bafna et al.'s method to solve the perfect phylogeny haplotyping problem?

9. Consider loci 1 and 2. Suppose the allele for locus 1 is either A or a and the allele for locus 2 is either B or b. Suppose A and B are associated such that (1) $p_{AB} = p_A p_B + D_1$, (2) $p_{Ab} = p_A p_b - D_2$, (3) $p_{aB} = p_a p_B - D_3$, and (4) $p_{ab} = p_a p_b + D_4$. Can you show that $D_1 = D_2 = D_3 = D_4$?

10. Consider the following two SNPs.

    - SNP1: Case — 40 A, 160 C, Control — 20 A, 180 C,
    - SNP2: Case — 20 A, 180 C, Control — 10 A, 190 C.

    Can you compute the relative risk and the odds ratio for the two SNPs? Which SNP has a higher risk?

11. Consider the following SNP with genotypes 0, 1, and 2. Suppose we have 200 case and 200 control samples.

- Case: 20 of genotype 0, 100 of genotype 1, and 80 of genotype 2
- Control: 50 of genotype 0, 100 of genotype 1, and 50 of genotype 2.

Can you perform associated study for the SNP using logistic regression?

References

[1] J. P. Abrahams, M. van den Berg, E. van Batenburg, and C. Pleij. Prediction of RNA secondary structure, including pseudoknotting, by computer simulation. *Nucleic Acids Research*, 18:3035–3044, 1990.

[2] E. N. Adams. Consensus techniques and the comparison of taxonomic trees. *Systematic Zoology*, 21(4):390–397, 1972.

[3] R. Agarwala and D. Fernandez-Baca. A polynomial-time algorithm for the perfect phylogeny when the number of character states is fixed. *SIAM Journal on Computing*, 23(6):1216–1224, 1994.

[4] A. V. Aho, D. S. Hirschberg, and J. D. Ullman. Bounds on the complexity of the longest common subsequence problem. *The Journal of ACM*, 23(1):1–12, 1976.

[5] Y. Akiyama and M. Kanehisa. NeuroFold: An RNA secondary structure prediction system using a Hopfield neural network. In *Genome Informatics Workshop*, 1992.

[6] T. Akutsu. Dynamic programming algorithm for RNA secondary structure prediction with pseudoknots. *Discrete Applied Mathematics*, 104 (1–3):45–62, 2000.

[7] N. Alon, B. Chor, F. Pardi, and A. Rapoport. Approximate maximum parsimony and ancestral maximum likelihood. *IEEE/ACM Transactions on Computational Biology and Bioinformatics*, 2008.

[8] S. F. Altschul. Generalized affine gap costs for protein sequence alignment. *Proteins*, 32(1):88–96, 1998.

[9] S. F. Altschul, W. Gish, W. Miller, E. W. Meyers, and D. J. Lipman. Basic local alignment search tool. *Journal of Molecular Biology*, 215: 403–410, 1990.

[10] S. F. Altschul, T. L. Madden, A. A. Schaffer, J. Zhang, Z. Zhang, W. Miller, and D. J. Lipman. Gapped BLAST and PSI-BLAST: A new generation of protein database search programs. *Nucleic Acids Research*, 25(17):3389–3402, 1997.

[11] N. Amenta, F. Clarke, and K. S. John. A linear-time majority tree algorithm. In *Proceedings of Third International Workshop on Algorithms in Bioinformatics*, pages 216–227, 2003.

[12] W. Ao, J. Gaudet, W. J. Kent, S. Muttumu, and S. E. Mango. Environmentally induced foregut remodeling by PHA-4/FoxA and DAF-12/NHR. *Science*, 305:1743–1746, 2004.

[13] D. A. Bader, B. M. E. Moret, and M. Yan. A linear-time algorithm for computing inversion distance between signed permutations with an experimental study. *Journal of Computational Biology*, 8(5):483–491, 2001.

[14] V. Bafna and P. Pevzner. Genome rearrangement and sorting by reversals. *SIAM Journal on Computing*, 25(2):272–289, 1996.

[15] V. Bafna, E. L. Lawler, and P. Pevzner. Approximation algorithms for multiple sequence alignment. *Theoretical Computer Science*, 182(1-2): 233–244, 1997.

[16] V. Bafna, D. Gusfield, G. Lancia, and S. Yooseph. Haplotyping as perfect phylogeny: A direct approach. *Journal of Computational Biology*, 10(3-4):323–340, 2003.

[17] T. Bailey and C. Elkan. Unsupervised learning of multiple motifs in biopolymers using expectation maximization. *Machine Learning*, 21: 51–80, 1995.

[18] T. L. Bailey and C. Elkan. Fitting a mixture model by expectation maximization to discover motifs in biopolymers. In *Proceedings of the Second International Conference on Intelligent Systems for Molecular Biology*, pages 28–36, 1994.

[19] Y. Barash, G. Elidan, T. Kaplan, and N. Friedman. Modeling dependencies in protein-DNA binding sites. In *Proceedings of the 7th Annual International Conference on Computational Molecular Biology*, pages 28–37, 2003.

[20] J. P. Barthelemy and F. McMorris. The median procedure for n-trees. *Journal of Classification*, 3(2):329–334, 1986.

[21] M. A. Bender and M. Farach-Colton. The LCA problem revisited. In *Latin American Theoretical Informatics*, pages 88–94, 2000.

[22] S. A. Benner, M. A. Cohen, and G. H. Gonnet. Empirical and structural models for insertions and deletions in the divergent evolution of proteins. *Journal of Molecular Biology*, 229(4):1065–1082, 1993.

[23] P. V. Benos, M. L. Bulyk, and G. D. Stormo. Additivity in protein-DNA interactions: How good an approximation is it? *Nucleic Acids Research*, 30(20):4442–4451, 2002.

[24] A. Bergeron. A very elementary presentation of the Hannenhalli-Pevzner theory. *Discrete Applied Mathematics*, 146(2):134–145, 2005.

[25] A. Bergeron, J. Mixtacki, and J. Stoye. Reversal distance without hurdles and fortresses. In *Proceedings of the 15th Annual Symposium on Combinatorial Pattern Matching*, pages 388–399, 2004.

[26] P. Berman and S. Hannenhalli. Fast sorting by reversal. In *Proceedings of the 7th Annual Symposium on Combinatorial Pattern Matching*, pages 168–185, 1996.

[27] P. Berman and M. Karpinski. On some tighter inapproximability results. In *Proceedings of the 26th International Colloquium on Automata, Languages and Programming*, pages 200–209, 1999.

[28] P. Berman, S. Hannenhalli, and M. Karpinski. 1.375-approximation algorithm for sorting by reversals. In *Proceedings of the 10th Annual European Symposium on Algorithms*, pages 200–210, 2002.

[29] V. Berry and O. Gascuel. Inferring evolutionary trees with strong combinatorial evidence. *Theoretical Computer Science*, 240(2):271–298, 2000.

[30] G. Blackshields, I. M. Wallace, M. Larkin, and D. G. Higgins. Analysis and comparison of benchmarks for multiple sequence alignment. *In Silico Biology*, 6(4):321–339, 2006.

[31] M. Blanchette and M. Tompa. Discovery of regulatory elements by a computational method for phylogenetic footprinting. *Genome Research*, 12(5):739–748, 2002.

[32] N. Blow. PCR's next frontier. *Nature Methods*, 4:869–875, 2007.

[33] L. A. Boyer, T. I. Lee, M. F. Cole, S. E. Johnstone, S. S. Levine, J. P. Zucker, M. G. Guenther, R. M. Kumar, H. L. Murray, R. G. Jenner, D. K. Gifford, D. A. Melton, R. Jaenisch, and R. A. Young. Core transcriptional regulatory circuitry in human embryonic stem cells. *Cell*, 122:947–956, 2005.

[34] N. Bray, I. Dubchak, and L. Pachter. AVID: A global alignment program. *Genome Research*, 13(1):97–102, 2003.

[35] L. Brocchieri and S. Karlin. A symmetric-iterated multiple alignment of protein sequences. *Journal of Molecular Biology*, 276(1):249–264, 1998.

[36] G. S. Brodal, R. Fagerberg, and C. N. S. Pedersen. Computing the quartet distance between evolutionary trees in time $O(n \log n)$. *Algorithmica*, 38:377–395, 2003.

[37] B. R. Brooks, R. E. Bruccokeri, B. D. Olafson, D. J. States, S. Swaminathan, and M. Karplus. CHARMM: A program for macromolecular energy, minimization, and dynamics calculations. *Journal of Computational Chemistry*, 4:187–217, 1983.

[38] M. Brow and C. Wilson. RNA pseudoknot modeling using intersections of stochastic context free grammars with applications to database search. In *Pacific Symposium on Biocomputing*, pages 109–125, 1996.

[39] M. Brudno, C. B. Do, G. M. Cooper, M. F. Kim, E. Davydov, N. I. S. C. Comparative Sequencing Program, E. D. Green, A. Sidow, and S. Batzoglou. LAGAN and Multi-LAGAN: Efficient tools for large-scale multiple alignment of genomic DNA. *Genome Research*, 13(4):721–731, 2003.

[40] D. Bryant. *DIMACS Series in Discrete Mathematics and Theoretical Computer Science*, chapter A classification of consensus methods for phylogenetics. 2001.

[41] D. Bryant, J. Tsang, P. Kearney, and M. Li. Computing the quartet distance between evolutionary trees. In *Proceedings of the 11th Annual ACM-SIAM Symposium on Discrete Algorithms*, pages 285–286, 2000.

[42] P. J. Buhler. Efficient large-scale sequence comparison by locality-sensitive hashing. *Bioinformatics*, 17(15):419–428, 2001.

[43] P. Buneman. A note on the metric properties of trees. *Journal of Combinatorial Theory (B)*, 17:48–50, 1974.

[44] S. Burkhardt, A. Crauser, P. Ferragina, H. P. Lenhof, E. Rivals, and M. Vingron. Q-gram based database searching using a suffix array. In *Proceedings of the 3rd Annual International Conference on Computational Molecular Biology*, pages 77–83, 1999.

[45] M. Burrows and D. J. Wheeler. A block-sorting lossless data compression algorithm. Technical report, Digital SRC Research Report, 1994.

[46] R. M. Bush, C. A. Bender, K. Subbarao, N. J. Cox, and W. M. Fitch. Predicting the evolution of human influenza A. *Science*, 286(5446): 1921–1925, 1999.

[47] H. J. Bussemaker, H. Li, and E. D. Siggia. Regulatory element detection using correlation with expression. *Nature Genetics*, 27:167 – 174, 2001.

[48] R. L. Cann, M. Stoneking, and A. C. Wilson. Mitochondrial DNA and human evolution. *Nature*, 325(6099):31–36, 1987.

[49] Y. Cao, R. M. Kumar, B. H. Penn, C. A. Berkes, C. Kooperberg, L. A. Boyer, R. A. Young, and S. J. Tapscott. Global and gene-specific analyses show distinct roles for myod and myog at a common set of promoters. *EMBO Journal*, 25:502–511, 2006.

[50] Alberto Caprara. Sorting by reversals is difficult. In *Proceedings of the 1st Annual International Conference on Computational Molecular Biology*, pages 75–83, 1997.

[51] C. S. Carlson, M. A. Eberle, M. J. Rieder, Q. Yi, L. Kruglyak, and D. A. Nickerson. Selecting a maximally informative set of single-nucleotide polymorphisms for association analyses using linkage disequilibrium. *American Journal of Human Genetics*, 74:106–120, 2004.

[52] H. Carroll, W. Beckstead, T. O'Connor, M. Ebbert, M. Clement, Q. Snell, and D. McClellan. DNA reference alignment benchmarks based on tertiary structure of encoded proteins. *Bioinformatics*, 23 (19):2648–2649, 2007.

[53] J. A. Cavender and J. Felsenstein. Invariants of phylogenies: Simple case with discrete states. *Journal of Classification*, 4:57–71, 1987.

[54] P. Chain, S. Kurtz, E. Ohlebusch, and T. Slezak. An applications-focused review of comparative genomics tools: Capabilities, limitations and future challenges. *Briefing in Bioinformatics*, 4(2):105–123, 2003.

[55] A. Chakravarty, J. M. Carlson, R. S. Khetani, and R. H. Gross. A parameter-free algorithm for improved *de novo* identification of transcription factor binding sites. *BMC Bioinformatics*, 8:29, 2007.

[56] H. L. Chan, T. W. Lam, W. K. Sung, P. W. H. Wong, S. M. Yiu, and X. Fan. The mutated subsequence problem and locating conserved genes. *Bioinformatics*, 21(10):2271–2278, 2005.

[57] W. I. Chang and E. L. Lawler. Sublinear expected time approximate string matching and biological applications. *Algorithmica*, 12:327–344, 1994.

[58] D. Che, S. Jensen, L. Cai, and J. S. Liu. BEST: Binding-site estimation suite tools. *Bioinformatics*, 21:2909–2911, 2005.

[59] T. Chen, M. Y. Kao, M. Tepel, J. Rush, and G. M. Church. A dynamic programming approach to de novo peptide sequencing via tandem mass spectrometry. *Journal of Computational Biology*, 8:325–337, 2001.

[60] D. Y. Chiang, P. O. Brown, and M. B. Eisen. Visualizing associations between genome sequences and gene expression data using genome-mean expression profiles. *Bioinformatics*, 17 Suppl 1:S49–S55, 2001.

[61] C. Christiansen, T. Mailund, C. N. S. Pedersen, and M. Randers. Computing the quartet distance between trees of arbitrary degree. In *Proceedings of the 5th International Workshop on Algorithms in Bioinformatics*, pages 77–88, 2005.

[62] D. A. Christie. A 3/2-approximation algorithm for sorting by reversals. In *Proceedings of the 9th Annual ACM-SIAM Symposium on Discrete Algorithms*, pages 244–252, 1998.

[63] A. G. Clark. Inference of haplotypes from PCR-amplified samples of diploid populations. *Molecular Biology and Evolution*, 7:111–122, 1990.

[64] E. M. Conlon, X. S. Liu, J. D. Lieb, and J. S. Liu. Integrating regulatory motif discovery and genome-wide expression analysis. *Proceedings of the National Academy of Sciences of USA*, 100(6):3339–3344, 2003.

[65] F. Corpet and B. Michot. RNAlign program: Alignment of RNA sequences using both primary and secondary structures. *Bioinformatics*, 10(4):389–399, 1994.

[66] R. Craig and R. C. Beavis. A method for reducing the time required to match protein sequences with tandem mass spectra. *Rapid Communcations in Mass Spectrometry*, 17:2310–2316, 2003.

[67] F. Crick. Central dogma of molecular biology. *Nature*, 227(5258):561–563, 1970.

[68] F. H. Crick. On protein synthesis. *The Symposia of the Society for Experimental Biology*, 12:138–163, 1958.

[69] E. Dam, K. Pleij, and D. Draper. Structural and functional aspects of RNA pseudoknots. *Biochemistry*, 31(47):11665–11676, 1992.

[70] V. Dančík, T. A. Addona, K. R. Clauser, J. E. Vath, and P. A. Pevzner. De novo peptide sequencing via tandem mass spectrometry. *Journal of Computational Biology*, 6:327–342, 1999.

[71] A. C. E. Darling, B. Mau, F. R. Blattner, and N. T. Perna. Mauve: Multiple alignment of conserved genomic sequence with rearrangements. *Genome Research*, 14(7):1394–1403, 2004.

[72] B. DasGupta, X. He, T. Jiang, M. Li, J. Tromp, and L. Zhang. On distances between phylogenetic trees. In *Proceedings of the 8th Annual ACM-SIAM Symposium on Discrete Algorithms*, pages 427–436, 1997.

[73] W. H. E. Day. Optimal algorithms for comparing trees with labeled leaves. *Journal of Classification*, 2:7–28, 1985.

[74] W. H. E. Day, D. S. Johnson, and D. Sankoff. The computational complexity of inferring rooted phylogenies by parsimony. *Mathematical Biosciences*, 81:33–42, 1986.

[75] M. O. Dayhoff, R. M. Schwartz, and B. C. Orcutt. A model of evolutionary change in proteins. *Atlas of Protein Sequence and Structure*, 5 (supplement 3):345–352, 1978.

[76] A. L. Delcher, S. Kasif, R. D. Fleischmann, J. Peterson, O. White, and S. L. Salzberg. Alignment of whole genomes. *Nucleic Acids Research*, 27(11):2369–2376, 1999.

[77] A. L. Delcher, A. Phillippy, J. Carlton, and S. L. Salzberg. Fast algorithms for large-scale genome alignment and comparison. *Nucleic Acids Research*, 30(11):2478–2483, 2002.

[78] Z. Ding, V. Filkov, and D. Gusfield. A linear-time algorithm for the perfect phylogeny haplotyping (PPH) problem. *Journal of Computational Biology*, 13(2):522–553, 2006.

[79] C. B. Do and K. Katoh. Protein multiple sequence alignment. *Functional Proteomics*, pages 379–413, 2008.

[80] C. B. Do, M. S. P. Mahabhashyam, M. Brudno, and S. Batzoglou. ProbCons: Probabilistic consistency-based multiple sequence alignment. *Genome Research*, 15(2):330–340, 2005.

[81] T. Dobzhansky and A. H. Sturtevant. Inversions in the chromosomes of drosophila pseudoobscura. *Genetics*, 23:28–64, 1928.

[82] R. F. Doolittle, M. Hunkapiller, L. E. Hood, S. Devare, K. Robbins, S. Aaronson, and H. Antoniades. Simian sarcoma virus, onc gene v-sis, is derived from the gene (or genes) encoding a platelet-derived growth factor. *Science*, 221:275–277, 1983.

[83] W. F. Doolittle. Phylogenetic classification and the universal tree. *Science*, 284(5423):2124–2128, 1999.

[84] T. A. Down and T. J. P. Hubbard. NestedMICA: Sensitive inference of over-represented motifs in nucleic acid sequence. *Nucleic Acids Res*, 33 (5):1445–1453, 2005.

[85] I. Dubchak, A. Poliakov, A. Kislyuk, and M. Brudno. Multiple whole-genome alignments without a reference organism. *Genome Research*, 19 (4):682–689, 2009.

[86] L. Duret and S. Abdeddaim. Multiple alignment for structural, functional or phylogenetic analyses of homologous sequences. In D. Higgins and W. Taylor, editors, *Bioinformatics sequence structure and databanks*, 2000.

[87] S. R. Eddy and R. Durbin. RNA sequence analysis using covariance models. *Nucleic Acids Research*, 22:2079–2088, 1994.

[88] R. C. Edgar. MUSCLE: A multiple sequence alignment method with reduced time and space complexity. *BMC Bioinformatics*, 5:113, 2004.

[89] R. C. Edgar. MUSCLE: Multiple sequence alignment with high accuracy and high throughput. *Nucleic Acids Research*, 32:1792–1797, 2004.

[90] R. C. Edgar and S. Batzoglou. Multiple sequence alignment. *Current Opinion in Structural Biology*, 16:368–373, 2006.

[91] R. C. Edgar and K. Sjöander. COACH: Profile-profile alignment of protein families using hidden markov models. *Bioinformatics*, 20(8): 1309–1318, 2004.

[92] R. C. Edgar and K. Sjölander. SATCHMO: Sequence alignment and tree construction using hidden Markov models. *Bioinformatics*, 19(11): 1404–1411, 2003.

[93] C. Ehresmann, F. Baudin, M. Mougel, P. Romby, J. P. Ebel, and B. Ehresmann. Probing the structure of RNAs in solution. *Nucleic Acids Research*, 15:9109–9112, 1987.

[94] J. K. Eng, A. L. McCormack, and J. R. Yates. An approach to correlate tandem mass spectral data of peptides with amino acid sequences in a protein database. *Journal of the American Society for Mass Spectrometry*, 5:976–989, 1994.

[95] E. Eskin and P. Pevzner. Finding composite regulatory patterns in DNA sequences. *Bioinformatics (Supplement 1)*, 18:S354–S363, 2002.

[96] E. Eskin, E. Halperin, and R. M. Karp. Large scale reconstruction of haplotypes from genotype data. In *Proceedings of the 7th Annual International Conference on Research in Computational Molecular Biology*, pages 104–113, 2003.

[97] G. Estabrook, F. McMorris, and C. Meacham. Comparison of undirected phylogenetic trees based on subtrees of four evolutionary units. *Systematic Zoology*, 34:193–200, 1985.

[98] L. Ettwiller, B. Paten, M. Ramialison, E. Birney, and J. Wittbrodt. Trawler: De novo regulatory motif discovery pipeline for chromatin immunoprecipitation. *Nature Methods*, 4(7):563–565, 2007.

[99] L. Excoffier and M. Slatkin. Maximum-likelihood estimation of molecular haplotype frequencies in a diploid population. *Molecular Biology and Evolution*, 12(5):921–927, 1995.

[100] M. Farach. Optimal suffix tree construction with large alphabets. In *Proceedings of the 38th IEEE Symposium on Foundations of Computer Science*, pages 137–143, 1997.

[101] M. Farach, S. Kannan, and T. Warnow. A robust model for finding optimal evolutionary trees. *Algorithmica*, 13(1):155–179, 1996.

[102] J. Felsenstein. Cases in which parsimony or compatibility methods will be positively misleading. *Systematic Zoology*, 27:401–410, 1978.

[103] J. Felsenstein. Evolutionary trees from DNA sequences: A maximum likelihood approach. *Journal of Molecular Evolution*, 17(6):368–376, 1981.

[104] D. F. Feng and R. F. Doolittle. Progressive sequence alignment as a prerequisite to correct phylogenetic trees. *Journal of Molecular Biology and Evolution*, 25:351–360, 1987.

[105] J. Fernandez-de-Cossio, J. Gonzalez, and V. Besada. A computer program to aid the sequencing of peptides in collision-activated decomposition experiments. *Bioinformatics*, 11(4):427–434, 1995.

[106] P. Ferragine and G. Manzini. Opportunistic data structures with applications. In *Proceedings of the 41st IEEE Symposium on Foundations of Computer Science*, pages 390–398, 2000.

[107] B. Fischer, V. Roth, F. Roos, J. Grossmann, S. Baginsky, P. Widmayer, W. Gruissem, and J. M. Buhmann. NovoHMM: A hidden Markov model for de novo peptide sequencing. *Analytical Chemistry*, 77:7265–7273, 2005.

[108] W. M. Fitch. Toward defining the course of evolution: Minimum change for a specified tree topology. *Systematic Zoology*, 20:406–416, 1971.

[109] W. M. Fitch and E. Margoliash. Construction of phylogenetic trees. *Science*, 155:279–284, 1987.

[110] L. R. Foulds and R. L. Graham. The Steiner problem in phylogeny is NP-complete. *Advances in Applied Mathematics*, 3:43–49, 1982.

[111] A. Frank and P. Pevzner. Pepnovo: De novo peptide sequencing via probabilistic network modeling. *Analytical Chemistry*, 77:964–973, 2005.

[112] E. Fratkin, B. T. Naughton, D. L. Brutlag, and S. Batzoglou. Motifcut: Regulatory motifs finding with maximum density subgraphs. *Bioinformatics*, 22(14):e150–e157, 2006.

[113] Z. Galil and R. Giancarlo. Speeding up dynamic programming with application to molecular biology. *Theoretical Computer Science*, 64(1): 107–118, 1989.

[114] F. Galisson. The fasta and BLAST programs. *Manuscript*, 2000.

[115] M. R. Garey and D. S. Johnson. *Computers and Intractability: A Guide to the Theory of NP-Completeness*. W. H. Freeman and Co., 1979.

[116] R. B. Gary and G. D. Stormo. Graph-theoretic approach to RNA modeling using comparative data. In *Proceedings of the 3rd International Conference on Intelligent Systems for Molecular Biology*, pages 75–80, 1995.

[117] W. H. Gates and C. H. Papadimitriou. Bound for sorting by prefix reversals. *Discrete Mathematics*, 27:47–57, 1979.

[118] L.Y. Geer, S. P. Markey, J. A. Kowalak, L. Wagner, M. Xu, D. M. Maynard, X. Yang, W. Shi, and S. H. Bryant. Open mass spectrometry search algorithm. *Journal of Proteome Research*, 3:958–964, 2004.

[119] D. B. Gordon, L. Nekludova, S. McCallum, and E. Fraenkel. TAMO: A flexible, object oriented framework for analyzing transcriptional regulation using DNA-sequences motifs. *Bioinformatics*, 21:3164–3165, 2005.

[120] O. Gotoh. An improved algorithm for matching biological sequences. *J Mol Biol*, 162(3):705–708, 1982.

[121] O. Gotoh. Optimal sequence alignment allowing for long gaps. *Bulletin of Mathematical Biology*, 52(3):359–373, 1990.

[122] O. Gotoh. Significant improvement in accuracy of multiple protein sequence alignments by iterative refinement as assessed by reference to structural alignments. *Journal of Molecular Biology*, 264(4):823–838, 1996.

[123] R. Grossi and J. S. Vitter. Compressed suffix arrays and suffix trees with applications to text indexing and string matching. In *Proceedings of the 32nd Annual ACM Symposium on Theory of Computing*, pages 397–406, 2000.

[124] X. Gu and W. H. Li. The size distribution of insertions and deletions in human and rodent pseudogenes suggests the logarithmic gap penalty for sequence alignment. *Journal of Molecular Evolution*, 40(4):464–473, 1995.

[125] A. P. Gultyaev, F. H. D. van Batenburg, and C. W. A. Pleij. The computer simulation of RNA folding pathways using a genetic algorithm. *Journal of Molecular Biology*, 250:37–51, 1995.

[126] S. K. Gupta, J. D. Kececioglu, and A. A. Schaffer. Improving the practical space and time efficiency of the shortest-paths approach to sum-of-pairs multiple sequence alignment. *Journal of Computational Biology: A Journal of Computational Molecular Cell Biology*, 2(3):459–472, 1995.

[127] D. Gusfield. Efficient methods for multiple sequence alignment with guaranteed error bounds. *Bulletin of Mathematical Biology*, 55(1):141–154, 1993.

[128] D. Gusfield. *Algorithms on Strings, Trees and Sequences: Computer Science and Computational Biology.* Cambridge University Press, 1997.

[129] D. Gusfield. Inference of haplotypes from samples of diploid populations: Complexity and algorithms. *Journal of Computational Biology*, 8(3):305–323, 2001.

[130] D. Gusfield. Haplotyping as perfect phylogeny: conceptual framework and efficient solutions. In *Proceedings of the 6th Annual International Conference on Computational Molecular Biology*, pages 166–175, 2002.

[131] B.D. Halligan, V. Ruotti, S. N. Twigger, and A. S. Greene. DeNovoID: a web-based tool for identifying peptides from sequence and mass tags deduced from de novo peptide sequencing by mass spectroscopy. *Nucleic Acids Research*, 33:376–381, 2005.

[132] A. M. Hamel and M. A. Steel. Finding a maximum compatible tree is NP-hard for sequences and trees. *Applied Mathematics Letters*, 9(2): 55–59, 1996.

[133] Y. Han, B. Ma, and K. Zhang. SPIDER: Software for protein identification from sequence tags with de novo sequencing error. *Journal of Bioinformatics and Computational Biolology*, 3:697–716, 2005.

[134] S. Hannenhalli and P. Pevzner. Transforming men into mice (polynomial algorithm for genomic distance problem). In *Proceedings of the 36th IEEE Symposium on Foundations of Computer Science*, pages 581–592, 1995.

[135] S. Hannenhalli and P. Pevzner. Transforming cabbage into turnip: Polynomial algorithm for sorting signed permutations by reversals. In *Proceedings of the 27th Annual ACM Symposium on Theory of Computing*, pages 178–189, 1995.

[136] C. T. Harbison, D. B. Gordon, T. I. Lee, N. J. Rinaldi, K. D. Macisaac, T. W. Danford, N. M. Hannett, J. B. Tagne, D. B. Reynolds, J. Yoo, E. G. Jennings, J. Zeitlinger, D. K. Pokholok, M. Kellis, P. A. Rolfe, K. T. Takusagawa, E. S. Lander, D. K. Gifford, E. Fraenkel, and R. A. Young. Transcription regulatory code of a eukaryotic genome. *Nature*, 431:99–104, 2004.

[137] D. Harel and R. E. Tarjan. Fast algorithms for finding nearest common ancestors. *SIAM Journal on Computing*, 13:338–355, 1984.

[138] J. A. Hartigan. Minimum mutation fits to a given tree. *Biometrics*, 20: 53–65, 1973.

[139] J. Hein. Reconstructing evolution of sequences subject to recombination using parsimony. *Mathematical Biosciences*, 98(2):185–200, 1990.

[140] M. D. Hendy and D. Penny. Branch and bound algorithms to determine minimal evolutionary trees. *Mathematical Biosciences*, 59:277–290, 1982.

[141] S. Henikoff and J. G. Henikoff. Automated assembly of protein blocks for database searching. *Nucleic Acids Res*, 19(23):6565–6572, 1991.

[142] S. Henikoff and J. G. Henikoff. Amino acid substitution matrices from protein blocks. *Proceedings of the National Academy of Sciences of USA*, 89(22):10915–10919, 1992.

[143] G. Z. Hertz and G. D. Stormo. Identifying DNA and protein patterns with statistically significant alignments of multiple sequences. *Bioinformatics*, 15(7-8):563–577, 1999.

[144] M. H. Heydari and I. H. Sudborough. On the diameter of the pancake network. *Journal of Algorithms*, 25(1):67–94, 1997.

[145] D. G. Higgins and P. M. Sharp. CLUSTAL: A package for performing multiple sequence alignment on a microcomputer. *Gene*, 73(1):237–244, 1988.

[146] D. M. Hillis, J. J. Bull, M. E. White, M. R. Badgett, and I. J. Molineux. Experimental phylogenetics: Generation of a known phylogeny. *Science*, 255(5044):589–592, 1992.

[147] D. S. Hirschberg. A linear space algorithm for computing maximal common subsequences. *Communications of the ACM*, 18(6):341–343, 1975.

[148] D. S. Hirschberg. An information-theoretic lower bound for the longest common subsequence problem. *Information Processing Letter*, 7(1):40–41, 1978.

[149] M. Hohl, S. Kurtz, and E. Ohlebusch. Efficient multiple genome alignment. *Bioinformatics*, 18 Suppl 1:S312–S320, 2002.

[150] I. Holmes. Using guide trees to construct multiple-sequence evolutionary HMMs. *Bioinformatics*, 19 Suppl 1:i147–i157, 2003.

[151] I. Holmes and W. J. Bruno. Evolutionary HMMs: A Bayesian approach to multiple alignment. *Bioinformatics*, 17(9):803–820, 2001.

[152] W. K. Hon, K. Sadakane, and W. K. Sung. Breaking a time and space barrier in constructing full-text indices. In *Proceedings of the 44th IEEE Symposium on Foundations of Computer Science*, 2003.

[153] P. Hong, X. S. Liu, Q. Zhou, X. Lu, J. S. Liu, and W. H. Wong. A boosting approach for motif modeling using chip-chip data. *Bioinformatics*, 21(11):2636–2643, 2005.

[154] J. Hu, Y. D. Yang, and D. Kihara. EMD: An ensemble algorithm for discovering regulatory motifs in DNA sequences. *BMC Bioinformatics*, 7:342, 2006.

[155] H. D. Huang, J. T. Horng, Y. M. Sun, A. P. Tsou, and S. L. Huang. Identifying transcriptional regulatory sites in the human genome using an integrated system. *Nucleic Acids Research*, 32:1948–1956, 2004.

[156] B. R. Huber and M. Bulyk. Meta-analysis discovery of tissue specific DNA sequence motifs from mammalian gene expression data. *BMC Bioinformatics*, 7:229–254, 2006.

[157] J. D. Hughes, P. W. Estep, S. Tavazoie, and G. M. Church. Computational identification of cis-regulatory elements associated with functionally coherent groups of genes in saccharomyces cerevisiae. *Journal of Molecular Biology*, 296:1205–1214, 2000.

[158] T. N. D. Huynh, W. K. Hon, T. W. Lam, and W. K Sung. Approximate string matching using compressed suffix arrays. *Theoretical Computer Science*, 352(1-3):240–249, 2006.

[159] S. Ieong, M. Y. Kao, T. W. Lam, W. K. Sung, and S. M. Yiu. Predicting RNA secondary structures with arbitrary pseudoknots by maximizing the number of stacking pairs. In *Proceedings of the 2nd IEEE International Symposium on Bioinformatics and Bioengineering*, pages 183–190, 2001.

[160] F. Jacob and J. Monod. Genetic regulatory mechanisms in the synthesis of proteins. *Journal of Molecular Biology*, 3:318–356, 1961.

[161] J. Jansson, H. K. Ng, K. Sadakane, and W. K. Sung. Rooted maximum agreement supertrees. *Algorithmica*, 43(4):293–307, 2005.

[162] S. Jensen and J. S. Liu. BioOptimizer: A Bayesian scoring function approach to motif discovery. *Bioinformatics*, 20:1557–1564, 2006.

[163] T. H. Jukes and C. R. Cantor. *Mammalian Protein Metabolism*, chapter Evolution of protein molecules, pages 21–132. Academic Press, 1969.

[164] W. Just. Computational complexity of multiple sequence alignment with SP-score. *Journal of Computational Biology*, 8(6):615–623, 2001.

[165] M. Y. Kao, T. W. Lam, W. K. Sung, T. M. Przytyccka, and H. F. Ting. General techniques for comparing unrooted evolutionary trees. In *Proceedings of the 29th Annual ACM Symposium on Theory of Computing*, pages 54–65, 1997.

[166] M. Y. Kao, T. W. Lam, W. K. Sung, and H. F. Ting. Cavity matchings, label compressions, and unrooted evolutionary trees. *SIAM Journal of Computing*, 30(2):602–624, 2000.

[167] M. Y. Kao, T. W. Lam, W. K. Sung, and H. F. Ting. Fast labeled tree comparisons via unbalanced bipartite matchings. *Journal of Algorithm*, 40(2):212–233, 2001.

[168] H. Kaplan, R. Shamir, and R. E. Tarjan. A faster and simpler algorithm for sorting signed permutations by reversals. *SIAM Journal on Computing*, 29(3):880–892, 1999.

[169] S. Karlin and G. Ghandour. Comparative statistics for DNA and protein sequences: Multiple sequence analysis. *Proceedings of the National Academy of Sciences of USA*, 82(18):6186–6190, 1985.

[170] K. Katoh, K. Misawa, K. Kuma, and T. Miyata. MAFFT: A novel method for rapid multiple sequence alignment based on fast Fourier transform. *Nucleic Acids Research*, 30(14):3059–3066, 2002.

[171] K. Katoh, K. Kuma, H. Toh, and T. Miyata. MAFFT version 5: Improvement in accuracy of multiple sequence alignment. *Nucleic Acids Research*, 33(2):511–518, 2005.

[172] J. Kececioglu and D. Sankoff. Exact and approximation algorithms for sorting by reversals, with application to genome rearrangement. *Algorithmica*, 13(1/2):180–210, 1995.

[173] U. Keich and P. A. Pevzner. Finding motifs in the twilight zone. *Bioinformatics*, 18(10):1374–1381, 2002.

[174] W. J. Kent. BLAT—the BLAST-like alignment tool. *Genome Research*, 12(4):656–664, 2002.

[175] J. Kim, S. Pramanik, and M. J. Chung. Multiple sequence alignment using simulated annealing. *Comput Appl Biosci*, 10(4):419–426, 1994.

[176] S. H. Kim, G. J. Suddath, G. J. Quigley, A. McPherson, J. L. Sussman, A. H. J. Wang, N. C. Seeman, and A. Rich. Three-dimensional tertiary structure of yeast phenylalanine transfer RNA. *Science*, 185:435–439, 1974.

[177] P. S. Klosterman, M. Tamura, S. R. Holbrook, and S. E. Brenner. SCOR: A structural classification of RNA database. *Nucleic Acids Research*, 30(1):392–394, 2002.

[178] D. A. Konings and R. R. Gutell. A comparison of thermodynamic foldings with comparatively derived structures of 16S and 16S-like rRNAs. *RNA*, 1(6):559–574, 1995.

[179] M. K. Kuhner and J. Felsenstein. A simulation comparison of phylogeny algorithms under equal and unequal evolutionary rates. *Molecular Biology and Evolution*, 11(3):459–468, 1994.

[180] S. Kurtz. Reducing the space requirement of suffix trees. *Software Practice and Experiences*, 29:1149–1171, 1999.

[181] S. Kurtz, A. Phillippy, A. L. Delcher, M. Smoot, M. Shumway, C. Antonescu, and S. L. Salzberg. Versatile and open software for comparing large genomes. *Genome Biology*, 5(2):R12, 2004.

[182] T. W. Lam, W. K. Sung, S. L. Tam, C. K. Wong, and S. M. Yiu. Compressed indexing and local alignment of DNA. *Bioinformatics*, 24 (6):791–797, 2008.

[183] C. E. Lawrence and A. A. Reilly. An expectation maximization (EM) algorithm for the identification and characterization of common sites in unaligned biopolymer sequences. *Proteins*, 7:41–51, 1990.

[184] C. E. Lawrence, S. F. Altschul, M. S. Boguski, J. S. Liu, A. F. Neuwald, and J. C. Wootton. Detecting subtle sequence signals: A Gibbs sampling strategy for multiple alignment. *Science*, 262:208–214, 1993.

[185] C. Lee, C. Grasso, and M. F. Sharlow. Multiple sequence alignment using partial order graphs. *Bioinformatics*, 18(3):452–464, 2002.

[186] T. Leitner, D. Escanilla, C. Franzen, M. Uhlen, and J. Albert. Accurate reconstruction of a known HIV-1 transmission history by phylogenetic tree analysis. *Proceedings of the National Academy of Sciences of USA*, 93(20):10864–10869, 1996.

[187] M. Li, B. Ma, D. Kisman, and J. Tromp. Patternhunter II: Highly sensitive and fast homology search. *Genome Informatics*, 14:164–175, 2003.

[188] S. Liang, M. P. Samanta, and B. A. Biegel. cWINNOWER algorithm for finding fuzzy DNA motifs. *Journal of Bioinformatics and Computational Biology*, 2(1):47–60, 2004.

[189] C. Linhart, Y. Halperin, and R. Shamir. Transcription factor and microRNA motif discovery: The Amadeus platform and a compendium of metazoan target sets. *Genome Research*, 18(7):1180–1189, 2008.

[190] D. J. Lipman and W. R. Pearson. Rapid and sensitive protein similarity searches. *Science*, 227:1435–1441, 1985.

[191] D. J. Lipman and W. R. Pearson. Improved tools for biological sequence comparison. *Proceedings of the National Academy of Sciences of USA*, 85:2444–2448, 1988.

[192] D. J. Lipman, S. F. Altschul, and J. D. Kececioglu. A tool for multiple sequence alignment. *Proceedings of the National Academy of Sciences of USA*, 86(12):4412–4415, 1989.

[193] X. Liu, D. L. Brutlag, and J. S. Liu. BioProspector: Discovering conserved DNA motifs in upstream regulatory regions of co-expressed genes. In *Proceedings of Pacific Symposium of Biocomputing*, pages 127–138, 2001.

[194] X. Liu, D. Brutlag, and J. Liu. An algorithm for finding protein-DNA binding sites with application to chromatin-immunoprecipitation microarray experiments. *Nature Biotechnology*, 20:835–839, 2002.

[195] A. Loria and T. Pan. Domain structure of the ribozyme from eubacterial ribonuclease P. *RNA*, 2(6):551–563, 1996.

[196] T. M. Lowe and S. R. Eddy. tRNAscan-SE: A program for improved detection of transfer RNA genes in genomic sequence. *Nucleic Acids Research*, 25(5):955–964, 1997.

[197] A. Loytynoja and N. Goldman. An algorithm for progressive multiple alignment of sequences with insertions. *Proceedings of the National Academy of Sciences of USA*, 102(30):10557–10562, 2005.

[198] A. Loytynoja and M. C. Milinkovitch. A hidden Markov model for progressive multiple alignment. *Bioinformatics*, 19(12):1505–1513, 2003.

[199] R. B. Lyngso and C. N. S. Pedersen. RNA pseudoknot prediction in energy-based models. *Journal of Computational Biology*, 7(3-4):409–427, 2000.

[200] R. B. Lyngso, M. Zuker, and C. N. S. Pedersen. Fast evaluation of internal loops in RNA secondary structure prediction. *Bioinformatics*, 15(6):440–445, 1999.

[201] B. Ma, J. Tromp, and M. Li. Patternhunter: Faster and more sensitive homology search. *Bioinformatics*, 18(3):440–445, Mar 2002.

[202] B. Ma, K. Zhang, and C. Liang. An effective algorithm for the peptide de novo sequencing from MS/MS spectrum. In *Proceedings of the 14th Annual Symposium on Combinatorial Pattern Matching*, pages 266–277, 2003.

[203] S. Mahony, P. E. Auron, and P. V. Benos. DNA familial binding profiles made easy: Comparison of various motif alignment and clustering strategies. *PLoS Computational Biology*, 3(3):e61+, 2007.

[204] U. Manber and G. Myers. Suffix arrays: A new method for on-line string searches. *SIAM Journal on Computing*, 22(5):935–948, 1993.

[205] T. Margush and F.R. McMorris. Consensus *n*-trees. *Bulletin of Mathematical Biology*, 43(2):239–244, 1981.

[206] L. Marsan and M. F. Sagot. Algorithms for extracting structured motifs using a suffix tree with an application to promoter and regulatory site consensus identification. *J Comput Biol*, 7(3-4):345–362, 2000.

[207] T. T. Marstrand, J. Frellsen, I. Moltke, M. Thiim, E. Valen, D. Retelska, and A. Krogh. Asap: A framework for over-representation statistics for transcription factor binding sites. *PLoS ONE*, 3(2):e1623, 2008.

[208] W. J. Masek and M. Paterson. A faster algorithm computing string edit distances. *Journal of Computer and System Science*, 20(1):18–31, 1980.

[209] D. M. Mathews, J. Sabina, M. Zuker, and D. H. Turner. Expanded sequence dependence of thermodynamic parameters improves prediction of RNA secondary structure. *Journal of Molecular Biology*, 288(5):911–940, 1999.

[210] S. McGinnis and T. L. Madden. BLAST: At the core of a powerful and diverse set of sequence analysis tools. *Nucleic Acids Research*, 32: W20–W25, 2004.

[211] E. M. McGreight. A space-economical suffix tree construction algorithm. *Journal of ACM*, 23:262–272, 1976.

[212] W. Miller and E. W. Myers. Sequence comparisons with concave weighting functions. *Bulletin of Mathematical Biology*, 50:97–120, 1988.

[213] G. W. Moore, M. Goodman, and J. Barnabas. An iterative approach from the standpoint of the additive hypothesis to the dendrogram problem posed by molecular data sets. *Journal of Theoretical Biology*, 38 (3):423–457, 1973.

[214] R. Mott. EST_GENOME: A program to align spliced DNA sequences to unspliced genomic DNA. *Comput Appl Biosci*, 13(4):477–478, 1997.

[215] T. Muller, R. Spang, and M. Vingron. Estimating amino acid substitution models: A comparison of Dayhoff's estimator, the resolvent approach and a maximum likelihood method. *Molecular Biology and Evolution*, 19:8–13, 2002.

[216] V. Narang, W.-K. Sung, and A. Mittal. Localmotif – An in-silico tool for detecting localized motifs in regulatory sequences. In *18th IEEE International Conference on Tools with Artificial Intelligence*, pages 791–799, 2006.

[217] National Center for Biotechnology Information: Comparative Maps. http://www.ncbi.nlm.nih.gov/Homology, 2004.

[218] S. B. Needleman and C. D. Wunsch. A general method applicable to the search for similarities in the amino acid sequence of two proteins. *Journal of Molecular Biology*, 48:443–453, 1970.

[219] C. R. Newton, A. Graham, L. E. Heptinstall, S. J. Powell, C. Summers, N. Kalsheker, J. C. Smith, and A. F. Markham. Analysis of any point mutation in DNA. The amplification refractory mutation system (ARMS). *Nucleic Acids Res*, 17(7):2503–2516, 1989.

[220] Z. Ning, A. J. Cox, and J. C. Mullikin. SSAHA: A fast search method for large DNA databases. *Genome Research*, 11(10):1725–1729, 2001.

[221] C. Notredame. Recent progress in multiple sequence alignment: A survey. *Pharmacogenomics*, 31(1):131–144, 2002.

[222] C. Notredame. Recent evolutions of multiple sequence alignment algorithms. *PLoS Comput Biology*, 3(8):e123, 2007.

[223] C. Notredame and D. G. Higgins. SAGA: Sequence alignment by genetic algorithm. *Nucleic Acids Research*, 24(8):1515–1524, 1996.

[224] R. Nussinov and A. B. Jacobson. Fast algorithm for predicting the secondary structure of single-stranded RNA. *Proceedings of the National Academy of Sciences of USA*, 77(11):6309–6313, 1980.

[225] D. T. Odom, N. Zizlsperger, D. B. Gordon, G. W. Bell, N. J. Rinaldi, H. L. Murray, T. L. Volkert, J. Schreiber, P. A. Rolfe, D. K. Gifford, E. Fraenkel, G. I. Bell, and R. A. Young. Control of pancreas and liver gene expression by HNF transcription factors. *Science*, 303:1378–1381, 2004.

[226] O. O'Sullivan, K. Suhre, C. Abergel, D. G. Higgins, and C. Notredame. 3DCoffee: Combining protein sequences and structures within multiple sequence alignments. *Journal of Molecular Biology*, 340(2):385–395, 2004.

[227] J. D. Palmer and L. A. Herbon. Plant mitochondrial DNA evolves rapidly in structure, but slowly in sequence. *Journal of Molecular Evolution*, 27:87–97, 1988.

[228] T. Palomero, W. K. Lim, D. T. Odom, M. L. Sulis, P. J. Real, A. Margolin, K. C. Barnes, J. O'Neil, D. Neuberg, A. P. Weng, J. C. Aster, F. Sigaux, J. Soulier, A. T. Look, R. A. Young, A. Califano, and A. A. Ferrando. NOTCH1 directly regulates c-MYC and activates a feed-forward-loop transcriptional network promoting leukemic cell growth. *Proceedings of the National Academy of Sciences of USA*, 103:18261–18266, 2006.

[229] C. Papanicolaou, M. Gouy, and J. Ninio. An energy model that predicts the correct folding of both the tRNA and 5S RNA molecules. *Nucleic Acids Research*, 12:31–44, 1984.

[230] G. Pavesi, G. Mauri, and G. Pesole. An algorithm for finding signals of unknown length in DNA sequencess. *Bioinformatics*, 17(90001):S207–S214, 2001.

[231] D. A. Pearlman, D. A. Case, J. W. Caldwell, W. S. Ross, III T. E. Cheatham, S. DeBolt, D. Ferguson, G. Seibel, and P. Kollman. AMBER, a package of computer programs for applying molecular mechanics, normal mode analysis, molecular dynamics and free energy calculations to simulate the structural and energetic properties of molecules. *Computer Physics Communications*, 91(1-3), 1995.

[232] J. Pei and N. V Grishin. MUMMALS: Multiple sequence alignment improved by using hidden markov models with local structural information. *Nucleic Acids Research*, 34(16):4364–4374, 2006.

[233] D. N. Perkins, D. J. C. Pappin, D. M. Creasy, and J. S. Cottrell. Probability-based protein identification by searching sequence databases using mass spectrometry data. *Electrophoresis*, 20(18):3551–3567, 1999.

[234] M. W. Perlin, M. B. Burks, R. C. Hoop, and E. P. Hoffman. Toward fully automated genotyping: Allele assignment, pedigree construction, phase determination, and recombination detection in Duchenne muscular dystrophy. *American Journal of Human Genetics*, 55(4):777–787, 1994.

[235] A. E. Pertiz, R. Kierzek, N. Sugimoto, and D. H. Turner. Thermodynamic study of internal loops in oligoribonucleotides: Symmetric loops are more stable than asymmetric loops. *Biochemistry*, 30:6428–6436, 1991.

[236] P. Pevzner. Multiple alignment, communication cost, and graph matching. *SIAM Journal on Applied Mathematics*, 52(6):1763–1779, 1992.

[237] P. A. Pevzner and S. H. Sze. Combinatorial approaches to finding subtle signals in DNA sequences. In *Proceedings of the 8th International Conference on Intelligent Systems for Molecular Biology*, pages 269–278, 2000.

[238] T. M. Phuong, C. B. Do, R. C. Edgar, and S. Batzoglou. Multiple alignment of protein sequences with repeats and rearrangements. *Nucleic Acids Research*, 34(20):5932–5942, 2006.

[239] C. W. Pleij. Pseudoknots: A new motif in the RNA game. *Trends Biochem Sci*, 15(4):143–147, 1990.

[240] C. P. Ponting, J. Schultz, F. Milpetz, and P. Bork. SMART: Identification and annotation of domains from signalling and extracellular protein sequences. *Nucleic Acids Research*, 27(1):229–232, 1999.

[241] G. P. S. Raghava, S. M. J. Searle, P. C. Audley, J. D. Barber, and G. J. Barton. OXBench: A benchmark for evaluation of protein multiple sequence alignment accuracy. *BMC Bioinformatics*, 4:47, 2003.

[242] B. Ren, H. Cam, Y. Takahashi, T. Volkert, J. Terragni, R. A. Young, and B. D. Dynlacht. E2F integrates cell cycle progression with DNA repair, replication and G2/M checkpoints. *Genes and Development*, 16:245–256, 2002.

[243] K. Rice and T. Warnow. Parsimony is hard to beat. In *Computing and Combinatorics*, pages 124–133, 1997.

[244] J. R. Riordan, J. M. Rommens, B. Kerem, N. Alon, R. Rozmahel, Z. Grzelczak, J. Zielenski, S. Lok, N. Plavsic, and J. L. Chou. Identification of the cystic fibrosis gene: Cloning and characterization of complementary DNA. *Science*, 245(4922):1066–1073, 1989.

[245] E. Rivas and S. Eddy. A dynamic programming algorithm for RNA structure prediction including pseudoknots. *Journal of Molecular Biology*, 285:2053–2068, 1999.

[246] D. F. Robinson and L. R. Foulds. Comparison of phylogentic trees. *Mathematical Biosciences*, 53:131–147, 1981.

[247] S. Roch. A short proof that phylogenetic tree reconstruction by maximum likelihood is hard. *IEEE/ACM Transactions on Computational Biology and Bioinformatics*, 3(1):92–94, 2006.

[248] K. A. Romer, G. R. Kayombya, and E. Fraenkel. WebMOTIFS: Automated discovery, filtering, and scoring of DNA sequence motifs using multiple programs and Bayesian approaches. *Nucleic Acids Research*, 35:W217–W220, 2007.

[249] M. Ronaghi, S. Karamohamed, B. Pettersson, M. Uhlen, and P. Nyren. Real-time DNA sequencing using detection of pyrophosphate release. *Anal Biochem*, 242(1):84–89, 1996.

[250] F. R. Roth, J. D. Hughes, P. E. Estep, and G. M. Church. Finding DNA regulatory motifs within unaligned non-coding sequences clustered by whole-genome mRNA quantitation. *Nature Biotechnology*, 16(10):939–945, 1998.

[251] G. Ruano, K. K. Kidd, and J. C. Stephens. Haplotype of multiple polymorphisms resolved by enzymatic amplification of single DNA molecules. *Proceedings of the National Academy of Sciences of USA*, 87 (16):6296–6300, 1990.

[252] M. F. Sagot. Spelling approximate repeated or common motifs using a suffix tree. In *LATIN*, pages 374–390, 1998.

[253] N. Saitou and M. Nei. The neighbor-joining method: A new method for reconstructing phylogenetic trees. *Molecular Biology and Evolution*, 4:405–425, 1987.

[254] H. Salgado, S. Gama-Castro, A. Martinez-Antonio, E. Diaz-Peredo, F. Sanchez-Solano, M. Peralta-Gil, D. Garcia-Alonso, V. Jimenez-Jacinto, A. Santos-Zavaleta, C. Bonavides-Martinez, and J. Collado-Vides. RegulonDB (version 4.0): Transcriptional regulation, operon organization and growth conditions in Escherichia coli K-12. *Nucleic Acids Research*, 32:D303, 2004.

[255] A. Sandelin, W. Alkema, P. Engstrom, W. W. Wasserman, and B. Lenhard. JASPAR: An open-access database for eukaryotic transcription factor binding profiles. *Nucleic Acids Research*, 32(Database issue):D91–D94, 2004.

[256] D. Sankoff. Minimal mutation trees of sequences. *SIAM Journal on Applied Mathematics*, 78:35–42, 1975.

[257] D. Sankoff. Simultaneous solution of RNA folding, alignment and protosequence problems. *SIAM Journal on Applied Mathematics*, 45(5): 810–825, 1985.

[258] D. Sankoff, G. Leduc, N. Antoine, B. Paquin, B. F. Lang, and R. Cedergren. Gene order comparisons for phylogenetic inference: Evolution ofthe mitochondrial genome. *Proceedings of the National Academy of Sciences of USA*, 22:448–450, 1992.

[259] B. Schieber and U. Vishkin. On finding lowest common ancestors: Simplifications and parallelization. *SIAM Journal on Computing*, 17:1253 1262, 1988.

[260] J. Schrieber, R. G. Jenner, H. L. Murray, G. K. Gerber, D. K. Gifford, and R. A. Young. Coordinated binding of NFKB family members in the response of human cells to lipopolysaccharide. *Proceedings of the National Academy of Sciences of USA*, 103:5899–5904, 2006.

[261] S. Schwartz, W. J. Kent, A. Smit, Z. Zhang, R. Baertsch, R. C. Hardison, D. Haussler, and W. Miller. Human-mouse alignments with BLASTZ. *Genome Research*, 13(1):103–107, Jan 2003.

[262] B. O. Searle, S. Dasari, M. Turner, A. P. Reddy, D. Choi, P. A. Wilmarth, A. L. McCormack, L. L. David, and S. B. Nagalla. High-throughput identification of proteins and unanticipated sequence modifications using a mass-based alignment algorithm for MS/MS de novo sequencing results. *Analytical Chemistry*, 76:2220–2230, 2004.

[263] A. Shevchenko, S. Sunyaev, A. Loboda, A. Shevchenko, P. Bork, W. Ens, and K. G. Standing. Charting the proteomes of organisms with unsequenced genomes by MALDI-quadrupole time-of-flight mass spectrometry and BLAST homology searching. *Analytical Chemistry*, 73:1917–1926, 2001.

[264] V. A. Simossis and J. Heringa. PRALINE: A multiple sequence alignment toolbox that integrates homology-extended and secondary structure information. *Nucleic Acids Research*, 33(Web Server issue):W289–W294, 2005.

[265] V. A. Simossis, J. Kleinjung, and J. Heringa. Homology-extended sequence alignment. *Nucleic Acids Research*, 33(3):816–824, 2005.

[266] S. Sinha and M. Tompa. A statistical method for finding transcription factor binding site. In *Proceedings of the Eighth International Conference on Intelligent Systems on Molecular Biology*, pages 344–354, 2000.

[267] T. F. Smith and M. S. Waterman. Identification of common molecular subsequences. *Journal of Molecular Biology*, 147:195–197, 1981.

[268] R. R. Sokal and F. J. Rohlf. Taxonomic congruence in the Leptopodomorpha re-examined. *Systematic Zoology*, 30(3):309–325, 1981.

[269] E. L. Sonnhammer, S. R. Eddy, E. Birney, A. Bateman, and R. Durbin. Pfam: multiple sequence alignments and HMM-profiles of protein domains. *Nucleic Acids Research*, 26(1):320–322, 1998.

[270] M. Steel. The complexity of reconstructing trees from qualitative characters and subtrees. *Journal of Classification*, 9(1):91–116, 1992.

[271] M. Steel and D. Penny. Distribution of tree comparison metrics – some new results. *Systematic Biology*, 42(2):126–141, 1993.

[272] M. Stephens and P. Donnelly. Inference in molecular population genetics. *J. R. Statist. Soc. B*, 62:605–635, 2000.

[273] M. Stephens, N. J. Smith, and P. Donnelly. A new statistical method for haplotype reconstruction from population data. *American Journal of Human Genetics*, 68(4):978–989, 2001.

[274] J. Stoye, V. Moulton, and A. W. Dress. DCA: An efficient implementation of the divide-and-conquer approach to simultaneous multiple sequence alignment. *Computer Applications in the Biosciences*, 13(6): 625–626, 1997.

[275] J. A. Studier and K. L. Keppler. A note on the neighbor-joining algorithm of Saitou and Nei. *Molecular Biology and Evolution*, 5:729–731, 1988.

[276] A. H. Sturtevant. Genetic factors affecting the strength of linkage in drosophila. *Proceedings of the National Academy of Sciences of USA*, 3:555–558, 1917.

[277] A. R. Subramanian, M. Kaufmann, and B. Morgenstern. DIALIGN-TX: Greedy and progressive approaches for segment-based multiple sequence alignment. *Algorithms for Molecular Biology*, 3:6, 2008.

[278] J. E. Tabaska, R. B. Cary, H. N. Gabow, and G. D. Stormo. An RNA folding method capable of identifying pseudoknots and base triples. *Bioinformatics*, 14(8):691–699, 1998.

[279] D. Tabb, A. Saraf, and J. R. Yates III. GutenTag: High-throughput sequence tagging via an empirically derived fragmentation model. *Analytical Chemistry*, 75:6415–6421, 2003.

[280] C. K. Tang and D. E. Draper. Unusual mRNA pseudoknot structure is recognized by a protein translational repressor. *Cell*, 57:531–536, 1989.

[281] S. Tanner, H. Shu, A. Frank, L. Wang, E. Zandi, M. Mumby, P. A. Pevzner, and V. Bafna. InsPecT: Identification of posttranslationally modified peptides from tandem mass spectra. *Analytical Chemistry*, 77 (14):4626–4639, 2005.

[282] S. Tanner, S. H. Payne, S. Dasari, Z. Shen, P. Wilmarth, L. David, W. F. Loomis, S. P. Briggs, and V. Bafna. Accurate annotation of peptide modifications through unrestrictive database search. *Journal of Proteome Research*, 7(1):170–181, 2008.

[283] E. Tannier, A. Bergeron, and M.-F. Sagot. Advances on sorting by reversals. *Discrete Applied Mathematics*, 155(6–7):881–888, 2007.

[284] J. A. Taylor and R. S. Johnson. Sequence database searches via de novo peptide sequencing by tandem mass spectrometry. *Rapid Communcations in Mass Spectrometry*, 11:1067–1075, 1997.

[285] J. A. Taylor and R. S. Johnson. Implementation and uses of automated de novo peptide sequencing by tandem mass spectrometry. *Analytical Chemistry*, 73:2594–2604, 2001.

[286] G. Thijs, M. Lescot, K. Marchal, S. Rombauts, B. De Moor, P. Rouze, and Y. Moreau. A higher-order background model improves the detection of promoter regulatory elements by Gibbs sampling. *Bioinformatics*, 17(12):1113–1122, 2001.

[287] J. D. Thompson, D. G. Higgins, and T. J. Gibson. CLUSTAL W: Improving the sensitivity of progressive multiple sequence alignment through sequence weighting, positions-specific gap penalties and weight matrix choice. *Nucleic Acids Research*, 22:4673–4680, 1994.

[288] J. D. Thompson, F. Plewniak, and O. Poch. A comprehensive comparison of multiple sequence alignment programs. *Nucleic Acids Research*, 27(13):12682–2690, 1999.

[289] J. D. Thompson, P. Koehl, R. Ripp, and O. Poch. BAliBASE 3.0: Latest developments of the multiple sequence alignment benchmark. *Proteins*, 61(1):127–136, 2005.

[290] I. Tinoco, O. C. Uhlenbeck, and M. D. Levine. Estimation of secondary structure in ribonucleic acids. *Nature*, 230:362–367, 1971.

[291] I. Tinoco, P. N. Borer, B. Dengler, M. D. Levine, O. C. Uhlenbeck, D. M. Crothers, and J. Gralla. Improved estimation of secondary structure in ribonucleic acids. *Nature New Biology*, 246:40–41, 1973.

[292] M. Tompa. An exact method for finding short motifs in sequences, with application to the ribosome binding site problem. In *Proceedings of the Seventh International Conference on Intelligent Systems on Molecular Biology*, pages 262–271, 1999.

[293] M. Tompa, N. Li, T. L. Bailey, G. M. Church, B. D. Moor, E. Eskin, A. V. Favorov, M. C. Frith, Y. Fu, W. J. Kent, V. J. Makeev, A. A. Mironov, W. S. Noble, G. Pavesi, G. Pesole, M. Regnier, N. Simonis, S. Sinha, G. Thijs, J. van Helden, M. Vandenbogaert, Z. Weng, C. Workman, C. Ye, and Z. Zhu. Assessing computational tools for the discovery of transcription factor binding sites. *Nature Biotechnology*, 23:137–144, 2005.

[294] D. Tsur, S. Tanner, E. Zandi, V. Bafna, and P. Pevzner. Identification of post-translational modifications via blind search of mass-spectra. *Nature Biotechnology*, 23:1562–1567, 2005.

[295] E. Ukkonen. On-line construction of suffix trees. *Algorithmica*, 14: 249–260, 1995.

[296] J. van Helden, B. Andre, and J. Collado-Vides. Extracting regulatory sites from the upstream region of yeast genes by computational analysis of oligonucleotide frequencies. *Journal of Molecular Biology*, 281:827–842, 1998.

[297] J. van Helden, A. F. Rios, and J. Collado-Vides. Discovering regulatory elements in non-coding sequences by analysis of spaced dyads. *Nucleic Acids Research*, 28:1808–1818, 2000.

[298] I. M. Wallace, G. Blackshields, and D. G. Higgins. Multiple sequence alignments. *Current Opinion in Structural Biology*, 15:261–266, 2005.

[299] I. M. Wallace, O. O'Sullivan, D. G. Higgins, and C. Notredame. M-Coffee: Combining multiple sequence alignment methods with t-coffee. *Nucleic Acids Research*, 34(6):1692–1699, 2006.

[300] I. V. Walle, I. Lasters, and L. Wyns. Align-M – a new algorithm for multiple alignment of highly divergent sequences. *Bioinformatics*, 20 (9):1428–1435, 2004.

[301] L. Wang and T. Jiang. On the complexity of multiple sequence alignment. *Journal of Computational Biology*, 1(4):337–348, 1994.

[302] M. S. Waterman. Efficient sequence alignment algorithms. *Journal of Theoretical Biology*, 108:333–337, 1984.

[303] M. S. Waterman, R. Arratia, and D. J. Galas. Pattern recognition in several sequences: Consensus and alignment. *Bulletin of Mathematical Biology*, 46(4):515–527, 1984.

[304] J. D. Watson and F. H. Crick. Molecular structure of nucleic acids: A structure for deoxyribose nucleic acid. *Nature*, 171(4356):737–738, 1953.

[305] P. Weiner. Linear pattern matching algorithms. *Switching and Automata Theory*, pages 1–11, 1973.

[306] E. Wijaya, K. Rajaraman, S. M. Yiu, and W. K. Sung. Detection of generic spaced motifs using submotif pattern mining. *Bioinformatics*, 23:1476–1485, 2007.

[307] E. Wijaya, S. M. Yiu, N. T. Son, R. Kanagasabai, and W. K. Sung. Motifvoter: A novel ensemble method for fine-grained integration of generic motif finders. *Bioinformatics*, 24(20):2288–2295, 2008.

[308] W. J. Wilbur and D. J. Lipman. The context dependent comparison of biological sequence. *SIAM J. Applied Maths*, 44(3):557–567, 1984.

[309] E. Wingender, X. Chen, E. Fricke, R. Geffers, R. Hehl, I. Liebich, M. Krull, V. Matys, H. Michael, R. Ohnhauser, M. Pruss, F. Schacherer, S. Thiele, and S. Urbach. The TRANSFAC system on gene expression regulation. *Nucleic Acids Research*, 29(1):281–283, 2001.

[310] C. R. Woese, O. Kandler, and M. L. Wheelis. Towards a natural system of organisms: Proposal for the domains Archaea, Bacteria, and Eucarya. *Proceedings of the National Academy of Sciences of USA*, 87:4576–4579, 1990.

[311] C. T. Workman and G. D. Stormo. ANN-Spec: A method for discovering transcription factor binding sites with improved specificity. In *Proceedings of Pacific Symposium of Biocomputing*, pages 467–478, 2000.

[312] D. Y. Wu, L. Ugozzoli, B. K. Pal, and R. B. Wallace. Allele-specific enzymatic amplification of beta-globin genomic DNA for diagnosis of sickle cell anemia. *Proceedings of the National Academy of Sciences of USA*, 86(8):2757–2760, 1989.

[313] E. P. Xing, M. I. Jordan, and W. Wu. Logos: A modular Bayesian model for de novo motif detection. *Journal of Bioinformatics and Computational Biology*, 2:127–154, 2003.

[314] J. R. Yates, J. K. Eng, and A. L. McCormack. Mining genomes: Correlating tandem mass spectra of modified and unmodified peptides to sequences in nucleotide databases. *Analytical Chemistry*, 67(18):3202–3210, 1995.

[315] K. Zhang, M. Deng, T. Chen, M. S. Waterman, and F. Sun. A dynamic programming algorithm for haplotype block partitioning. *Proceedings of the National Academy of Sciences of USA*, 99:7335–7339, 2002.

[316] Y. Zhang and M. S. Waterman. An Eulerian path approach to local multiple alignment for DNA sequences. *Proceedings of the National Academy of Sciences of USA*, 102(5):1285–1290, 2005.

[317] Z. Zhang, S. Schwartz, L. Wagner, and W. Miller. A greedy algorithm for aligning DNA sequences. *Journal of Computational Biology*, pages 203–214, 2000.

[318] H. Zhou and Y. Zhou. SPEM: Improving multiple sequence alignment with sequence profiles and predicted secondary structures. *Bioinformatics*, 21(18):3615–3621, 2005.

[319] Q. Zhou and J. S. Liu. Modeling within-motif dependence for transcription factor binding site predictions. *Bioinformatics*, 20(6):909–916, 2004.

[320] M. Zuker. On finding all suboptimal foldings of an RNA molecule. *Science*, 244:48–52, 1989.

[321] M. Zuker and P. Stiegler. Optimal computer folding of large RNA sequences using thermodynamics and auxiliary information. *Nucleic Acids Research*, 9:133–148, 1981.

Index